Encyclopaedia of Mathematical Sciences

Volume 62

Editor-in-Chief: R.V. Gamkrelidze

A.N. Parshin I.R. Shafarevich (Eds.)

Number Theory II

Algebraic Number Theory

Springer-Verlag
Berlin Heidelberg New York
London Paris Tokyo
Hong Kong Barcelona
Budapest

Consulting Editors of the Series:
A.A. Agrachev, A.A. Gonchar, E.F. Mishchenko,
N.M. Ostianu, V.P. Sakharova, A.B. Zhishchenko

Title of the Russian edition:
Itogi nauki i tekhniki, Sovremennye problemy matematiki,
Fundamental'nye napravleniya, Vol. 62, Teoriya chisel 2
Publisher VINITI, Moscow 1990

Mathematics Subject Classification (1991):
11Rxx, 11Sxx

ISBN 3-540-53386-9 Springer-Verlag Berlin Heidelberg New York
ISBN 0-387-53386-9 Springer-Verlag New York Berlin Heidelberg

Library of Congress Cataloging-in-Publication Data
Teoriia chisel 2. English
Number theory II:
algebraic number theory / A.N. Parshin, I.R. Shafarevich (eds.).
p. cm.—(Encyclopaedia of mathematical sciences; v. 62)
Translation of: Teoriia chisel 2, issued as v. 62 of the serial: Itogi nauki i tekhniki.
Seriia Sovremennye problemy matematiki. Fundamental'nye napravleniia.
Includes bibliographical references and index.
ISBN 3-540-53386-9 (Berlin).—ISBN 0-387-53386-9 (New York)
1. Algebraic number theory. I. Parshin, A.N. II. Shafarevich, I.R. (Igor' Rostislavovich), 1923–
III. Title. IV. Title: Number theory 2. V. Title: Number theory two. VI. Series.
QA247.T45513 1992 512'.74—dc20 92-2200

Printed in the United States of America
Typesetting: Asco Trade Typesetting Ltd., Hong Kong
41/3140-543210—Printed on acid-free paper

List of Editors and Author

Editor-in-Chief

R.V. Gamkrelidze, Russian Academy of Sciences, Steklov Mathematical Institute, ul. Vavilova 42, 117966 Moscow, Institute for Scientific Information (VINITI), ul. Usievicha 20a, 125219 Moscow, Russia

Consulting Editors

A.N. Parshin, Steklov Mathematical Institute, ul. Vavilova 42, 117966 Moscow, Russia

I.R. Shafarevich, Steklov Mathematical Institute, ul. Vavilova 42, 117966 Moscow, Russia

Author

H. Koch, Max-Planck-Gesellschaft zur Förderung der Wissenschaften e.V., Arbeitsgruppe "Algebraische Geometrie und Zahlentheorie" an der Humboldt-Universität zu Berlin, Mohrenstr. 39, 1086 Berlin, FRG

Algebraic Number Fields

H. Koch

Contents

Preface

The purpose of the present article is the description of the main structures and results of algebraic number theory. The main subjects of interest are the algebraic number fields. We included in this article mostly properties of general character of algebraic number fields and of structures related to them. Special results appear only as examples which illustrate general features of the theory.

On the other hand important parts of algebraic number theory such as class field theory have their analogies for other types of fields like functions fields of one variable over finite fields. More general class field theory is part of a theory of fields of finite dimension, which is now in development. We mention these analogies and generalizations only in a few cases. From the application of algebraic number theory to problems about Diophantine equations we explain the application to the theory of complete forms, in particular binary quadratic forms (Chap. 1.1.5–7), and to Fermat's last theorem (Chap. 1.3.10, Chap. 2.6.4, Chap. 4.2.6).

A part of algebraic number theory serves as basis science for other parts of mathematics such as arithmetic algebraic geometry and the theory of modular forms. This concerns our Chap. 1, Basic number theory, Chap. 2, Class field theory, and parts of Chap. 3, Galois cohomology. These domains are presented in more detail than other parts of the theory.

In accordance with the principles of this encyclopedia we give in general no reference to the history of results and refer only to the final or most convenient source for the result in question. Many paragraphs and sections contain after the title a main reference which served as basis for the presentation of the material of the paragraph or section.

Each chapter and some paragraphs contain a detailed introduction to their subject. Therefore we give here only some technical remarks: At the end of Theorems, Propositions, Lemmas, Proofs, or Exercises stands the sign □ or ⊠. The sign □ means that the Theorem, Proposition, Lemma, or Exercise is easy to prove or that we gave a full sketch of the proof. The sign ⊠ means that one needs for the proof ideas which are not or only partially explained in this article.

We use Bourbaki's standard notation such as $\mathbb{Z}$ for the ring of integers, $\mathbb{Q}$ for the field of rational numbers, $\mathbb{R}$ for the field of real numbers, and $\mathbb{C}$ for the field of complex numbers. For any ring Λ with unit element, $\Lambda^\times$ denotes the group of units, i.e. the group of invertible elements and Λ^* denotes the set of elements distinct from 0. Furthermore μ_n denotes the group of roots of unity with order dividing n. For any set M we denote by $|M|$ the cardinality of M.

I am grateful to I.R. Shafarevich and A.N. Parshin who read a preliminary version of this book and proposed several important alterations and additions, to U. Bellack and B. Wüst for their help in the production of the final manuscript and to the staff of Springer-Verlag for their cooperation.

Helmut Koch

Chapter 1
Basic Number Theory

The first goal of algebraic number theory is the generalization of the theorem on the unique representation of natural numbers as products of prime numbers to algebraic numbers. Gauss considered the ring $\mathbb{Z}[\sqrt{-1}]$ of all numbers of the form $a + \sqrt{-1}b$ with $a, b \in \mathbb{Z}$ and showed that $\mathbb{Z}[\sqrt{-1}]$ is a ring with unique factorization in prime elements (see § 2.1). He introduced these numbers for the development of his theory of biquadratic residues. Another motivation for the study of the arithmetic of algebraic numbers comes from the theory of Tiophantine equations. For example, the quadratic form $f(x_1, x_2) = x_1^2 - Dx_2^2$ with $D \in \mathbb{Z}$, $\sqrt{D} \notin \mathbb{Z}$ can be written in the form $(x_1 - \sqrt{D}x_2)(x_1 + \sqrt{D}x_2)$. Hence the question about the representation of integers by $f(a_1, a_2)$ with $a_1, a_2 \in \mathbb{Z}$ can be reformulated as a question of factorization of algebraic numbers of the form $a_1 + \sqrt{D}a_2$. These numbers form a module in the field $\mathbb{Q}(\sqrt{D})$.

Beginning with the year 1840 Kummer (1975) considered the ring of numbers of the form $a_0 + a_1\zeta + \cdots + a_{p-1}\zeta^{p-1}$, $a_0, \ldots, a_{p-1} \in \mathbb{Z}$, where p is a prime and ζ a primitive p-th root of unity, i.e. $\zeta^p = 1$ and $\zeta \neq 1$. Since in general this ring has not unique factorization in prime numbers, Kummer introduced "ideal numbers" and showed the unique factorization in ideal prime numbers. With this concept he was able to prove Fermat's last theorem in many new cases using the identity

$$x^p - y^p = \prod_{i=0}^{p-1} (x - \zeta^i y).$$

Up to now Kummer's deep and beautiful results about cyclotomic fields $\mathbb{Q}(\zeta)$ serve as paradigmas for research in algebraic number theory (see Chap. 4) but it needed some 30 years until Kronecker and Dedekind found the right generalization of Kummer's ideal numbers for arbitrary number fields K: One has to define the notion of integral algebraic numbers. The integral algebraic numbers contained in K form a ring $\mathfrak{O}_K$ (§ 1.1), which is the natural realm for the generalization of the unique factorization in prime numbers.

There are three methods to establish arithmetic in $\mathfrak{O}_K$. Kronecker considers polynomials with coefficients in $\mathfrak{O}_K$ (§ 2.3). Dedekind introduces ideals in $\mathfrak{O}_K$, defining so one of the most important notions of algebra, and the third method due to Zolotarev and Hensel uses what is nowadays called localization (§ 4).

An important part of the theory, in particular the Dirichlet unit theorem, is valid for rings $\mathfrak{O}$ in K with $K = \mathbb{Q}(\mathfrak{O})$, $1 \in \mathfrak{O}$ and $\mathfrak{O} \subseteq \mathfrak{O}_K$, called orders of K. Orders and modules of elements in K appear in connection with decomposible forms (§ 1.5–6). Therefore we begin Chap. 1 with the theory of modules and orders in algebraic number fields (§ 1). In § 2 we define the notion of a ring with divisor theory formulating axiomatically what one awaits from an arithmetic theory of rings. For a domain such a divisor theory is always unique. We indicate its existence for the ring $\mathfrak{O}_K$ by means of Kronecker's method. In §3 we develop systematically the ideal theory of Dedekind rings including the theory of the different and discriminant and the theory of ramification groups. Nowadays algebraic number fields are mostly studied by a mixture of ideal and valuation theory. The latter, being the third method mentioned above, is presented in §4. $\mathfrak{p}$-adic analysis, which is an essential part of the valuation theoretic method, is developed in this article only so far as it is needed in the following. §5 explains harmonic analysis in local and global fields and §6 is devoted to the study of L-series and their application to the arithmetic of algebraic number fields.

Chapter 1 of this article covers the content of most of the numerous books on algebraic number theory which serve as an introduction to the field. We mention only the following: Artin (1951), Borevich, Shafarevich (1985), Eichler (1963), Hasse (1979), Hecke (1923), Lang (1970), Narkiewicz (1990), Weiss (1963), Weyl (1940).

Artin (1951), Eichler (1963), and Hasse (1979) are introductions to the theory of algebraic numbers and algebraic functions as well. Narkiewicz (1990) contains a voluminous review about results in the elementary and analytical theory of algebraic numbers and a complete bibliography of this domain.

§1. Orders in Algebraic Number Fields

(Main reference: Borevich, Shafarevich (1985), Chap. 2)

The subject of this article are the *algebraic numbers*. A complex number α is called *algebraic* if it satisfies an equation of the form

$$\alpha^n + a_1\alpha^{n-1} + \cdots + a_n = 0$$

with $a_1, \ldots, a_n \in \mathbb{Q}$.

An *algebraic number field* K is a finite field extension of $\mathbb{Q}$ lying in $\mathbb{C}$. It is always of the form $K = \mathbb{Q}(\alpha)$ with an algebraic number α. The degree $[K:\mathbb{Q}]$ of K over $\mathbb{Q}$ is called the *degree* of K.

The first goal of algebraic number theory is the extension of the arithmetic in $\mathbb{Z}$ and $\mathbb{Q}$ to algebraic number fields.

By arithmetic in $\mathbb{Z}$ we mean the unique factorization of natural numbers in the product of prime numbers. The arithmetic in $\mathbb{Q}$ is given by the arithmetic in $\mathbb{Z}$: Any $r \in \mathbb{Q} - \{0\}$ has a uniquely determined representation

$$r = (-1)^\nu \prod_p p^{\nu_p(r)}$$

where the product runs over all prime numbers p and $\nu \in \{0, 1\}$, $\nu_p(r) \in \mathbb{Z}$ are uniquely determined by r.

If we want to generalize the arithmetic of $\mathbb{Q}$ to an algebraic number field K, the first question we have to answer is, what will be the right generalization of $\mathbb{Z}$? This should be a ring $\mathfrak{O}$ in K with the following properties:

1. K is the quotient field of $\mathfrak{O}$.
2. $\mathfrak{O} \cap \mathbb{Q} = \mathbb{Z}$.
3. The additive group of $\mathfrak{O}$ is finitely generated.

A ring in K with these properties is called an *order* of K.

It is easy to see (compare Example 2) that for $K \neq \mathbb{Q}$ there are infinitely many orders of K. But we shall show in § 1.1 that there is one maximal order $\mathfrak{O}_K$ containing all orders of K, an element α of K belongs to $\mathfrak{O}_K$ if and only if there are integers $a_1, \ldots, a_s$ for some s such that

$$\alpha^s + a_1\alpha^{s-1} + \cdots + a_s = 0. \tag{1.1}$$

Furthermore we shall see in § 2 that $\mathfrak{O}_K$ is the natural generalization of $\mathbb{Z}$ in the sense that it is possible to develop arithmetic in $\mathfrak{O}_K$. The elements of $\mathfrak{O}_K$ are called the *integers of* K, and any complex number α satisfying an equation of the form (1.1) is called an *integral algebraic number.*

Orders in a number field K arise in a natural way in connection with *modules* in K: In this paragraph a module $\mathfrak{m}$ in K means a finitely generated subgroup of K^+. Since the group K^+ has no torsion, $\mathfrak{m}$ is a free $\mathbb{Z}$-module of rank $\leqslant [K:\mathbb{Q}]$. The module $\mathfrak{m}$ is called *complete* (or a lattice in K) if its rank is equal to $[K:\mathbb{Q}]$.

Modules play an important role in the arithmetic of K and in number theoretical questions connected with algebraic number fields (§ 1.5–7). Therefore we start with the study of modules and orders.

1.1. Modules and Orders. Let $\mathfrak{m}_1$ and $\mathfrak{m}_2$ be modules in K. Then the *product* $\mathfrak{m}_1\mathfrak{m}_2$ is the abelian group generated by the elements $\mu_1\mu_2$ with $\mu_1 \in \mathfrak{m}_1$, $\mu_2 \in \mathfrak{m}_2$. Obviously $\mathfrak{m}_1\mathfrak{m}_2$ is finitely generated, i.e. $\mathfrak{m}_1\mathfrak{m}_2$ is a module in K.

Proposition 1.1. *A number α in an algebraic number field K is an algebraic integer if and only if there exists a module $\mathfrak{m} \neq \{0\}$ in K with $\alpha\mathfrak{m} \subset \mathfrak{m}$.*

Proof. Let α be an algebraic integer in K. Then we can take for $\mathfrak{m}$ the module generated by $1, \alpha, \alpha^2, \ldots, \alpha^{n-1}$ where $n := [K:\mathbb{Q}]$. On the other hand if we have $\alpha\mathfrak{m} \subset \mathfrak{m}$ for some $\alpha \in K$ and $\mu_1, \ldots, \mu_s$ is a basis of $\mathfrak{m}$, then there are $a_{ij} \in \mathbb{Z}$ with

$$\alpha\mu_i = \sum_{j=1}^{s} a_{ij}\mu_j \qquad \text{for } i = 1, \ldots, s.$$

Hence $\det(\alpha\delta_{ij} - a_{ij}) = 0$.

Let $\mathfrak{m}$ be a complete module in K. Then

$$\mathfrak{O}(\mathfrak{m}) := \{\alpha \in K \mid \alpha\mathfrak{m} \subset \mathfrak{m}\}$$

is called the *order of* $\mathfrak{m}$.

Proposition 1.2. *The order of a complete module $\mathfrak{m}$ in K is an order in K. For every order $\mathfrak{O}$ in K there exists a complete module $\mathfrak{m}$ in K with $\mathfrak{O} = \mathfrak{O}(\mathfrak{m})$,* e.g. $\mathfrak{m} = \mathfrak{O}$. □

Example 1. Let $K := \mathbb{Q}(\alpha)$ with an algebraic integer α. Then $\mathbb{Z}[\alpha]$ is an order in K. □

Let $\omega_1, \ldots, \omega_n$ be a basis of the complete module $\mathfrak{m}$ in K. Then the discriminant $d_{K/\mathbb{Q}}(\omega_1, \ldots, \omega_n)$ (see App. 1) is independent of the choice of $\omega_1, \ldots, \omega_n$. It is called the *discriminant of* $\mathfrak{m}$ and is denoted by $d(\mathfrak{m})$.

Let $\beta_1, \ldots, \beta_n$ be a basis of $\mathfrak{O}(\mathfrak{m})$ and let $A \in GL_n(\mathbb{Q})$ such that

$$(\omega_1, \ldots, \omega_n)^T = A(\beta_1, \ldots, \beta_n)^T.$$

Then the absolute value $|\det A|$ of the determinant of A is independent of the choice of the bases. It is called the *norm of* $\mathfrak{m}$ and is denoted by $N(\mathfrak{m})$. It is easy to see that

$$d(\mathfrak{m}) = d(\mathfrak{O}(\mathfrak{m}))N(\mathfrak{m})^2 \tag{1.2}$$

and

$$N(\alpha\mathfrak{m}) = N_{K/\mathbb{Q}}(\alpha)N(\mathfrak{m}) \qquad \text{for } \alpha \in K^\times. \tag{1.3}$$

$d(\mathfrak{m})$ is a rational number, distinct from 0. The sign of $d(\mathfrak{m})$ is determined by the following proposition.

Proposition 1.3 (Brill's discriminant Theorem). *Let r_1 be the number of real embeddings and $2r_2$ the number of complex embeddings of K in $\mathbb{C}$. Then $d(\mathfrak{m})$ is positive if and only if r_2 is even.* □ (Narkiewicz (1974), p. 59)

Now let $\mathfrak{O}$ be an order in K. Beside Prop. 1.3 one has the following simple proposition about $d(\mathfrak{O})$.

Proposition 1.4 (Stickelberger's discriminant Theorem). *$d(\mathfrak{O})$ is an integer with $d(\mathfrak{O}) \equiv 1$ or $d(\mathfrak{O}) \equiv 0 \pmod 4$.* □ (Narkiewicz (1974), p. 59)

From the theory of finitely generated abelian groups follows

Proposition 1.5. *Let $\mathfrak{m}_1$, $\mathfrak{m}_2$ be complete modules with the same order $\mathfrak{O}$ and $\mathfrak{m}_1 \supseteq \mathfrak{m}_2$. Then $N(\mathfrak{m}_2)/N(\mathfrak{m}_1)$ is equal to the index $[\mathfrak{m}_1 : \mathfrak{m}_2]$.* □

The main result of the present section is

Theorem 1.6. *The set $\mathfrak{O}_K$ of all algebraic integers in K is an order in K, called the maximal order.*

Proof. Let $\alpha_1, \alpha_2 \in \mathfrak{O}_K$ and let $\mathfrak{m}_1 := \mathbb{Z}[\alpha_1]$, $\mathfrak{m}_2 := \mathbb{Z}[\alpha_2]$. Then

$$(\alpha_1 \pm \alpha_2)\mathfrak{m}_1\mathfrak{m}_2 \subseteq \mathfrak{m}_1\mathfrak{m}_2 \qquad \text{and} \qquad \alpha_1\alpha_2\mathfrak{m}_1\mathfrak{m}_2 \subseteq \mathfrak{m}_1\mathfrak{m}_2.$$

Hence $\mathfrak{O}_K$ is a ring by Prop. 1.1. It remains to show that $\mathfrak{O}_K$ is finitely generated. Let m be a complete module contained in $\mathfrak{O}_K$. If $\mathfrak{m} \neq \mathfrak{O}_K$ take $\alpha_1 \notin \mathfrak{O}_K - \mathfrak{m}$, and let $(\mathfrak{m}, \alpha_1)$ be the module in K generated by $\mathfrak{m}$ and α_1. Then $D((\mathfrak{m}, \alpha_1))$ is a proper divisor of $D(\mathfrak{m})$. If $(\mathfrak{m}, \alpha_1) \neq \mathfrak{O}_K$ take $\alpha_2 \in \mathfrak{O}_K - (\mathfrak{m}, \alpha_1)$ and so forth. After finitely many steps we end up with an α_s such that $(\mathfrak{m}, \alpha_1, \ldots, \alpha_s) = \mathfrak{O}_K$. □

The following two propositions are easy to prove.

Proposition 1.7 *An algebraic number α is integral if and only if the minimal polynomial of α with respect to $\mathbb{Q}$ has integral coefficients.* □

Proposition 1.8. *Let $f(x) = x^m + \alpha_1 x^{m-1} + \cdots + \alpha_m \in \mathfrak{O}_K[x]$ and let $\alpha \in K$ with $f(\alpha) = 0$. Then α is an integer of K.* □

Example 2. A *quadratic number field* K is by definition an algebraic number field of degree two. There exists a unique square free $d \in \mathbb{Z}$ such that $K = \mathbb{Q}(\sqrt{d})$. Let

$$\omega = \begin{cases} (1 + \sqrt{d})/2 & \text{if } d \equiv 1 \pmod 4 \\ \sqrt{d} & \text{if } d \equiv 2, 3 \pmod 4). \end{cases}$$

Then $\{1, \omega\}$ is a basis of $\mathfrak{O}_K$.

An arbitrary order in K is of the form $\mathfrak{O}_f := \mathbb{Z}[f\omega]$ with a positive rational integer f called the conductor of $\mathfrak{O}_f$. □

Example 3. For number fields K of degree > 2 the maximal order $\mathfrak{O}_K$ is not always of the form $\mathbb{Z}[\alpha]$. For instance let $K = \mathbb{Q}(\beta)$ with $\beta^3 + \beta^2 - 2\beta + 8 = 0$. Then $\gamma = (\beta + \beta^2)/2 \in \mathfrak{O}_K$ and $1, \beta, \gamma$ is a basis of the module $\mathfrak{O}_K$, but there is no $\alpha \in \mathfrak{O}_K$ with $\mathfrak{O}_K = \mathbb{Z}[\alpha]$ (Hasse (1979), Chap. 25.7, Weiss (1963), p. 170). ⊠

The discriminant $d_K := d(\mathfrak{O}_K)$ is called the *discriminant* of K. Beside the degree $[K : \mathbb{Q}]$ the discriminant is the most important invariant of an algebraic number field K. By the discriminant theorem of Hermite (Theorem 1.12) there are only finitely many algebraic number fields with given discriminant. Hence one can classify algebraic number fields by means of their degree and their discriminant, of course conjugate fields have the same discriminant. By Example 2 the discriminant of the quadratic field $K = \mathbb{Q}(\sqrt{d})$ is given as follows: $d_K = d$ if $d \equiv 1 \pmod 4$, $d_K = 4d$ if $d \equiv 2, 3 \pmod 4$.

This shows that a quadratic field is uniquely determined by its in discriminant. This is not true in general for fields of higher degree (Example 19). See Table 5 for non-conjugate totally real cubic fields with the same discriminant < 500000.

See § 6.11 and Chap. 3.2.8 for discriminant estimation and for the smallest discriminant for fields with given degree.

1.2. Module Classes. Let $\mathfrak{O}$ be an order in the algebraic number field K. Obviously for any $\alpha \in K^\times$ the complete module $\alpha\mathfrak{O}$ has order $\mathfrak{O}$. If all complete modules with order $\mathfrak{O}$ have the form $\alpha\mathfrak{O}$ for some $\alpha \in K^\times$, then $\mathfrak{O}$ is a principal ideal ring and there is an arithmetic in $\mathfrak{O}$ which is quite similar to the arithmetic in $\mathbb{Z}$ (see § 2.1). In general K acts on the set $\mathfrak{M}(\mathfrak{O})$ of complete modules with order $\mathfrak{O}$ by multiplication. The orbits of this action are called *module classes*, i.e. a module class of $\mathfrak{O}$ consists of all modules of the form $\alpha\mathfrak{m}$, $\alpha \in K^\times$, for some fixed module $\mathfrak{m} \in \mathfrak{M}(\mathfrak{O})$. Two modules lying in the same class are called *equivalent*. The number $h(\mathfrak{O})$ of module classes of $\mathfrak{O}$ measures the deviation of $\mathfrak{O}$ from being a principal ideal domain. We show in this section that $h(\mathfrak{O})$ is finite.

In the application to the theory of quadratic forms (§ 1.6) we need also a finer classification of modules:

An algebraic number is called *totally positive* if all its real conjugates are positive. The modules $\mathfrak{m}_1$ and $\mathfrak{m}_2$ are called *equivalent in the narrow sense* if $\mathfrak{m}_2 = \alpha\mathfrak{m}_1$ with a totally positive $\alpha \in K^\times$.

The set of classes of equivalent modules with order $\mathfrak{O}$ will be denoted by $CL(\mathfrak{O})$.

Theorem 1.9. *$CL(\mathfrak{O})$ is a finite set.*

By means of Minkowski's geometry of numbers (Borevich, Shafarevich (1985), Chap. 2, § 3) one gets a more precise result:

Theorem 1.10. *Let K be an algebraic number field with r_1 real and $2r_2$ complex conjugates* (App. 1.2), $n := [K : \mathbb{Q}] = r_1 + 2r_2$, *and let $\mathfrak{m}$ be a complete module in K with order $\mathfrak{O}$. Then there is an $\alpha \in \mathfrak{m}$ with $\alpha \neq 0$ and*

$$|N_{K/\mathbb{Q}}(\alpha)| \leqslant \left(\frac{4}{\pi}\right)^{r_2} \frac{n!}{n^n} \sqrt{|d(\mathfrak{m})|}.$$

Theorem 1.10 implies Theorem 1.9: Taking in account $\mathfrak{O} \subseteq \alpha^{-1}\mathfrak{m}$, Prop. 1.5 and (1.2), (1.3) one finds

$$[\alpha^{-1}\mathfrak{m} : \mathfrak{O}] = N(\alpha^{-1}\mathfrak{m})^{-1} = \frac{|N_{K/\mathbb{Q}}(\alpha)|}{N(\mathfrak{m})} \leqslant \left(\frac{4}{\pi}\right)^{r_2} \frac{n!}{n^n} \sqrt{|d(\mathfrak{O})|}. \tag{1.4}$$

Hence in each class of $CL(\mathfrak{O})$ we have a module $\mathfrak{m}'$ containing $\mathfrak{O}$ such that $[\mathfrak{m}' : \mathfrak{O}]$ is limited. It follows from the theory of finitely generated abelian groups that there are only finitely many such modules in $CL(\mathfrak{O})$. □

For the proof of Theorem 1.10 we consider complete modules as lattices in $\mathbb{R}^n$ in the following way: Let $g_1, \ldots, g_{r_1}$ be the real isomorphisms of K and let g_{r_1+1}, $g_{r+1}, \ldots, g_r, g_n$ be the pairs of conjugate complex isomorphisms of K, $r = r_1 + r_2$. Then

$$\phi: \alpha \to (g_1\alpha, \ldots, g_{r_1}\alpha, \operatorname{Re} g_{r_1+1}\alpha, \ldots, \operatorname{Re} g_r\alpha, \operatorname{Im} g_{r_1+1}\alpha, \ldots, \operatorname{Im} g_r\alpha)$$

defines an injection of K in $\mathbb{R}^n$. It is easy to see that for $\mathfrak{m} \in \mathfrak{M}(\mathfrak{O})$ the image $\phi(\mathfrak{m})$ is a lattice in $\mathbb{R}^n$ with discriminant $2^{-r_2}\sqrt{|d(\mathfrak{m})|}$.

We are going to apply the following geometrical theorem to $\phi(\mathfrak{m})$: *Minkowski's convex body theorem.* Let L be a lattice in $\mathbb{R}^n$ with discriminant l. Furthermore let M be a convex closed set in $\mathbb{R}^n$ which is central symmetric with respect to the origin and has volume $V(M)$ with $V(M) \geqslant 2^n l$. Then M contains at least one point of L distinct from the origin. ⊠ (Hasse (1979), Chap. 30.2)

The application of Minkowski's convex body theorem is based on the following

Lemma. *Let $\mathfrak{m} \in \mathfrak{M}(\mathfrak{O})$ and let M be a convex closed set which is central symmetric with respect to the origin and contained in*

$$N = \left\{(\lambda_1, \ldots, \lambda_n) \in \mathbb{R}^n \,\middle|\, \prod_{\nu=1}^{r_1} |\lambda_\nu| \prod_{\nu=r_1+1}^{r} |(\lambda_\nu + i\lambda_{\nu+r_2})|^2 \leqslant 1\right\}.$$

Then there is a number $\alpha \neq 0$ *in* $\mathfrak{m}$ *with*

$$|N_{K/\mathbb{Q}}(\alpha)| \leqslant \frac{2^r}{V(M)}\sqrt{|d(\mathfrak{m})|}. \tag{1.5}$$

Proof. We apply Minkowski's convex body theorem to an M and the lattice $t\phi(\mathfrak{m})$ where $t > 0$ is determined by

$$V(M) = (2t)^n 2^{-r_2}\sqrt{|d(\mathfrak{m})|}.$$

Then the assumptions of the theorem are fulfilled. Hence there is an element $t\phi(\alpha) \neq 0$ in $t\phi(\mathfrak{m})$ with $\alpha \in \mathfrak{m}$. Now $t\phi(\alpha) \in N$ implies (1.5).

It remains to choose M. We put

$$M = \left\{(\lambda_1, \ldots, \lambda_n) \in \mathbb{R}^n \,\middle|\, \sum_{\nu=1}^{r_1} |\lambda_\nu| + 2 \sum_{\nu=r_1+1}^{r} |\lambda_\nu + i\lambda_{\nu+r_2}| \leqslant n\right\}.$$

This is obviously a convex closed set in $\mathbb{R}^n$ which is central symmetric with respect to the origin. Since the geometrical mean is smaller or equal the arithmetical mean, we have $M \subseteq N$. Now Theorem 1.10 follows from the computation of the volume of M:

$$V(M) = \frac{2^{r_1-r_2}\pi^{r_2}n^n}{n!}. \quad \square$$

Example 4. Let $K = \mathbb{Q}(\sqrt{d})$ as in Example 2. Then

$$d(\mathfrak{O}_K) = \begin{cases} d \text{ if } d \equiv 1 \pmod 4 \\ 4d \text{ if } d \equiv 2, 3 \pmod 4). \end{cases}$$

According to (1.4) for the computation of $|CL(\mathfrak{O}_K)|$ we have to consider the complete modules $\mathfrak{m}' \supseteq \mathfrak{O}_K$ with

$$[\mathfrak{m}' : \mathfrak{O}_K] \leqslant \left(\frac{4}{\pi}\right)^{r_2} \frac{1}{2}\sqrt{|d(\mathfrak{O}_K)|}.$$

a) Let $d < 0$ and therefore $r_2 = 1$. Then $\left(\frac{4}{\pi}\right)\frac{1}{2}\sqrt{|d(\mathfrak{O}_K)|} < 2$ if $|d(\mathfrak{O}_K)| < 10$. Hence $|CL(\mathfrak{O}_K)| = 1$ if $d = -1, -2, -3, -7$.

b) Let $d > 0$ and therefore $r_2 = 0$. Then $\frac{1}{2}\sqrt{|d(\mathfrak{O}_K)|} < 2$ if $|d(\mathfrak{O}_K)| < 16$. Hence $|CL(\mathfrak{O}_K)| = 1$ if $d = 2, 3, 5$. $\square$

Applying Theorem 1.10 to $\mathfrak{m} = \mathfrak{O} = \mathfrak{O}_K$ we find

$$1 \leqslant |N_{K/\mathbb{Q}}(\alpha)| \leqslant \left(\frac{4}{\pi}\right)^{r_2} \frac{n!}{n^n}\sqrt{|d(\mathfrak{O}_K)|}.$$

This proves the following *discriminant theorem of Minkowski*:

Theorem 1.11. *Let K be an algebraic number field with* r_1 *real and* $2r_2$ *complex conjugates. Then*

$$|d(\mathfrak{O}_K)| \geqslant \left(\frac{\pi}{4}\right)^{2r_2} \cdot \frac{n^{2n}}{(n!)^2}$$

and in particular $|d(\mathfrak{O}_K)| > 1$ *for* $n > 1$. □

Furthermore by means of Stirling's formula we get the estimation

$$|d(\mathfrak{O}_K)| \geqslant \left(\frac{\pi}{4}\right)^{2r_2} \frac{1}{2\pi n} \exp\left(2n - \frac{1}{6n}\right)$$

for $n > 1$, called *Minkowski's discriminant bound* (see § 6.15 and Chap. 3.2.8 for further results on discriminant estimation).

Minkowski's convex body theorem is also useful for the proof of the following *discriminant theorem of Hermite*:

Theorem 1.12. *Let N be a natural number. There are only finitely many algebraic number fields K with* $|d(\mathfrak{O}_K)| \leqslant N$. □ (Narkiewicz (1974), Theorem 2.11)

1.3. The Unit Group of an Order. Let $\mathfrak{O}$ be an order in the algebraic number field K with r_1 real and $2r_2$ complex conjugates. The structure of the group $\mathfrak{O}^\times$ of invertible element in $\mathfrak{O}$ is given by *Dirichlet's unit theorem*:

Theorem 1.13. $\mathfrak{O}^\times$ *is the direct product of the finite cyclic group of roots of unity in* $\mathfrak{O}$ *and a free abelian group of rank* $r_1 + r_2 - 1$.

The proof of Theorem 1.13 proceeds in several steps.

a) Let $g_1, \ldots, g_n$ be as in the proof of Theorem 1.10 and put $l_\nu = 1$ for $\nu = 1, \ldots, r_1$ and $l_\nu = 2$ for $\nu = r_1 + 1, \ldots, r$. The logarithmical components of $\alpha \in K^\times$ are defined as $l_\nu(\alpha) := l_\nu \log|g_\nu\alpha|$ for $\nu = 1, \ldots, r$. Furthermore we put $l(\alpha) := (l_1(\alpha), \ldots, l_r(\alpha))$. Then l is a homomorphism of $K^\times$ into the additive group of $\mathbb{R}^r$. Moreover $\mathfrak{O}^\times$ is mapped into the subspace U of $\mathbb{R}^r$ given by

$$U = \{(\lambda_1, \ldots, \lambda_r) \in \mathbb{R}^r \mid \lambda_1 + \cdots + \lambda_r = 0\}.$$

b) The kernel of the map $l\colon \mathfrak{O}^\times \to \mathbb{R}^r$ is finite. This follows from

Lemma 1.14. *Let c be a positive real number. Then there are only finitely many* $\alpha \in \mathfrak{O}$ *with* $|g_\nu\alpha| \leqslant c$ *for* $\nu = 1, \ldots, n$.

Proof. Let $\omega_1, \ldots, \omega_n$ be a basis of $\mathfrak{O}$ and $\kappa_1, \ldots, \kappa_n$ the complementary basis (Appendix 1.1). For $\alpha = h_1\omega_1 + \cdots + h_n\omega_n$, $h_\nu \in \mathbb{Z}$, we have

$$h_\nu = \mathrm{tr}_{K/\mathbb{Q}}(\alpha\kappa_\nu) = g_1\alpha g_1\kappa_\nu + \cdots + g_n\alpha g_n\kappa_\nu,$$

$$|h_\nu| \leqslant c(|g_1\kappa_\nu| + \cdots + |g_n\kappa_\nu|) \qquad \text{for } \nu = 1, \ldots, n. \; \square$$

Since the kernel of l is a finite group, it is a group of roots of unity, and since all roots of unity in $\mathfrak{O}$ are mapped by l to 0 the kernel of l is the group of all roots of unity in $\mathfrak{O}$.

c) From Lemma 1.14 it follows also that $l(\mathfrak{O}^\times)$ is a discrete subgroup of U, and since the vector space U has dimension $r - 1$ the abelian group $l(\mathfrak{O}^\times)$ has rank $\leqslant r - 1$.

d) For the proof of Theorem 1.13 it remains to show that the rank of $l(\mathfrak{O}^\times)$ is $\geqslant r-1$. This is the main problem. Dirichlet solved it by means of his famous pigeon hole principle, Minkowski applied his geometry of numbers (see Borevich, Shafarevich (1985), Chap. 2, § 4 for the last approach). We explain here the method of Dirichlet.

The *pigeon hole principle* consists in the remark that if a number of pigeons is distributed in a smaller number of holes, then there is one hole containing at least two pigeons. Minkowski's convex body theorem (§ 1.2) can be considered as a refinement of Dirichlet's pigeon hole principle.

We restrict to the case of a real-quadratic field K, which shows the idea of the proof in the simplest non trivial situation (see also Koch (1986), 19.7, for the proof in the general situation).

Lemma 1.15. *Let g be the non trivial automorphism of K and let ω_1, ω_2 be a basis of $\mathfrak{O}$. We put*

$$u := \max\{|\omega_1| + |\omega_2|, |g\omega_1| + |g\omega_2|.\}$$

Then for an arbitrary positive real number c there is a number $\gamma \in \mathfrak{O}$ such that

$$|\gamma| < c, \qquad |N_{K/\mathbb{Q}}(\gamma)| < 2u^2.$$

Proof. Let k be a natural number and

$$\Omega(k) = \{h_1\omega_1 + h_2\omega_2 \mid h_\nu \in \mathbb{Z}, 0 \leqslant h_\nu \leqslant k \text{ for } \nu = 1, 2\}.$$

By definition of u we have

$$|\delta| \leqslant uk, \qquad |g\delta| \leqslant uk \qquad \text{for } \delta \in \Omega(k).$$

We apply the pigeon hole principle as follows: For $\delta \in \Omega(k)$ we have $-uk \leqslant \delta \leqslant uk$. Put $d := 2uk/(k^2+1)$, hence $d < 2u/k$. We divide the interval $[-uk, uk]$ in subintervals of length d. These subintervals are our pigeon holes. δ belongs to one of the subintervals. If δ lies on the border of two adjacent intervals, we put it into one of both. Since we have $(k+1)^2$ numbers in $\Omega(k)$ and k^2+1 subintervals, there is one subinterval containing two different numbers α, β in $\Omega(k)$. We put $\gamma := \alpha - \beta$. Then

$$|\gamma| = |\alpha - \beta| \leqslant d < 2u/k$$

and

$$|N_{K/\mathbb{Q}}(\gamma)| = |\gamma||g\gamma| < 2u^2.$$

Now we see that we can satisfy the claim of Lemma 1.15 choosing k big enough. □

By means of Lemma 1.15 we construct a unit $\varepsilon \in \mathfrak{O}^\times$ such that $|\varepsilon| < 1$ as follows: There is an infinite sequence $\gamma_1, \gamma_2, \ldots$ of numbers in $\mathfrak{O}$ such that

$$N_{K/\mathbb{Q}}(\gamma_\nu) < 2u^2 \qquad \text{for } \nu = 1, 2, \ldots$$

and

$$|\gamma_1| > |\gamma_2| > \cdots. \tag{1.6}$$

Since the natural numbers $|N(\gamma_\nu)|$ are bounded we may assume without loss of generality that they are equal. But there are only finitely many pairwise non associated numbers $\gamma \in 0$ with fixed $|N(\gamma)|$. This follows from

Lemma 1.16. *Let* $\alpha, \beta \in \mathfrak{O}$ *such that* $|N(\alpha)| = |N(\beta)| =: a$ *and* $\alpha - \beta \in a\mathfrak{O}$. *Then* α *and* β *are associated, i.e.* α/β *is a unit in* $\mathfrak{O}$.

Proof. Since $\alpha - \beta = a\delta$ with $\delta \in \mathfrak{O}$

$$\frac{\alpha}{\beta} = 1 + \frac{|N(\beta)|\delta}{\beta} \in \mathfrak{O}, \qquad \frac{\beta}{\alpha} = 1 - \frac{|N(\alpha)|\delta}{\alpha} \in \mathfrak{O}. \ \square$$

Now we take associated numbers $\gamma_1, \gamma_2 \in \mathfrak{O}$ with (1.6). Then $\varepsilon = \gamma_2\gamma_1^{-1}$ satisfies $|\varepsilon| < 1$. Hence the rank of $l(\mathfrak{O}^\times)$ is $\geqslant 1$. This ends the proof of Theorem 1.13 in the case that K is a real quadratic field. $\square$

We now come back to the general situation of an algebraic number field K of degree n, with r_1 real and $2r_2$ complex conjugates, $r := r_1 + r_2$.

A set $\{\varepsilon_1, \ldots, \varepsilon_{r-1}\}$ of units in $\mathfrak{O}$ is called a *fundamental system of units of* $\mathfrak{O}$ if the vectors $l(\varepsilon_1), \ldots, l(\varepsilon_{r-1})$ generate the abelian group $l(\mathfrak{O}^\times)$. The number

$$R(\varepsilon_1, \ldots, \varepsilon_{r-1}) := |\det(l_\nu(\varepsilon_\mu)_{\nu,\mu=1,\ldots,r-1})|$$

is independent of the choice of the sequence of isomorphisms $g_1, \ldots, g_{r-1}$. Furthermore $R(\varepsilon_1, \ldots, \varepsilon_{r-1})$ has the same value for all fundamental systems of units of $\mathfrak{O}$. It is called the regulator $R(\mathfrak{O})$ of $\mathfrak{O}$. In the case $r = 1$ we put $R(\mathfrak{O}) = 1$. The linear independence of $l(\varepsilon_1), \ldots, l(\varepsilon_{r-1})$ implies $R(\mathfrak{O}) \neq 0$.

1.4. The Unit Group of a Real-Quadratic Number Field. In the case of a real-quadratic field $K = \mathbb{Q}(\sqrt{d})$ one has $r = 2$ and therefore a fundamental system of units consists of one unit. There is a nice method to compute this fundamental unit of an order $\mathfrak{O}_f = \mathbb{Z}[\omega f]$, which will be explained in this section. We keep the notation of Example 2. This method is related to the continued fraction algorithm. We begin with some definitions and facts about this algorithm.

Let $\alpha > 1$ be an irrational real number. We associate to α an infinite sequence $[a_1, a_2, \ldots]$ of natural numbers defined by induction: We put $a_1 := [\alpha]$ (the integral part of α), $\alpha_2 := (\alpha - a_1)^{-1}$ and $a_n := [\alpha_n]$, $\alpha_{n+1} := (\alpha_n - a_n)^{-1}$, $n = 2, 3, \ldots$, $[a_1, a_2, \ldots]$ is called the *continued fraction of* α.

The sequence $[a_1, a_2, \ldots]$ is periodic if and only if $\mathbb{Q}(\alpha)$ is a real-quadratic field and purely periodic if and only if $\alpha > 1$ is *reduced*, i.e. $-1/\alpha' > 1$ where α' denotes the conjugate of α. If $[a_1, a_2, \ldots]$ has the period $a_i, \ldots, a_{i+s}$ we write

$$[a_1, \ldots, a_{i-1}, \overline{a_i, \ldots, a_{i+s}}] := [a_1, a_2, \ldots, a_{i+s}, \ldots].$$

The *n-th approximation of* $[a_1, a_2, \ldots]$ is defined by

$$p_n/q_n = [a_1, a_2, \ldots, a_n] = a_1 + 1/(a_2 + 1/(\ldots + 1/(a_{n-1} + 1/a_n)\ldots))$$

with $(p_n, q_n) = 1$. We put $p_0 = 1$, $q_0 = 0$, $p_{-1} = 0$, $q_{-1} = 1$. It is easy to see that $p_n = a_n p_{n-1} + p_{n-2}$ and $q_n = a_n q_{n-1} + q_{n-2}$ for $n = 1, 2, \ldots$.

Now we are able to formulate the main result of this section:

Theorem 1.17. *Let $\alpha > 1$ be a reduced number in K with $d_{K/\mathbb{Q}}(\alpha) = d(\mathfrak{O}_f)$, e.g. $\alpha := (f\omega - [f\omega])^{-1}$. Let $[\overline{a_1, \ldots, a_k}]$ be the continued fraction of α with smallest possible period. Then $\varepsilon = q_k\alpha + q_{k-1}$ is a fundamental unit of $\mathfrak{O}_f$.* ⊠ (Hasse (1964), § 16.5, Koch, Pieper (1976), Kap. 9.3)

Remark. ε is the unique fundamental unit of $\mathfrak{O}_f$ with $\varepsilon > 1$.

Example 5. $K = \mathbb{Q}(\sqrt{5})$, $f = 1$. Then $\alpha = (1 + \sqrt{5})/2$ is reduced and $\alpha = [\overline{1}]$. Therefore $q_1 = 1$, $\varepsilon = \alpha$. □

Example 6. $K = \mathbb{Q}(\sqrt{19})$, $f = 1$, $\alpha = (\sqrt{19} - 4)^{-1}$. Then $\alpha = [\overline{2, 1, 3, 1, 2, 8}]$. $q_1 = 1, q_2 = 1, q_3 = 4, q_4 = 5, q_5 = 14, q_6 = 117$. Therefore $\varepsilon = 117(\sqrt{19} - 4)^{-1} + 14 = 170 + 39\sqrt{19}$. □

1.5. Integral Representations of Rational Numbers by Complete Forms. From Dirichlet's unit theorem one gets results about certain Tiophantine equations as will be explained in this section. Let $x_1, \ldots, x_n$ be variables and $F(x_1, \ldots, x_n)$ a form in $x_1, \ldots, x_n$ of degree n with rational coefficients. Then F is called *complete* if there are an algebraic number field K of degree n, a complete module $\mathfrak{m} = (\mu_1, \ldots, \mu_n)$ in K, and a rational number $a \neq 0$ such that

$$F(x_1, \ldots, x_n) = aN_{K/\mathbb{Q}}(x_1\mu_1 + \cdots + x_n\mu_n). \tag{1.7}$$

If $\mu_1', \ldots, \mu_n'$ is another basis of $\mathfrak{m}$, then

$$(\mu_1, \ldots, \mu_n)^T = A(\mu_1', \ldots, \mu_n')^T \text{ with } A \in GL_n(\mathbb{Z}). \text{ We put}$$

$$(x_1', \ldots, x_n') = (x_1, \ldots, x_n)A.$$

Then

$$G(x_1', \ldots, x_n') := aN_{K/\mathbb{Q}}(x_1'\mu_1' + \cdots + x_n'\mu_n') = F(x_1, \ldots, x_n).$$

The forms F and G are called *equivalent*.

Let b be a rational number. We are interested in this section in solutions of the Diophantine equation

$$F(x_1, \ldots, x_n) = b. \tag{1.8}$$

in rational integers $b_1, \ldots, b_n$. By means of (1.7) this is equivalent to the solution of the equation $N_{K/\mathbb{Q}}(\mu) = b/a$ with $\mu = b_1\mu_1 + \cdots + b_n\mu_n \in \mathfrak{m}$.

Some insight in the solutions of (1.8) is given by the following proposition in connection with Dirichlet's unit theorem.

Proposition 1.18. *There are finitely many numbers $\alpha_1, \ldots, \alpha_k \in \mathfrak{m}$ such that every $\mu \in \mathfrak{m}$ with $N_{K/\mathbb{Q}}(\mu) = b/a$ has a unique representation in the form $\mu = \alpha_\nu\varepsilon$ with $\nu = 1, \ldots, k$, $\varepsilon \in \mathfrak{O}(\mathfrak{m})^\times$, $N_{K/\mathbb{Q}}(\varepsilon) = 1$.*

For the proof see Lemma 1.16. □

Remark. $E_0 = \{\varepsilon \in \mathfrak{O}^\times \mid N_{K/\mathbb{Q}}(\varepsilon) = 1\}$ is a subgroup of $\mathfrak{O}^\times$ of index 1 or 2. This index is called the unit index. If K has no real conjugate then of course every unit has norm 1. In general no simple criterion is known for the determination of the unit index. □

Example 7. Let $K = \mathbb{Q}(\sqrt{d})$ with d squarefree, $d > 0$. If d is divisible by a prime $p \equiv 3 \pmod 4$, then $[\mathfrak{O}^\times : E_0] = 1$ since $N(\varepsilon) = -1$ implies the existence of $x, y \in \mathbb{Z}$ with $x^2 - dy^2 = -4$, but

$$\left(\frac{x^2 - dy^2}{p}\right) = \left(\frac{x^2}{p}\right) = 1, \qquad \left(\frac{-4}{p}\right) = -1.$$

where $\left(\frac{a}{p}\right)$ denotes the Legendre symbol. On the other hand $[\mathfrak{O}^\times : E_0] = 2$ for $d = 5$ (Example 5). □

With the notation above let $a = b = 1$ and let $\mathfrak{m} = \mathfrak{O}$. Then Proposition 1.18 gives a one to one correspondence between the solutions of (1.8) and the units of $\mathfrak{O}$ with norm 1.

Example 8. Let D be a natural number which is not a square. The equation

$$x^2 - Dy^2 = 1$$

is called Pell's equation. Its solutions correspond to the units with norm 1 of the order $(1, \sqrt{D})$ in the field $\mathbb{Q}(\sqrt{D})$. □

Example 9. Let a be a rational integer which is not a cube. The equation

$$x^3 + ay^3 + a^2z^3 - 3axyz = 1 \tag{1.9}$$

can be written in the form

$$N_{K/\mathbb{Q}}(x + \sqrt[3]{a}y + \sqrt[3]{a^2}z) = 1$$

with $K := \mathbb{Q}(\sqrt[3]{a})$. Hence the solutions of (1.9) correspond to the units with norm 1 of the order $(1, \sqrt[3]{a}, \sqrt[3]{a^2})$ in K. □

1.6. Binary Quadratic Forms and Complete Modules in Quadratic Number Fields. Our knowledge about *complete binary quadratic forms* F, i.e.

$$F(x, y) = ax^2 + bxy + cy^2, \qquad a, b, c \in \mathbb{Z},$$

where $d(F) := b^2 - 4ac$ is not a square number, is much richer than our knowledge about forms of higher degree. In fact such forms are a main topic of the number theoretical investigations of Euler, Lagrange, and Gauss (Weil (1983)).

Let $\mathfrak{O}$ be an order in the quadratic number field $K = \mathbb{Q}(\sqrt{d})$. The nontrivial automorphism of K will be denoted by g. First of all one has the following theorem about the set $\mathfrak{M}(\mathfrak{O})$ of complete modules with order $\mathfrak{O}$.

Theorem 1.19. *$\mathfrak{M}(\mathfrak{O})$ is a group with respect to the multiplication of modules. The unit element of this group is $\mathfrak{O}$ and the inverse of $\mathfrak{m} \in \mathfrak{M}(\mathfrak{O})$ is $N(\mathfrak{m})^{-1} g\mathfrak{m}$*

where $g\mathfrak{m} = \{g\mu | \mu \in \mathfrak{m}\}$. *Furthermore*

$$N(\mathfrak{m}_1\mathfrak{m}_2) = N(\mathfrak{m}_1)N(\mathfrak{m}_2) \qquad \text{for } \mathfrak{m}_1, \mathfrak{m}_2 \in \mathfrak{M}(\mathfrak{O}).$$

Proof. First of all one shows $\mathfrak{m}g\mathfrak{m} = N(\mathfrak{m})\mathfrak{O}$. Furthermore if $\mathfrak{m}_1, \mathfrak{m}_2 \in \mathfrak{M}(\mathfrak{O})$, then

$$(\mathfrak{m}_1\mathfrak{m}_2)g(\mathfrak{m}_1\mathfrak{m}_2) = \mathfrak{m}_1 g\mathfrak{m}_1\mathfrak{m}_2 g\mathfrak{m}_2 = N(\mathfrak{m}_1)N(\mathfrak{m}_2)\mathfrak{O}. \tag{1.10}$$

If $\mathfrak{O}_1$ is the order of $\mathfrak{m}_1\mathfrak{m}_2$, then

$$(\mathfrak{m}_1\mathfrak{m}_2)g(\mathfrak{m}_1\mathfrak{m}_2) = N(\mathfrak{m}_1\mathfrak{m}_2)\mathfrak{O}_1. \tag{1.11}$$

(1.10) and (1.11) imply $\mathfrak{O} = \mathfrak{O}_1$ and $N(\mathfrak{m}_1)N(\mathfrak{m}_2) = N(\mathfrak{m}_1\mathfrak{m}_2)$. □

In general, Theorem 1.19 is not true for orders in fields of degree $\geqslant 3$. See Dade, Taussky, Zassenhaus (1962) for a study of the general situation.

The *principal modules* $\alpha\mathfrak{O}$ for $\alpha \in K^*$ form a subgroup $\mathfrak{H}(\mathfrak{O})$ of $\mathfrak{M}(\mathfrak{O})$, and $CL(\mathfrak{O})$ can be identified with the factor group $\mathfrak{M}(\mathfrak{O})/\mathfrak{H}(\mathfrak{O})$.

We want to connect the set of classes of binary quadratic forms with $CL(\mathfrak{O})$. We have already associated with a module $\mathfrak{m}$ an equivalence class of complete forms. We want to get a one to one correspondence between classes of modules and classes of forms. For this purpose we have to change slightly the notions of equivalence of modules and forms:

Instead of $\mathfrak{H}(\mathfrak{O})$ we take the subgroup $\mathfrak{H}_0(\mathfrak{O})$ of principal modules $\alpha\mathfrak{O}$ with $\alpha \in K^\times$ which are totally positive. $CL_0(\mathfrak{O}) := \mathfrak{M}(\mathfrak{O})/\mathfrak{H}_0(\mathfrak{O})$ is called the *group of classes in the narrow sense.* If $d < 0$, then $\mathfrak{H}_0(\mathfrak{O}) = \mathfrak{H}(\mathfrak{O})$. If $d > 0$, then $[\mathfrak{H}(\mathfrak{O}) : \mathfrak{H}_0(\mathfrak{O})] | 2$ and $\mathfrak{H}_0(\mathfrak{O}) = \mathfrak{H}(\mathfrak{O})$ if and only if there is a unit $\varepsilon \in \mathfrak{O}^\times$ with $N(\varepsilon) = -1$.

On the other hand we consider *primitive forms* $F(x, y) = ax^2 + bxy + cy^2$, i.e. $(a, b, c) = 1$, with *discriminant* $D(F) = D(\mathfrak{O})$. Two such forms F_1, F_2 are called *properly equivalent* if there is a matrix $A \in SL_2(\mathbb{Z})$ with

$$F_2(x, y) = F_1((x, y)A).$$

A basis μ_1, μ_2 of a module $\mathfrak{m} \in \mathfrak{M}(\mathfrak{O})$ is called admissible if $\det\begin{pmatrix} \mu_1 \mu_2 \\ g\mu_1 g\mu_2 \end{pmatrix} > 0$ for $d > 0$, $i \det\begin{pmatrix} \mu_1 \mu_2 \\ g\mu_1 g\mu_2 \end{pmatrix} < 0$ for $d < 0$.

One has the following fundamental correspondence.

Theorem 1.20. *Let* $\mathfrak{m} \in \mathfrak{M}(\mathfrak{O})$ *and* μ_1, μ_2 *an admissible basis of* $\mathfrak{m}$. *Then we associate to* $\mathfrak{m}$ *the class* $\phi(\mathfrak{m})$ *of properly equivalent forms which contains* $N(\mu_1 x + \mu_2 y)/N(\mathfrak{m})$.

ϕ *induces a one to one correspondence* $\overline{\psi}$ *between the classes in* $CL_0(\mathfrak{O})$ *and the classes of properly equivalent primitive binary quadratic forms with discriminant* $D(\mathfrak{O})$ *which are positive definite if* $d < 0$. □

We can transfer the group structure of $CL_0(\mathfrak{O})$ to the set of form classes appearing in the theorem. The resulting product of form classes is Gauss' famous *composition of form classes.*

1.7. Representatives for Module Classes in Quadratic Number Fields. Let $\mathfrak{O}$ be an order in the field $K = \mathbb{Q}(\sqrt{d})$. In this section we determine representatives of $CL(\mathfrak{O})$ in $\mathfrak{M}(\mathfrak{O})$.

We begin with the case $d < 0$. A number $\alpha \in K$ is called *reduced* if the following conditions are fulfilled:

$$\operatorname{Im} \alpha > 0, \qquad -\tfrac{1}{2} < \operatorname{Re} \alpha \leqslant \tfrac{1}{2}, \qquad |\alpha| > 1 \text{ if } -\tfrac{1}{2} < \operatorname{Re} \alpha < 0,$$
$$|\alpha| \geqslant 1 \text{ if } 0 \leqslant \operatorname{Re} \alpha \leqslant \tfrac{1}{2}.$$

Theorem 1.21. *Let $\mathfrak{O}$ be an order in an imaginary-quadratic number field. Every class of $CL(\mathfrak{O})$ contains one and only one module with basis* 1, α *where α is a reduced number.* □

$M = (1, \alpha)$ has order $\mathfrak{O}$ if and only if $a\alpha^2 + b\alpha + c = 0$ with $a, b, c \in \mathbb{Z}$, $a > 0$, $(a, b, c) = 1$ and $D(\mathfrak{O}) = b^2 - 4ac$. $\alpha = (-b + i\sqrt{|D(\mathfrak{O})|})/2a$ is reduced if and only if $-a \leqslant b < a$, $c \geqslant a$ if $b \leqslant 0$, $c > a$ if $b > 0$. Hence the classes in $CL(\mathfrak{O})$ correspond to the triples a, b, c of integers such that $D(\mathfrak{O}) = b^2 - 4ac$, $a > 0$, $(a, d, c) = 1$, $-a \leqslant b < a$, and $c \geqslant a$ if $b \leqslant 0$, $c > a$ if $b > 0$.

Example 10. $\mathfrak{O} = \mathbb{Z}[(1 + \sqrt{-19})/2]$. Then $|b| \leqslant a < \sqrt{|D(\mathfrak{O})|/3} < 3$. There is only one possible triple: $1, -1, 5$. □

Example 11. $\mathfrak{O} = \mathbb{Z}[(1 + \sqrt{-47})/2]$. Then there are five possible triples: $3, \pm 1, 4$; $2, \pm 1, 6$; $1, -1, 12$. □

Now let $d > 0$. We keep the notation of § 1.4.

Theorem 1.22. *Let $\alpha = \alpha_1$, β be reduced numbers of a real-quadratic number field. Then the modules $(1, \alpha)$ and $(1, \beta)$ belong to the same order $\mathfrak{O}$ and the same class in $CL(\mathfrak{O})$ if and only if $\beta = \alpha_\nu$ for some $\nu = 1, 2, \ldots$.* ⊠

Every class in $CL(\mathfrak{O})$ contains a module $(1, \alpha)$ where α is a reduced number.

As in the case $d < 0$ a module $\mathfrak{m} = (1, \alpha)$ has order $\mathfrak{O}$ if and only if $a\alpha^2 + b\alpha + c = 0$ with $a, b, c \in \mathbb{Z}$, $a > 0$, $(a, b, c) = 1$ and $D(\mathfrak{O}) = b^2 - 4ac$. $\alpha = (-b + D(\mathfrak{O}))/2a$ is reduced if and only if $-b + \sqrt{D(\mathfrak{O})} > 2a > b + \sqrt{D(\mathfrak{O})} > 0$.

Example 12. $\mathfrak{O} = \mathbb{Z}[\sqrt{6}]$. Then there are two reduced numbers $\alpha = (\sqrt{6} + 2)/2$ and $\beta = \sqrt{6} + 2$, but $\beta = \alpha_2$ and therefore $|CL(\mathfrak{O})| = 1$. □

The reduction theory as explained above contains a new proof for the finiteness of the number of module classes with given order in a quadratic field.

In the case of imaginary-quadratic fields two modules $(1, \alpha)$ and $(1, \beta)$ are equivalent if and only if

$$\beta = \frac{k\alpha + l}{m\alpha + n} \tag{1.12}$$

with $k, l, m, n \in \mathbb{Z}$ and $kn - lm = \pm 1$. More generally we call two complex numbers α and β equivalent if (1.12) is satisfied for some $k, l, m, n \in \mathbb{Z}$ with $kn - lm = \pm 1$. There is one and only one reduced number in every equivalence class of complex, non-real numbers. This equivalence of complex numbers plays an important role in the theory of elliptic functions and modular forms (see Serre (1970), Chap. 7).

§ 2. Rings with Divisor Theory

(Main reference: Borevich, Shafarevich (1985), Chap. 3)

Let K be an algebraic number field and $\mathfrak{O}$ an order in K. In this paragraph we consider the problem of generalization of the unique factorization of natural numbers in products of prime numbers. We will see that this is possible in a satisfying manner if and only if $\mathfrak{O}$ is the maximal order $\mathfrak{O}_K$ of K.

2.1. Unique Factorization in Prime Elements. A domain R is a commutative ring with unit element which has no zero divisors. The concept of unique factorization in prime numbers can be formulated for a domain R as follows:

An element $\alpha \neq 0$ of R is called a *prime element* of R if α is not a unit and any divisor β of α in R is a unit or β is *associated* to α i.e. α/β is a unit. R is called a *ring with unique factorization in prime elements* if every $\alpha \in R - R^\times$, $\alpha \neq 0$, can be written as a product of prime elements $\pi_1, \ldots, \pi_u$ and if $\alpha = \pi'_1, \ldots, \pi'_t$ in another representation of α as product of prime elements, then $t = u$ and there is a permutation $\varphi \in S_u$ such that π_i is associated to $\pi'_{\varphi(i)}$ for $i = 1, \ldots, u$.

A ring with unique factorization in prime elements is denoted by UFD (unique factorization domain) for short.

In a more abstract manner the condition for a domain R to be a UFD can be formulated as follows. Let $\mathfrak{D}$ be the factor semigroup of the multiplicative semigroup R^* of elements in R distinct from 0 by the group $R^\times$ of units of R. Then R is a UFD if and only if $\mathfrak{D}$ is a free abelian semigroup.

One proves the unique factorization in prime numbers in $\mathbb{Z}$ by means of the Euclidean algorithm. In general a domain R is called *Euclidean* if there exists a map h, called the *height*, from R in the set of non-negative rational integers such that the following conditions are fulfilled:

1) For all $\alpha, \beta \in R$, $\beta \neq 0$, there exists a $\gamma \in R$ with $\alpha = \beta\gamma$ or $h(\alpha - \beta\gamma) < h(\beta)$.
2) $h(\alpha) = 0$ if and only if $\alpha = 0$.

A ring R is called *principal ideal ring* (PIR) if every ideal $\mathfrak{a}$ of R is principal, i.e. $\mathfrak{a} = \alpha R$ for some $\alpha \in R$.

Proposition 1.23. *An Euclidean ring is a principal ideal ring. A principal ideal domain is a* UFD. □

An algebraic number field K is called *Euclidean* if the maximal order $\mathfrak{O}_K$ is Euclidean with height $h(\alpha) := N_{K/\mathbb{Q}}(\alpha)$ for $\alpha \in \mathfrak{O}_K$.

Example 13. An imaginary-quadratic number field K is Euclidean if and only if $D(\mathfrak{O}_K) = -3, -4, -7, -8, -11$. □

Example 14. A real-quadratic number field K is Euclidean if and only if $D(\mathfrak{O}_K) = 5, 8, 12, 13, 17, 21, 24, 28, 29, 33, 37, 41, 44, 57, 73, 76$ (Chatland, Davenport (1950)). ⊠

Example 15. Let p be a prime with $p \leqslant 13$ and let ζ_p be a primitive p-th root of unity. Then $\mathbb{Q}(\zeta_p + \zeta_p^{-1})$ is Euclidean (Lenstra (1977)). ⊠

Example 13 shows together with Example 10 that not every UFD is an Euclidean ring, see also Chap. 5.5.3–4.

If the maximal order $\mathfrak{O}_K$ of an algebraic number field K is a PIR, one gets particularly simple results about the solution of the Tiophantine equations connected with the normform of a basis $\omega_1, \ldots, \omega_n$ of $\mathfrak{O}_K$:

$$f(x_1, \ldots, x_n) := N_{K/Q}(\omega_1 x_1 + \cdots + \omega_n x_n) = a, \qquad a \in \mathbb{Z},$$

(compare § 1.5).

Example 16. Let $K = \mathbb{Q}(\sqrt{-1})$, $\omega_1 = 1$, $\omega_2 = \sqrt{-1}$. Then

$$N_{K/\mathbb{Q}}(x + \sqrt{-1}y) = x^2 + y^2.$$

Each prime element of $\mathfrak{O}_K$ divides a natural prime p. If $p = 2$, then $2 = \sqrt{-1}(1 - \sqrt{-1})^2$, if $p \equiv 1 \pmod 4$, then $p = (a + \sqrt{-1}b)(a - \sqrt{-1}b)$, $a, b \in \mathbb{N}$, where $1 - \sqrt{-1}$, $a + \sqrt{-1}b$ and $a - \sqrt{-1}b$ are prime elements in $\mathfrak{O}_K$. If $p \equiv 3 \pmod 4$, then p is a prime element in $\mathfrak{O}_K$. This implies the following theorem of Euler: A prime p can be represented as the sum of two squares of natural numbers if and only if $p = 2$ or $p \equiv 1 \pmod 4$. There is only one such representation. □

Below we will show that among orders of algebraic number fields only maximal orders can be UFD. The following example shows that not every maximal order is a UFD.

Example 17. In the ring $\mathbb{Z}[\sqrt{-5}]$ the number 6 has the representations $2 \cdot 3$ and $(1 + \sqrt{-5})(1 - \sqrt{-5})$ as product of prime elements which are all non-associated. □

2.2. The Concept of a Domain with Divisor Theory. Example 17 shows that the concept of unique factorization in prime elements is not useful for a generalization of the arithmetic in $\mathbb{Z}$. Kummer found the way out of this situation in the case of the rings $\mathbb{Z}[\zeta]$, where ζ is a root of unity of prime order, introducing the concept of ideal numbers, which one can formulate for arbitrary domains R in the following way:

A *divisor theory for a domain* R is given by a free abelian semigroup $\mathfrak{D}$ and a homomorphism ( ) of the semigroup $R^* := R - \{0\}$ into $\mathfrak{D}$ with the following properties (called axioms in the following):

1) $\alpha \in R^*$ divides $\beta \in R^*$ if and only if (α) divides (β). We say that $\mathfrak{a} \in \mathfrak{D}$ divides $\alpha \in R$ if $\alpha = 0$ or $\mathfrak{a}$ is a divisor of (α), notation: $\mathfrak{a} | \alpha$.
2) If $\mathfrak{a} \in \mathfrak{D}$ divides $\alpha \in R$ and $\beta \in R$, then $\mathfrak{a}$ divides $\alpha \pm \beta$.

3) Let $\mathfrak{a}, \mathfrak{b} \in \mathfrak{D}$ such that

$$\{\alpha \in R \mid \mathfrak{a}|\alpha\} = \{\beta \in R \mid \mathfrak{b}|\beta\},$$

then $\mathfrak{a} = \mathfrak{b}$.

The concept of a domain with divisor theory was first formulated in Borevich, Shafarevich (1985), Chap. 3, § 3. Skula (1970) showed that 2) follows from 1) and 3). It is easy to see that there exists a divisor theory for a domain R if and only if R is a Krull domain (Bourbaki (1965), Močkoř (1983), Chap. 10).

The elements of $\mathfrak{D}$ are called divisors of R, and the prime elements of $\mathfrak{D}$ are called *prime divisors.* The quotient group of $\mathfrak{D}$ will be denoted by $Q(\mathfrak{D})$. The *principal divisors* of R are the elements of $\mathfrak{D}$ which are in the image of ( ). The *greatest common divisor* (g.c.d.) and the *least common multiple* (l.c.m.) of sets of divisors in $\mathfrak{D}$ are defined in the obvious manner.

Let $\mathfrak{a}$ be a divisor in $\mathfrak{D}$. The set

$$\tilde{\mathfrak{a}} := \{\alpha \in R \mid \mathfrak{a}|\alpha\}$$

is not empty. Moreover one can find an $\alpha \in R$ with $a|\alpha$ such that $(\alpha)/\mathfrak{a}$ is prime to a given divisor $\mathfrak{b}$:

$$\text{g.c.d.}((\alpha)/\mathfrak{a}, \mathfrak{b}) = (1). \tag{1.13}$$

α can be constructed as follows: Let $\mathfrak{p}_1, \ldots, \mathfrak{p}_s$ be the prime divisors of $\mathfrak{b}$. By axiom 3) there is an $\alpha_i \in R$ such that

$$\mathfrak{a}\mathfrak{p}_1 \ldots \mathfrak{p}_s/\mathfrak{p}_i|\alpha_i, \qquad \mathfrak{a}\mathfrak{p}_1 \ldots \mathfrak{p}_s \nmid \alpha_i, \qquad i = 1, \ldots, s.$$

$\alpha := \alpha_1 + \cdots + \alpha_s$ has the property (1.13) by axiom 2).

If there exists a divisor theory for a domain R then it is uniquely determined by R:

Proposition 1.24. *Let $R^* \to \mathfrak{D}$ and $R^* \to \mathfrak{D}'$ be two divisor theories for R. Then there is a unique isomorphism $\mathfrak{D} \to \mathfrak{D}'$ such that the following diagram is commutative:*

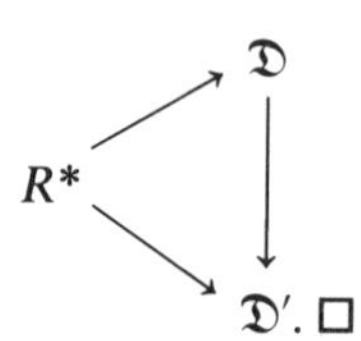

Proposition 1.25. *R is a* UFD *if and only if R has a divisor theory such that all divisors are principal divisors.* □

Proposition 1.26. *If there exists a divisor theory for a domain R, then R is integrally closed* (Appendix 1.3).

Proof. Let ξ be an element of the quotient field of R such that $\xi \notin R$ and that there are elements $a_1, \ldots, a_s$ in R with

$$\xi^s + a_1\xi^{s-1} + \cdots + a_s = 0.$$

Let $\xi = a/b$ with $a, b \in R$. Then $(b) \nmid (a)$. Hence there is a prime divisor $\mathfrak{p}$ in $\mathfrak{D}$ and an integer k such that $\mathfrak{p}^{k+1} | (b)$, $\mathfrak{p}^{k+1} \nmid (a)$. The right side of the equation

$$a^s = -a_1 b a^{s-1} - \cdots - a_s b^s$$

is divisible by $\mathfrak{p}^{ks+1}$ but $\mathfrak{p}^{ks+1} \nmid a^s$. □

In view of Prop. 1.24 it would be desirable to have a canonical realization of the divisor theory of R. This can be done in the following way: Let z be a variable. For $\alpha \in R[z]^*$ we define the content $I(\alpha) \in \mathfrak{D}$ as the g.c.d. of the coefficients of α. The proof of the following *Lemma of Gauss* is easy. One needs only the axioms 1) and 2).

Lemma 1.27. $I(\alpha\beta) = I(\alpha)I(\beta)$ *for* $\alpha, \beta \in R[z]^*$. □

Let K be the quotient field of R. By means of Lemma 1.27 we extend I to $K(z)^*$ putting

$$I(\gamma) := I(\alpha)/I(\beta) \in Q(\mathfrak{D}) \text{ for } \gamma \in K(z)^*, \gamma = \alpha/\beta, \alpha, \beta \in R[z]^*.$$

Now we put

$$R(z) := \{\alpha \in K(z)^* | I(\alpha) \in \mathfrak{D}\} \cup \{0\}.$$

By Prop. 1.24 the definition of $R(z)$ is independent of the choice of the divisor theory of R.

Proposition 1.28. $R(z)$ *is a* UFD. *The elements* α *of* $R(z)$ *have the form*

$$\alpha = \beta/\gamma \text{ with } \beta, \gamma \in R[z], \qquad I(\gamma) = (1).$$

Proof. It is easy to see that $R(z)$ is a ring with divisor theory defined by I. Since I maps *onto* the divisor semigroup $\mathfrak{D}$, it follows from Prop. 1.25 that $R(z)$ is a UFD. Now let $\alpha = \beta'/\gamma' \in R(z)$ with $\beta', \gamma' \in R[z]$ and $I(\gamma') \neq (1)$. We take an $\mathfrak{a} \in \mathfrak{D}$ such that $I(\gamma')\mathfrak{a} = (c)$, $c \in R$ and a $\delta \in R[z]$ such that $I(\delta) = \mathfrak{a}$. Then $\beta := \beta'\delta c^{-1}$, $\gamma := \gamma'\delta c^{-1} \in R[z]$, $I(\gamma) = (1)$ and $\alpha = \beta/\gamma$. □

It follows from Prop. 1.28 that the natural map $R^* \to R(z)^*/R(z)^\times$ defines a divisor theory for R. It is called *Kronecker's divisor theory.*

2.3. Divisor Theory for the Maximal Order of an Algebraic Number Field. Proposition 1.26 shows that with respect to orders in algebraic number fields K one can hope to have a divisor theory only if the order is integrally closed, i.e. if it is the maximal order $\mathfrak{O}_K$ of K. Dedekind (1894), Kronecker (1882), and Zolotarev (1880) proved first that in fact $\mathfrak{O}_K$ has a divisor theory. The ideal theoretical method of Dedekind will be explained in the next paragraph. Zolotarev's approach (see §3.3) represents the beginning of local algebra, which was further developed by Hensel's valuation theory, which is the subject to §4. The next theorem, which includes the existence of a divisor theory for $\mathfrak{O}_K$, will be proved by Kronecker's method of adjunction of variables.

Theorem 1.29. *Let R be a domain with divisor theory, let K be the quotient field of R and S the integral closure of R in a finite extension L of K. Then S is a domain with divisor theory.*

Proof. (Eichler (1963), Chap. 2, § 1). Let $R^* \to \mathfrak{D}$ be the divisor theory of R, z a variable, $R(z)$ the UFD defined in § 2.2, and $S(z)$ the integral closure of $R(z)$ in $L(z)$. In the following we show, that $S(z)$ is a principal ideal ring and that the natural map $S^* \to S(z)^*/S(z)^\times$ defines a divisor theory for S. It follows that this divisor theory is Kronecker's divisor theory in the sense defined in § 2.2.

We divide the proof of Theorem 1.23 into three lemmas.

Lemma 1.30. *The elements α of $S(z)$ have the form $\alpha = \beta/c$ with $\beta \in S[z]$, $c \in R[z]$, $I(c) = (1)$.*

Proof. α satisfies an equation

$$\alpha^s + a_1\alpha^{s-1} + \cdots + a_s = 0$$

with $a_i \in R(z)$. By Prop. 1.28 the coefficients a_i have the form b_i/c with b_i, $c \in R[z]$, $I(c) = (1)$, $i = 1, \ldots, s$. It follows that $c\alpha$ satisfies an equation

$$(c\alpha)^s + a_1'(c\alpha)^{s-1} + \cdots + a_s' = 0$$

with $a_i' \in R[z]$. Then $c\alpha \in S[z]$ since $S[z]$ is the integral closure of $R[z]$ in $L(z)$ (Appendix 1.3). □

Lemma 1.31. *$S(z)$ is a principal ideal ring.*

Proof. First let $B = (\beta_1, \ldots, \beta_s)$ be a finitely generated ideal of $S(z)$. We are going to show that B is generated by the linear combination $\beta_1 x_1 + \cdots + \beta_s x_s$, where the x_i, $i = 1, \ldots, s$, are certain powers of z which will be fixed below. We consider the characteristic polynomials

$$f_k(\xi) = N_{L/K}\left(\xi - \frac{\beta_k}{\sum_{\nu=1}^s \beta_\nu x_\nu}\right) = \frac{N_{L/K}(\sum_{\nu=1}^s \beta_\nu(x_\nu\xi - \delta_{\nu k}))}{N_{L/K}(\sum_{\nu=1}^s \beta_\nu x_\nu)}$$

of $\dfrac{\beta_k}{\sum_{\nu=1}^s \beta_\nu x_\nu}$, $k = 1, \ldots, s$, where $N_{L/K}$ means the norm from $L(z, \xi)$ to $K(z, \xi)$. We have to show that the coefficients of f_k lie in $R(z)$. Let

$$N_{L/K}\left(\sum_{\nu=1}^s \beta_\nu x_\nu\right) =: \sum \alpha^{-1}\alpha_{i_1\ldots i_s} x_1^{i_1} \ldots x_s^{i_s} \text{ with } \alpha, \alpha_{i_1\ldots i_s} \in R[z]$$

where the sum runs over all s-tuples $i_1, \ldots, i_s$ with

$$\sum_{\nu=1}^s i_\nu = [L:K] =: n. \tag{1.14}$$

We put $y_\nu := x_\nu\xi - \delta_{\nu k}$. Then

$$f_k(\xi) = \frac{\sum a_{i_1\ldots i_s} y_1^{i_1} \ldots y_s^{i_s}}{\sum a_{i_1\ldots i_s} x_1^{i_1} \ldots x_s^{i_s}}.$$

Now we fix the z-powers x_i such that

$$I(\sum \alpha_{i_1 \ldots i_s} x_1^{i_1} \ldots x_s^{i_s}) = \text{g.c.d.}(I(\alpha_{i_1 \ldots i_s})): \qquad (1.15)$$

Let $t - 1$ be the maximum of the degrees of the coefficients $a_{i_1 \ldots i_s}$ as polynomials in z. We put $x_i = z^{t(n+1)^i}$. Then the z-powers $x_1^{i_1} \ldots x_s^{i_s}$ are distinct for all s-tuples $i_1, \ldots, i_s$ with (1.14) and are powers of z^t. This implies (1.15). It follows that $f_k(\xi)$ is a polynomial with coefficients in $R(z)$ and highest coefficient 1. It follows that the root $\beta_k / \sum_{v=1}^{s} \beta_v x_v$ of $f_k(\xi)$ belongs to $S(z)$, i.e. $B = (\sum_{v=1}^{s} \beta_v x_v)$.

Now let B be an arbitrary ideal of $S(z)$ and let $\beta_1, \beta_2, \ldots$ be elements of B such that

$$(\beta_1) \subsetneqq (\beta_1, \beta_2) \subsetneqq \cdots \subsetneqq (\beta_1, \beta_2, \ldots, \beta_s) \subsetneqq \ldots. \qquad (1.16)$$

We know already that the ideals in this chain are principal. Let $(\beta_1, \beta_2, \ldots, \beta_s) = (\gamma_s)$, $s = 1, 2, \ldots$. Then $N_{L/K}\gamma_{i+1}$ is a proper divisor of $N_{L/K}\gamma_i$ for $i = 1, 2, \ldots$. It follows that the chain can not be infinite. □

Remark. Kronecker adjoints infinitely many variables. This simplifies the proof of Lemma 1.31 since then we can choose $x_1, \ldots, x_s$ as variables. On the other hand a function field in one variable appears to the mathematicians of our time as a nicer object and it is a simpler object than a function field in infinitely many variables. □

It remains to prove that the map $S^* \to S(z)^*/S(z)^\times$ satisfies the axioms of a divisor theory. Axioms 1) and 2) are easily verified. By means of Lemma 1.30 axiom 3) is equivalent to the following

Lemma 1.32. $(\alpha) = I(\alpha)$ *for all* $\alpha \in S[z]$.

Proof. Since $S[z]$ is the integral closure of $R[z]$ in $L(z)$, we have $N_{L/K}\alpha = \alpha\beta$ with $\beta \in S[z]$ (Appendix 1.3). Furthermore since the natural map $R(z)^*/R(z)^\times \to S(z)^*/S(z)^\times$ of the divisor semigroup of R in the divisor semigroup of S is an injection, we have $I(N_{L/K}\alpha) = (N_{L/K}\alpha)$ hence $I(\alpha)I(\beta) = (\alpha)(\beta)$. This implies $(\alpha) = I(\alpha)$ since $I(\alpha)|\alpha$ and $I(\beta)|\beta$. □

§ 3. Dedekind Rings

(Main reference: Borevich, Shafarevich (1985), Chap. 3)

Let K be an algebraic number field and $\mathfrak{O}_K$ the maximal order of K. Dedekind (1894) showed that the ideals $\mathfrak{a} \neq 0$ of $\mathfrak{O}_K$ form a free abelian semigroup under multiplication. This leads to a realization of a divisor theory for $\mathfrak{O}_K$ which is perhaps the most natural one. More general we consider in this paragraph Dedekind rings and explain their basic arithmetical properties.

The *product* $\mathfrak{ab}$ of ideals $\mathfrak{a}$, $\mathfrak{b}$ in an arbitrary ring R is defined as the ideal of R generated by the products $\alpha\beta$, $\alpha \in \mathfrak{a}$, $\beta \in \mathfrak{b}$. The ideals $\neq \{0\}$ of a domain form a semigroup under this multiplication. In the following an ideal of a domain always means an ideal $\neq \{0\}$.

3.1. Definition of Dedekind Rings. Let R be a ring with divisor theory $R^* \to \mathfrak{D}$. To every divisor $\mathfrak{a} \in \mathfrak{D}$ there corresponds the ideal

$$\tilde{\mathfrak{a}} = \{\alpha \in R \mid \mathfrak{a} | (\alpha)\}$$

of R. If $\mathfrak{p}$ is a prime divisor, then $\tilde{\mathfrak{p}}$ is a prime ideal. In general $\tilde{\mathfrak{p}}$ is not a maximal ideal of R:

Example 18. Let F be an arbitrary field and $R = F[x_1, x_2]$ with variables x_1, x_2. Lemma 1.27 implies that $F[x_1, x_2]$ is a UFD. But the ideal $x_1 R$ is a prime ideal which is not maximal. □

A domain R with divisor theory is called a *Dedekind ring* if all prime ideals of R are maximal.

Theorem 1.33. *Let R be a Dedekind ring and $R^* \to \mathfrak{D}$ a divisor theory of R. Then the correspondence $\mathfrak{a} \to \tilde{\mathfrak{a}}$ defines an isomorphism between the semigroup $\mathfrak{D}$ and the semigroup of ideals of R.* □

Theorem 1.33 implies that every ideal of a Dedekind ring can be uniquely represented as a product of prime ideals.

Theorem 1.34. *Let K be an algebraic number field. Then $\mathfrak{O}_K$ is a Dedekind ring.*

Proof. Taking in account Theorem 1.29, one has only to prove that every prime ideal $\mathfrak{P}$ of $\mathfrak{O}_K$ is maximal. Let p be the prime number with $(p) = \mathfrak{P} \cap \mathbb{Z}$. Then the natural injection $\mathbb{Z}/(p) \to \mathfrak{O}_K/\mathfrak{P}$ shows that the domain $\mathfrak{O}_K/\mathfrak{P}$ is a $\mathbb{Z}/(p)$-vector space. This implies that $\mathfrak{O}_K/\mathfrak{P}$ is a field, hence $\mathfrak{P}$ is maximal. □

The following theorem contains the more traditional definition of Dedekind rings and the main theorem about these rings due to E. Noether.

Theorem 1.35. *Let R be a Noetherian domain. The following properties are equivalent:*
1) *R is a Dedekind ring.*
2) *The ideals of R form a free semigroup.*
3) *R is integrally closed and all prime ideals of R are maximal.* ⊠ (van der Waerden (1967), § 137)

We recall that a *ring* is called *Noetherian* if every ascending chain of distinct ideals is finite.

In the proof that 3) implies 2) one uses the concept of a fractional ideal of R. This is a finitely generated R-module in the quotient field $Q(R)$. If $R = \mathfrak{O}_K$, a *fractional ideal* is the same as a complete module with order $\mathfrak{O}_K$.

The product $\mathfrak{a}\mathfrak{b}$ of two fractional ideals $\mathfrak{a}$, $\mathfrak{b}$ is the R-module generated by the elements $\alpha\beta$ with $\alpha \in \mathfrak{a}$, $\beta \in \mathfrak{b}$.

Let $\mathfrak{M}(R)$ be the semigroup of fractional ideals $\neq \{0\}$. If $\mathfrak{p}$ is a prime ideal of R a ring with 3) one proves that the fractional ideal

$$\mathfrak{p}^{-1} := \{\alpha \in Q(R) | \alpha\mathfrak{p} \subset R\}$$

is the inverse of $\mathfrak{p}$ in $\mathfrak{M}(R)$. It follows that $\mathfrak{M}(R)$ is a group. More exactly $\mathfrak{M}(R)$ is a free abelian group with the prime ideals as a set of free generators.

In commutative algebra one considers more generally Noetherian domains R with finite Krull dimension. The *Krull dimension* is defined as follows: The height of a prime ideal $\mathfrak{p}$ of R is a equal to h if there is a chain

$$\mathfrak{p}_1 \subsetneqq \cdots \subsetneqq \mathfrak{p}_h = \mathfrak{p}$$

of prime ideals and if there is no such chain with $h + 1$ prime ideals. The dimension of R is defined as the maximum of the heights of the prime ideals of R. If R has no prime ideals (beside $\{0\}$), i.e. if R is a field, then the dimension of R is 0 by definition.

A typical example for a ring with Krull dimension n is the polynomial ring $K[x_1, \ldots, x_n]$ in n variables $x_1, \ldots, x_n$ over a field K. By Theorem 1.35 Dedekind rings are characterized as integrally closed Noetherian rings of dimension 1. In particular $K[x]$ is a Dedekind ring. In fact $K[x]$ is even an Euclidean ring (§ 2.1) with the degree as the height of a polynomial in $K[x]$.

In this paragraph from now on R denotes always a Dedekind ring. Every fractional ideal $\mathfrak{a} \neq \{0\}$ has a unique representation

$$\mathfrak{a} = \prod_{\mathfrak{p}} \mathfrak{p}^{v_{\mathfrak{p}}(\mathfrak{a})}$$

where the sum runs over all prime ideals $\mathfrak{p}$ of R and almost all $v_{\mathfrak{p}}(\mathfrak{a})$ are zero. $v_{\mathfrak{p}}(\mathfrak{a})$ is called the $\mathfrak{p}$-*exponent* of $\mathfrak{a}$. The *support* $\operatorname{supp}(\mathfrak{a})$ is the set of prime ideals $\mathfrak{p}$ with $v_{\mathfrak{p}}(\mathfrak{a}) \neq 0$. Two fractional ideals $\mathfrak{a}, \mathfrak{b} \in \mathfrak{M}(R)$ are called prime to each other if $\operatorname{supp}(\mathfrak{a}) \cap \operatorname{supp}(\mathfrak{b}) = \varnothing$. Furthermore we define the *greatest common divisor* and the *least common multiple of* $\mathfrak{a}, \mathfrak{b}$ by g.c.d.$(\mathfrak{a}, \mathfrak{b}) = \mathfrak{c}$, l.c.m.$(\mathfrak{a}, \mathfrak{b}) = \mathfrak{d}$ with

$$v_{\mathfrak{p}}(\mathfrak{c}) = \min\{v_{\mathfrak{p}}(\mathfrak{a}), v_{\mathfrak{p}}(\mathfrak{b})\}, \qquad v_{\mathfrak{p}}(\mathfrak{d}) = \max\{v_{\mathfrak{p}}(\mathfrak{a}), v_{\mathfrak{p}}(\mathfrak{b})\}.$$

Proposition 1.36. g.c.d.$(\mathfrak{a}, \mathfrak{b}) = \mathfrak{a} + \mathfrak{b}$, l.c.m.$(\mathfrak{a}, \mathfrak{b}) = \mathfrak{a} \cap \mathfrak{b}$. □

If $\alpha \in R^*$, we write $(\alpha) := \alpha R$, $v_{\mathfrak{p}}(\alpha) := v_{\mathfrak{p}}((\alpha))$, a.s.o. We extent the notion of g.c.d. and l.c.m. to $\mathfrak{M}(R) \cup \{0\}$ according to the right side of the formulas in Prop. 1.36. For convenience of notation we put $v_{\mathfrak{p}}(0) = \infty$, where the symbol ∞ is considered as being greater than any real number.

3.2. Congruences. Let R be a Dedekind ring and $\mathfrak{a}$ an ideal of R. Two elements $\alpha, \beta \in R$ are called *congruent* modulo $\mathfrak{a}$ if $\alpha - \beta \in \mathfrak{a}$. In this case one writes $\alpha \equiv \beta \pmod{\mathfrak{a}}$. Congruences mod $\mathfrak{a}$ are equivalent to equalities in the quotient ring $R/\mathfrak{a}$.

We quote the main facts about congruences.

Proposition 1.37. *Let* $\alpha, \beta \in R$ *and* $\mathfrak{a}$ *an ideal of* R. *The congruence*

$$\alpha\xi \equiv \beta \pmod{\mathfrak{a}} \tag{1.17}$$

has a solution if and only if β *is divisible by* g.c.d.$((\alpha), \mathfrak{a})$. *If* ξ_0 *is a solution of* (1.17), *then an arbitrary solution has the form* $\xi_0 + \gamma$ *with* $\gamma \in \mathfrak{a}((\alpha), \mathfrak{a})^{-1}$. □

Proposition 1.38 (Chinese remainder theorem). *Let* $\mathfrak{a}_1, \ldots, \mathfrak{a}_s$ *be ideals of* R *which are pairwise prime to each other. Furthermore let* $\alpha_1, \ldots, \alpha_s$ *be arbitrary* *elements of R. Then there is an* $\alpha \in R$ *such that*

$$\alpha \equiv \alpha_i \ (\mathrm{mod}\ \mathfrak{a}_i) \ \textit{for}\ i = 1, \ldots, s.$$

The class of $\alpha \bmod \mathfrak{a}_1 \mathfrak{a}_2 \ldots \mathfrak{a}_s$ *is uniquely determined.* □

Now we specialize to the case that $R = \mathfrak{O}_K$ is the ring of integers of an algebraic number field and $\mathfrak{a}$ is an ideal of $\mathfrak{O}_K$. In § 1.1 we have defined the norm $N(\mathfrak{a})$ of $\mathfrak{a}$, it is equal to the number of elements of the finite ring $\mathfrak{O}_K/\mathfrak{a}$. We define the *generalized Euler function* $\varphi(\mathfrak{a})$ as the number of classes of $\mathfrak{O}_K/\mathfrak{a}$ which are prime to $\mathfrak{a}$, i.e. $\varphi(\mathfrak{a}) = |(\mathfrak{O}_K/\mathfrak{a})^\times|$.

We have the following generalizations of theorems of Fermat and Euler.

Proposition 1.39. *Let K be an algebraic number field,* $\mathfrak{p}$ *a prime ideal and* $\mathfrak{a}$ *an arbitrary ideal of* $\mathfrak{O}_K$. *Then*
1) $\alpha^{N(\mathfrak{p})} \equiv \alpha \ (\mathrm{mod}\ \mathfrak{p})$ *for all* $\alpha \in \mathfrak{O}_K$.
2) $\alpha^{\varphi(\mathfrak{a})} \equiv 1 \ (\mathrm{mod}\ \mathfrak{a})$ *for all* $\alpha \in \mathfrak{O}_K$ *which are prime to* $\mathfrak{a}$.
3) $\varphi(\mathfrak{a}) = \prod_{\mathfrak{p} \in \mathrm{supp}\, \mathfrak{a}} N(\mathfrak{p})^{v_\mathfrak{p}(\mathfrak{a})-1}(N(\mathfrak{p}) - 1)$. □

3.3. Semilocalization. Let R be a Dedekind ring and let $\mathfrak{S}$ be a set of prime ideals in R. Then $R_\mathfrak{S}$ denotes the ring of elements α/β, $\alpha, \beta \in R$, in the quotient field of R such that $\beta \notin \mathfrak{p}$ for all $\mathfrak{p} \in \mathfrak{S}$.

$R_\mathfrak{S}$ is a Dedekind ring with prime ideals $\mathfrak{p}' = \mathfrak{p}R_\mathfrak{S}$ for $\mathfrak{p} \in \mathfrak{S}$. If $\mathfrak{S}$ is finite, then $R_\mathfrak{S}$ is a principal ideal ring by the following

Proposition 1.40. *Let R be a Dedekind ring with finitely many prime ideals. Then R is a principal ideal ring.*

Proof. Let $\mathfrak{p}_1, \ldots, \mathfrak{p}_s$ be the prime ideals of R and $\mathfrak{a}$ an arbitrary ideal of R. There are $\alpha_i \in R$ with $\mathfrak{a}\mathfrak{p}_1 \ldots \mathfrak{p}_s \mathfrak{p}_i^{-1} | \alpha_i$ and $\mathfrak{a}\mathfrak{p}_1 \ldots \mathfrak{p}_s \nmid \alpha_i$ for $i = 1, \ldots, s$. Hence $\mathfrak{a} = (\alpha_1 + \cdots + \alpha_s)$. □

Proposition 1.40 allows to reduce many questions about ideals in Dedekind rings to the case of principal ideal rings. This method is called *semilocalization.* If $\mathfrak{S}$ consists of one prime ideal $\mathfrak{p}$, then $R_\mathfrak{S} = R_\mathfrak{p}$ is called the *localization* of R at $\mathfrak{p}$ and $R_\mathfrak{p}$ is called a local ring.

Example 19. Let p be a prime number. The localization $\mathbb{Z}_{(p)}$ of $\mathbb{Z}$ at (p) consists of all rational numbers a/b, $a, b \in \mathbb{Z}$, with $p \nmid b$. The integral closure $\mathfrak{O}_{K,p}$ of $\mathbb{Z}_{(p)}$ in an algebraic number field K is a principal ideal ring (Theorem 1.29, Proposition 1.40). The prime elements of $\mathfrak{O}_{K,p}$ correspond to the prime divisors of p in $\mathfrak{O}_K$. Zolotarev's foundation of the arithmetic in K (§ 2.3) is based on the study of the rings $\mathfrak{O}_{K,p}$ (see Borevich, Shafarevich (1985) for more details). □

3.4. Extensions of Dedekind Rings. In this and in the next section we consider the following situation. Let R be a Dedekind ring with quotient field K and let S be the integral closure of R in a finite extension L of K.

Then also S is a Dedekind ring (Theorem 1.29). We are interested in the connections between the ideal groups $\mathfrak{M}(R)$ and $\mathfrak{M}(S)$.

First of all we have a homomorphism ι of $\mathfrak{M}(R)$ into $\mathfrak{M}(S)$:

$$\iota(\mathfrak{a}) = \mathfrak{a}S \text{ for } \mathfrak{a} \in \mathfrak{M}(R).$$

If $\mathfrak{a} \in \operatorname{Ker} \iota$, then $\mathfrak{a} \subseteq R$ and $\mathfrak{a}^{-1} \in \operatorname{Ker} \iota$. Hence $\mathfrak{a} = R$. Therefore ι is an injection. In the following we identify $\mathfrak{M}(R)$ with its image in $\mathfrak{M}(S)$ if there is no danger of confusion.

Now let g be an isomorphism of L onto the extension gL of K and let $\mathfrak{a} \in \mathfrak{M}(S)$. Then

$$g\mathfrak{a} := \{g\alpha | \alpha \in \mathfrak{a}\}$$

is a fractional ideal of the Dedekind ring gS, and g induces an isomorphism of $\mathfrak{M}(S)$ onto $\mathfrak{M}(gS)$.

Furthermore we have a homomorphism $N_{L/K}$ of $\mathfrak{M}(S)$ into $\mathfrak{M}(R)$, called the *ideal norm*. $N_{L/K}$ is uniquely determined by the property

$$N_{L/K}(\alpha S) = (N_{L/K}\alpha)R \text{ for } \alpha \in L^{\times}.$$

We construct $N_{L/K}\mathfrak{A}$ for $\mathfrak{A} \in \mathfrak{M}(S)$ as follows. Take $\beta \in \mathfrak{A}$ such that $(\beta)\mathfrak{A}$ is prime to $\mathfrak{a} := \mathfrak{A} \cap K$ and put $(N_{L/K}(\beta))R = \prod_{\mathfrak{p}} \mathfrak{p}^{b_{\mathfrak{p}}}$. Then

$$N_{L/K}\mathfrak{A} := \prod_{\mathfrak{p} \in \operatorname{supp} \mathfrak{a}} \mathfrak{p}^{b_{\mathfrak{p}}}.$$

By means of semilocalization one proves that $N_{L/K}$ is a well defined homomorphism and that it has the following basic properties.

Proposition 1.41. 1) *Let* $n := [L:K]$ *and* $\mathfrak{a} \in \mathfrak{M}(R)$. *Then* $N_{L/K}\mathfrak{a} = \mathfrak{a}^n$.

2) *For a field tower* $M \supseteq L \supseteq K$ *one has* $N_{M/K} = N_{L/K}N_{M/L}$. *If* M/K *is normal and* G *denotes the set of isomorphisms of* L *in* M *which let* K *unchanged, then*

$$N_{L/K}\mathfrak{A} = \prod_{g \in G} g\mathfrak{A} \qquad \text{for } \mathfrak{A} \in \mathfrak{M}(S).$$

3) *Let* $\mathfrak{A}$ and $\mathfrak{B}$ *be integral ideals in* $\mathfrak{M}(S)$. *Then* $\mathfrak{A}|\mathfrak{B}$ *implies* $N_{L/K}\mathfrak{A}|N_{L/K}\mathfrak{B}$.

4) *If* $R = \mathbb{Z}$, *then* $N_{L/K}$ *coincides with the norm defined in* §1.1. □

Now we consider the relation between a prime ideal $\mathfrak{P}$ in $\mathfrak{M}(S)$ and the corresponding prime ideal $\mathfrak{p} := \mathfrak{P} \cap R$ in $\mathfrak{M}(R)$.

We define the *ramification index* $e = e_{L/K}(\mathfrak{P})$ as the highest exponent e such that $\mathfrak{P}^e$ divides $\mathfrak{p}$, i.e. $e = v_{\mathfrak{P}}(\mathfrak{p})$. Furthermore the *inertia degree* $f = f_{L/K}(\mathfrak{P})$ is defined by $N_{L/K}\mathfrak{P} = \mathfrak{p}^f$ ($\mathfrak{P}|\mathfrak{p}$ implies $N_{L/K}\mathfrak{P}|\mathfrak{p}^n$).

From these definitions it follows immediately that

$$[L:K] = \sum_{\mathfrak{P}|\mathfrak{p}} e_{L/K}(\mathfrak{P})f_{L/K}(\mathfrak{P}). \tag{1.18}$$

If L/K is a separable extension, then

$$f_{L/K}(\mathfrak{P}) = [S/\mathfrak{P} : R/\mathfrak{p}].$$

Since the inclusion $R \subseteq S$ induces an injection $R/\mathfrak{p} \hookrightarrow S/\mathfrak{P}$, we may consider $S/\mathfrak{P}$ as a field extension of $R/\mathfrak{p}$. More generally $S/\mathfrak{A}$ can be considered as $R/\mathfrak{p}$-vector space for all ideals of S with $\mathfrak{A}|\mathfrak{p}$. (1.18) can be interpreted as an equation about the dimensions of the vector spaces in the following special case of the chinese remainder theorem (Proposition 1.38):

$$S/\mathfrak{p} \cong \sum_{\mathfrak{P}|\mathfrak{p}} S/\mathfrak{P}^{e_{L/K}(\mathfrak{P})} \tag{1.19}$$

(see also Proposition 1.74) (By means of semilocalization (§ 3.3) we may assume that R and S are principal ideal rings. Then $\dim S/\mathfrak{p} = [L:K]$ is easily proved.)

The following theorem of Kummer allows in many cases to determine the decomposition of $\mathfrak{p}$ in prime ideals in S:

Theorem 1.42. *Let $\alpha \in S$ be a generating element of the extension L/K and let $f(x)$ be the minimal polynomial of α over K. Moreover let $\mathfrak{p}$ be a prime ideal of R such that $\mathfrak{p}S \cap R[\alpha] = \mathfrak{p}[\alpha]$. If $\bar{f}$ has the decomposition*

$$\bar{f} = \mathfrak{f}_1^{a_1} \dots \mathfrak{f}_s^{a_s}$$

in irreducible polynomials in $(R/\mathfrak{p})[x]$, then

$$\mathfrak{p} = \mathfrak{P}_1^{a_1} \dots \mathfrak{P}_s^{a_s}$$

is the prime ideal decomposition of $\mathfrak{p}$ in S, $f_{L/K}(\mathfrak{P}_i)$ equals the degree of $\mathfrak{f}_i$, and

$$\mathfrak{P}_i = (\mathfrak{p}, f_i(\alpha))$$

where f_i is a polynomial in $R[x]$ with $\bar{f}_i = \mathfrak{f}_i$, $i = 1, \dots, s$. ⊠ (Norkiewicz (1974), Chap. 4.3.1)

Example 20. Let $K_1 = \mathbb{Q}(\alpha_1)$, $\alpha_1^3 - 18\alpha_1 - 6 = 0$, $K_2 = \mathbb{Q}(\alpha_2)$, $\alpha_2^3 - 36\alpha_2 - 78 = 0$, $K_3 = \mathbb{Q}(\alpha_3)$, $\alpha_3^3 - 54\alpha_3 - 150 = 0$. Then the ring of integers in K_i is $\mathbb{Z}[\alpha_i]$, $i = 1, 2, 3$. The three fields have the same discriminant $2^2 \cdot 3^5 \cdot 23$, but they are distinct, since (5) is prime ideal in K_1, K_2 but $(5) = \mathfrak{p}_1\mathfrak{p}_2\mathfrak{p}_3$ in K_3, and (11) is prime ideal in K_2 but $(11) = \mathfrak{p}'_1\mathfrak{p}'_2\mathfrak{p}'_3$ in K_1. □

Let $M \supseteq L \supseteq K$ be a field tower, $\mathfrak{P}$ a prime ideal of the integral closure of R in M and $\mathfrak{P}_L := \mathfrak{P} \cap S$. Then

$$e_{M/K}(\mathfrak{P}) = e_{M/L}(\mathfrak{P})e_{L/K}(\mathfrak{P}_L), \tag{1.20}$$

$$f_{M/K}(\mathfrak{P}) = f_{M/L}(\mathfrak{P})f_{L/K}(\mathfrak{P}_L). \tag{1.21}$$

Let $\mathfrak{p}$ be a prime ideal of R. A polynomial

$$f(x) = x^n + a_1x^{n-1} + \dots + a_n \in R[x]$$

is called *Eisenstein polynomial* (for $\mathfrak{p}$) if $a_1, \dots, a_n \in \mathfrak{p}$ and $v_{\mathfrak{p}}(a_n) = 1$.

Proposition 1.43. 1) (Eisenstein's irreducibility criterion) *Let $f(x)$ be an Eisenstein polynomial for $\mathfrak{p}$ and α a root of $f(x)$ in some extension of R. Then $[K(\alpha):K] = n$, i.e. $f(x)$ is irreducible over K.*

2) *Let S be the integral closure of R in $K(\alpha)$ and $\mathfrak{P}$ the ideal of S generated by α and $\mathfrak{p}$. Then $\mathfrak{P}$ is a prime ideal and $\mathfrak{p} = \mathfrak{P}^n$.*

3) *Let L/K be a finite extension of degree n and let S be the integral closure of R in L. Assume that $\mathfrak{P}^n = \mathfrak{p}$ for a certain prime ideal $\mathfrak{P}$ of S, $\mathfrak{p} = \mathfrak{P} \cap R$ and let*

α be an element of S with $v_{\mathfrak{P}}(\alpha) = 1$. Then $L = K(\alpha)$ and α is the root of an Eisenstein polynomial for $\mathfrak{p}$.

Proof. 1), 2) Let $\mathfrak{P}$ be a prime divisor of $\mathfrak{p}$ in S. Then

$$\alpha^n = -a_1\alpha^{n-1} - \cdots - a_n$$

implies $nv_{\mathfrak{P}}(\alpha) = v_{\mathfrak{P}}(a_n) = v_{\mathfrak{P}}(\mathfrak{p})$, hence $\mathfrak{p} = \mathfrak{P}^n$ by (1.18) and $\mathfrak{P} = (\alpha, \mathfrak{p})$.

3) Let $f(x) = x^m + b_1x^{m-1} + \cdots + b_m \in R[x]$ be the irreducible polynomial with $f(\alpha) = 0$. Then $v_{\mathfrak{P}}(\alpha^m) = m$, $v_{\mathfrak{P}}(b_i\alpha^{m-i}) \equiv m - i \pmod n$ for $i = 1, \ldots, m$. Therefore $m = v_{\mathfrak{P}}(\alpha^m) = v_{\mathfrak{P}}(b_m) \equiv 0 \pmod n$ and $v_{\mathfrak{P}}(b_i) > 0$ for $i = 1, \ldots, m - 1$.

□

3.5. Different and Discriminant. (Main reference: Lang (1970), Chap. 3). Again let R be a Dedekind ring with quotient field K and S the integral closure of R in a finite separable extension L of K. Let $\mathfrak{A} \in \mathfrak{M}(S)$. The *complementary ideal* $\mathfrak{A}^\wedge$ of $\mathfrak{A}$ (with respect to R) is defined by

$$\mathfrak{A}^\wedge := \{\alpha \in L \mid \mathrm{tr}_{L/K}(\alpha\mathfrak{A}) \subseteq R\}.$$

From the definition it is clear that $\mathfrak{A}^\wedge$ is indeed a fractional ideal in $\mathfrak{M}(S)$. It has the following properties, $S^\wedge \supseteq S$, $\mathfrak{A}^\wedge = S^\wedge\mathfrak{A}^{-1}$, $(\mathfrak{A}^\wedge)^\wedge = \mathfrak{A}$.

The integral ideal $(S^\wedge)^{-1}$ is called the *different* of S/R and is denoted by $\mathfrak{D}_{L/K}$. The following theorem states the main properties of the different.

Theorem 1.44. *Let R and S be as above and let the residue field extensions $(S/\mathfrak{P})/(R/\mathfrak{p})$ be separable for all prime ideals $\mathfrak{P}$ of S and $\mathfrak{p} = \mathfrak{P} \cap R$.*

1) *$\mathfrak{D}_{L/K}$ is the* g.c.d. *of the element differents $D_{L/K}(\alpha)$ for $\alpha \in S$* (Appendix 1.1).

2) *(Dedekind's different theorem) Let $\mathfrak{P}$ be a prime ideal of S and $\mathfrak{p} := \mathfrak{P} \cap R$, $e := v_{\mathfrak{P}}(\mathfrak{p})$. Let p be the characteristic of the field $S/\mathfrak{P}$. Then*

$$v_{\mathfrak{P}}(\mathfrak{D}_{L/K}) = e - 1 \qquad \text{if } p \nmid e$$

and

$$v_{\mathfrak{P}}(\mathfrak{D}_{L/K}) > e - 1 \qquad \text{if } p \mid e.$$

3) *(Different tower theorem) Let T be the integral closure of S in a finite separable extension M of L. Then*

$$\mathfrak{D}_{M/K} = \mathfrak{D}_{M/L}\mathfrak{D}_{L/K}. \boxtimes$$

Remark. The assumptions of Theorem 1.44 are of course fulfilled in the case of algebraic number fields K, L and $R = \mathfrak{O}_K$, $S = \mathfrak{O}_L$ since in this case the residue class fields are finite. □

A prime ideal $\mathfrak{P}$ of S is called ramified over $\mathfrak{p} = \mathfrak{P} \cap R$ if $e = v_{\mathfrak{P}}(\mathfrak{p}) > 1$. This notation comes from the theory of algebraic functions: Let $K = \mathbb{C}(z)$ be the field of rational functions with complex coefficients, $R = \mathbb{C}[z]$ and $L = K(f)$. Then the prime ideals of R are in a one to one correspondence with the complex numbers. To $\alpha \in \mathbb{C}$ corresponds the prime ideal $(z - \alpha)R$. The prime ideals of S

can be identified with the points of the Riemann surface of f lying above $\mathbb{C}$. Since $\mathbb{C}$ is algebraically closed, the natural map $\mathbb{C} \to S/\mathfrak{P}$ is an isomorphism of $\mathbb{C}$ onto $S/\mathfrak{P}$. If $f \in S$, the value of f at the point $\mathfrak{P}$ of the Riemann surface is the uniquely determined complex number in the class $f + \mathfrak{P}$. The prime ideal $\mathfrak{P}$ is ramified over $\mathfrak{p}$ if and only if the point $\mathfrak{P}$ is ramified in the sense of coverings of surfaces, i.e. in a neighborhood of $\mathfrak{P}$ the Riemann surface of f has the form $\sqrt[e]{z-\alpha}$.

Theorem 1.44.2) implies that $\mathfrak{P}$ is ramified if and only if $e > 1$. Hence we have the following

Theorem 1.45. *Let the assumptions be as in Theorem* 1.44. *Then there are only finitely many prime ideals $\mathfrak{P}$ of S which are ramified over $\mathfrak{p} = \mathfrak{P} \cap R$.* □

We define the *discriminant* $\mathfrak{d}_{L/K}$ of S/R as the ideal norm $N_{L/K}\mathfrak{D}_{L/K}$ of the different. In the case $R = \mathbb{Z}$ this is the ideal generated by the discriminant d_L (§ 1.1). From Theorem 1.44 we get the main properties of the discriminant:

Theorem 1.46. *Let the assumptions be as in Theorem* 1.44 *and let $\mathfrak{p} = \mathfrak{P}_1^e \dots \mathfrak{P}_s^e$ be the decomposition of the prime ideal $\mathfrak{p}$ of R in prime ideals in S. Furthermore let p be the characteristic of $R/\mathfrak{p}$ and f_i the inertia degree of $\mathfrak{P}_i$, $i = 1, \dots, s$. Then*

1) (Dedekind's discriminant theorem)

$$v_{\mathfrak{p}}(\mathfrak{d}_{L/K}) = (e_1 - 1)f_1 + \cdots + (e_s - 1)f_s \text{ if } p \nmid e_i, i = 1, \dots, s,$$

$$v_{\mathfrak{p}}(\mathfrak{d}_{L/K}) > (e_1 - 1)f_1 + \cdots + (e_s - 1)f_s \text{ if } p | e_i \text{ for some } i.$$

2) (Discriminant tower theorem) *Let T be the integral closure of S in a finite separable extension M of L. Then*

$$\mathfrak{d}_{M/K} = N_{L/K}(\mathfrak{d}_{M/L})\mathfrak{d}_{L/K}^{[M:L]}.$$

3) $\mathfrak{d}_{L/K}$ *is the* g.c.d. *of all discriminants $d_{L/K}(\alpha_1, \dots, \alpha_n)$ for $\alpha_1, \dots, \alpha_n \in S$. In particular if R is a* PIR *and $\omega_1, \dots, \omega_n$ a basis of S over R, then $\mathfrak{d}_{L/K} = (d_{L/K}(\omega_1, \dots, \omega_n))$.* ⊠

A prime ideal $\mathfrak{p}$ of R is called ramified in S/R if one of its prime divisor $\mathfrak{P}$ in S is ramified over $\mathfrak{p}$. Theorem 1.46, 1) shows that $\mathfrak{p}$ is ramified in S/R if and only if $\mathfrak{p} | \mathfrak{d}_{L/K}$.

If the residue characteristic of $\mathfrak{P}$, i.e. the characteristic of $S/\mathfrak{P}$, is prime to $e_{L/K}(\mathfrak{P})$ and $e_{L/K}(\mathfrak{P}) > 1$ we say that $\mathfrak{P}$ is tamely ramified. In this case by means of Theorem 1.44, 2) we can compute the exact power of $\mathfrak{P}$ in $\mathfrak{D}_{L/K}$ if we know $e_{L/K}(\mathfrak{P})$. If the residue characteristic of $\mathfrak{P}$ divides $e_{L/K}(\mathfrak{P})$, we say that $\mathfrak{P}$ is wildly ramified. This is motivated by the fact that this is by far the most complicated case as we will see in the following.

Theorem 1.11 implies the following theorem of Minkowski.

Theorem 1.47. *For any number field $K \neq \mathbb{Q}$ there exists a prime number which is ramified in K (i.e. in $\mathfrak{O}_K/\mathbb{Z}$).* □

Corresponding to Theorem 1.45 we reformulate Theorem 1.12 as follows:

Theorem 1.48. *Let $\mathfrak{S}$ be a finite set of prime numbers. There are only finitely many algebraic number fields with given degree which are ramified only in the primes in $\mathfrak{S}$.*

The proof of Theorem 1.48 will be given later on (§4.6). □

Theorem 1.48 is also valid for function fields over $\mathbb{C}$. Moreover we have a very good insight in the structure of the (infinite) extension of $\mathbb{C}(z)$ consisting of all algebraic functions which are ramified in a fixed set of points $\mathfrak{S} \subset \mathbb{C}$ (Chap. 3, Example 13). We know much less about algebraic number fields with given ramification (Chap. 3.2.5–6).

Let $\mathfrak{p}$ be a prime ideal of R. We define the $\mathfrak{p}$-*different* $\mathfrak{D}_\mathfrak{p}$ of S over R by

$$\mathfrak{D}_\mathfrak{p} := \prod_{\mathfrak{P}|\mathfrak{p}} \mathfrak{P}^{v_\mathfrak{P}(\mathfrak{D}_{L/K})}$$

By means of Theorem 1.44.1) $\mathfrak{D}_\mathfrak{p}$ can be identified with the different of $S_\mathfrak{p}/R_\mathfrak{p}$ (§3.3). Hence we get the "global different" $\mathfrak{D}_{L/K}$ as the product of the "(semi)local differents" for all prime ideals $\mathfrak{p}$ (see also §4.5). The same is true for the discriminant of S/R.

We conclude the section with an application of Theorem 1.46 to the compositum of fields.

Theorem 1.49. *Let R be a principal ideal ring and $\omega_1, \ldots, \omega_n$ a basis of S as R-module, let M be as above and F an intermediate field of M/K such that $[FL:K] = [F:K][L:K]$. Furthermore let $\mathfrak{O}$ be the integral closure of R in F and $\theta_1, \ldots, \theta_m$ a basis of $\mathfrak{O}$ as R-module. We assume that the discriminants $d_{L/K}$ and $d_{F/K}$ are relatively prime. Then*

$$B := \{\omega_i\theta_j | i = 1, \ldots, n, j = 1, \ldots, m\}$$

is a basis of the integral closure of R in FL as R-module.

Proof. The discriminant tower theorem implies $\mathfrak{d}_{FL/K} = \mathfrak{d}_{F/K}^n \mathfrak{d}_{L/K}^m$. On the other hand B is a basis of FL/K with discriminant $\mathfrak{d}_{F/K}^n \mathfrak{d}_{L/K}^m$. □

Example 21 (Lang (1970), Chap. 5, §4). Let $K = \mathbb{Q}(\beta)$ with $\beta^5 - \beta + 1 = 0$. Since the polynomial $f(x) := x^5 - x + 1$ is irreducible (because it is irreducible mod 5), K has degree 5. The discriminant of a polynomial of the form $x^5 + ax + b$ is $5^5b^4 + 2^8a^5$. Hence $f(x)$ has the discriminant $2869 = 19 \cdot 151$. Since this number is square-free, it is equal to the discriminant d of $\mathfrak{O}_K$ (§1.1). Hence $\mathfrak{O}_K = \mathbb{Z}[\beta]$. According to Theorem 1.46 the ramified primes in $K/\mathbb{Q}$ are 19 and 151. By (1.3) in every ideal class of K there is an integral ideal $\mathfrak{a}$ with $N(\mathfrak{a}) < 4$. Since 2 and 3 have no prime factor of degree 1 in K (Theorem 1.42), there is no integral ideal $\mathfrak{a}$ with $N(\mathfrak{a}) = 2$ or $N(\mathfrak{a}) = 3$. Hence K has class number 1.

One can show that the normal closure N of K has degree 120 over $\mathbb{Q}$ and that $N/\mathbb{Q}(\sqrt{d})$ is unramified. □

Example 22 (Lang (1970), Chap. 5, §4). Let K be an arbitrary algebraic number field and L/K a finite normal extension. Then there exist infinitely many finite extensions F of K such that $L \cap F = K$ and FL/FK is unramified. □

3.6. Inessential Discriminant Divisors (Main reference: Hasse (1979), Chap. 25.6). The reader may ask whether there corresponds to Theorem 1.44, 1) a statement about the discriminant. One should hope that $\mathfrak{d}_{L/K}$ is the g.c.d. of the element discriminants

$$d_{L/K}(\alpha) := N_{L/K} f'_\alpha(\alpha) = \pm d_{L/K}(1, \alpha, \ldots, \alpha^{n-1}) \qquad \text{for } \alpha \in S$$

where f_α denotes the characteristic polynomial of α, $n := [L:K]$ (Appendix 1.1). But in general this is not so even if $K = \mathbb{Q}$ as was already known to Dedekind.

For $\alpha \in S$ we have

$$d_{L/K}(\alpha)R = \mathfrak{m}(\alpha)^2 \mathfrak{d}_{L/K} \tag{1.22}$$

with an integral ideal $\mathfrak{m}(\alpha)$, called the *inessential discriminant divisor* of α or the index of α. If $R = \mathbb{Z}$, (1.22) is a special case of (1.1). In general (1.22) is proved by the method of semilocalization.

An ideal $\mathfrak{a}$ of R is called a common inessential discriminant divisor of S/R if $\mathfrak{a}$ divides $\mathfrak{m}(\alpha)$ for all $\alpha \in S$.

Theorem 1.50. *Let K be an algebraic number field and L a finite extension of K. A prime ideal $\mathfrak{p}$ of $\mathfrak{O}_K$ fails to be a common inessential discriminant divisor of $\mathfrak{O}_L/\mathfrak{O}_K$ if and only if for every natural number f, the number $r_\mathfrak{p}(f)$ of prime divisors $\mathfrak{P}$ of $\mathfrak{p}$ in $\mathfrak{O}_L$ with $f_{L/K}(\mathfrak{P}) = f$ satisfies the inequality*

$$r_\mathfrak{p}(f) \leqslant \frac{1}{f} \sum_{d|f} \mu\left(\frac{f}{d}\right) N_{K/\mathbb{Q}}(\mathfrak{p})^d. \boxtimes$$

In the formula the ideal $N_{K/\mathbb{Q}}(\mathfrak{p})$ is identified with the natural number which generates it and μ denotes the Moebius function.

It follows from Theorem 1.50 that for a prime ideal $\mathfrak{p}$ of $\mathfrak{O}_K$ to be a common inessential discriminant divisor of $\mathfrak{O}_L/\mathfrak{O}_K$, it is necessary that $N_{K/\mathbb{Q}}(\mathfrak{p}) < [L:K]$. This condition is also sufficient if $\mathfrak{p}$ decomposes in L into $[L:K]$ distinct prime divisors.

Example 23 (Dedekind). 2 is a common inessential discriminant divisor in the extension $\mathbb{Q}(\beta)/\mathbb{Q}$ with $\beta^3 + \beta^2 - 2\beta + 8 = 0$, considered in example 3. □

3.7. Normal Extensions (Main reference: Serre (1962), Chap. 1. §§ 7, 8, Chap. 4). Let R be a Dedekind ring with quotient field K and S the integral closure of R in a finite separable normal extension N of K. Let G be the Galois group of N/K.

By means of Galois theory one gets a finer description of the ramification behavior of the prime ideals in the extension S/R: To every prime ideal $\mathfrak{P}$ of S one associates a series of subgroups of G, called the ramification groups. They determine not only the ramification behavior of $\mathfrak{P}$ over $\mathfrak{p} := R \cap \mathfrak{P}$ but also of the corresponding prime ideals in the intermediate fields of L/K.

We have already seen in § 3.4 that G acts as a group of automorphisms on $\mathfrak{M}(S)$. Let $\mathfrak{P}$ be a prime ideal of S. Then the subgroup

$$G_\mathfrak{P} := \{g \in G | g\mathfrak{P} = \mathfrak{P}\}$$

of G is called *decomposition group* of $\mathfrak{P}$ and the fixed field of $G_{\mathfrak{P}}$ is called *decomposition field* of $\mathfrak{P}$. We assume that the residue field extension $(S/P)/(R/P \cap R)$ is separable.

Proposition 1.51. *Let R and S be as above and let $\mathfrak{P}$ be a prime ideal of S.*

1) The prime divisors of $\mathfrak{p} := \mathfrak{P} \cap R$ in S are the prime ideals $g\mathfrak{P}$ for $g \in G$. These prime ideals have common ramification index e and inertia degree f.

2) The decomposition group $G_{\mathfrak{P}}$ has the order ef.

3) Let $K_{\mathfrak{P}}$ be the decomposition field of $\mathfrak{P}$. In the extension $N/K_{\mathfrak{P}}$, the prime ideal $\mathfrak{P}$ has ramification index e and inertia degree f. In the extension $K_{\mathfrak{P}}/K$, the prime ideal $\mathfrak{P} \cap K_{\mathfrak{P}}$ of $S \cap K_{\mathfrak{P}}$ has ramification index 1 and inertia degree 1. ☒

For $n = 0, 1, \ldots$ the *n-th ramification group* $V_n = V_n(\mathfrak{P}) = V_n(\mathfrak{P}, N/K)$ of the prime ideal $\mathfrak{P}$ of S is defined by

$$V_n := \{g \in G_{\mathfrak{P}} | g\alpha \equiv \alpha \ (\mathrm{mod}\ \mathfrak{P}^{n+1}) \text{ for } \alpha \in S\}.$$

$T_{\mathfrak{P}} := V_0(\mathfrak{P})$ is called *inertia group* of $\mathfrak{P}$ and the fixed field of $T_{\mathfrak{P}}$ is called the *inertia field* of $\mathfrak{P}$. It is easy to see that $V_0, V_1, \ldots$ is a descending sequence of invariant subgroups of $G_{\mathfrak{P}}$. The group $T_{\mathfrak{P}}$ is characterized as the maximal subgroup of $G_{\mathfrak{P}}$ which acts trivially on $S/\mathfrak{P}$. Hence we may identify $G_{\mathfrak{P}}/T_{\mathfrak{P}}$ with the corresponding subgroup of the Galois group $\mathfrak{G}_{\mathfrak{P}}$ of the residue field extension $(S/\mathfrak{P})/(\mathfrak{P} \cap R)$.

We put $V_{-1} := G_{\mathfrak{P}}$. Simple facts about the ramification groups are collected in the following proposition.

Proposition 1.52. *Let R and S be as above, let $\mathfrak{P}$ be a prime ideal of S such that $(S/\mathfrak{P})/(R/\mathfrak{P} \cap R)$ is a separable extension, and let e be the ramification index and f the inertia degree of $\mathfrak{P}$ with respect to R.*

1) $G_{\mathfrak{P}}/T_{\mathfrak{P}} = \mathfrak{G}_{\mathfrak{P}}$.

2) Let F be an intermediate field of N/K and $H := G(N/F)$. Then the n-th ramification group of $\mathfrak{P}$ with respect to N/F is equal to $V_n(\mathfrak{P}, N/K) \cap H$, $n = -1, 0, \ldots$.

3) Let $I_{\mathfrak{P}}$ be the inertia field of $\mathfrak{P}$ with respect to K. Then $\mathfrak{P}$ has ramification index $e = [N : I_{\mathfrak{P}}]$ with respect to $I_{\mathfrak{P}}$ and the prime ideal $\mathfrak{P} \cap I_{\mathfrak{P}}$ of $S \cap I_{\mathfrak{P}}$ has inertia degree $f = [I_{\mathfrak{P}} : F_{\mathfrak{P}}]$ with respect to $K_{\mathfrak{P}}$.

4) Let π be an element of $\mathfrak{P}$ with $\pi \notin \mathfrak{P}^2$. Then the ramification group V_n is characterized as the group of all $g \in T_{\mathfrak{P}}$ such that

$$g\pi \equiv \pi \ (\mathrm{mod}\ \mathfrak{P}^{n+1}).$$

5) $V_n = \{1\}$ for n sufficiently large.

6) There are monomorphisms

$$V_0/V_1 \to (S/\mathfrak{P})^\times \tag{1.23}$$

and

$$V_n/V_{n+1} \to (S/\mathfrak{P})^+ \text{ for } n = 1, 2, \ldots. \tag{1.24}$$

7) Let $\mathfrak{p} := \mathfrak{P} \cap R$ and $[R : \mathfrak{p}] = N(\mathfrak{p})$ finite. Then there is one and only one

$\bar{g} \in G_{\mathfrak{P}}/T_{\mathfrak{P}}$ with

$$g\alpha \equiv \alpha^{N(\mathfrak{p})} \pmod{\mathfrak{P}} \text{ for all } \alpha \in S.$$

$\bar{g}$ generates the group $G_{\mathfrak{P}}/T_{\mathfrak{P}}$. It is called the Frobenius automorphism of $\mathfrak{P}$ with respect to N/K and it will be denoted by $\left(\frac{N/K}{\mathfrak{P}}\right)$.

8) *Let h be an isomorphism of N onto hN. Then $h\mathfrak{P}$ is a prime ideal of hS. Furthermore*

$$V_n(h\mathfrak{P}) = hV_n(\mathfrak{P})h^{-1}, \qquad n = -1, 0, \ldots,$$

$$\left(\frac{hN/hK}{h\mathfrak{P}}\right) = h\left(\frac{N/K}{\mathfrak{P}}\right)h^{-1}. \boxtimes$$

If $G(N/K)$ is abelian and $h \in G(N/K)$, then $\left(\frac{N/K}{\mathfrak{P}}\right)$ only depends on $\mathfrak{p} = \mathfrak{P} \cap R$. In this case one writes $\left(\frac{N/K}{\mathfrak{P}}\right) =: \left(\frac{N/K}{\mathfrak{p}}\right)$.

3) implies that any class in $S/\mathfrak{P}$ has a representative in $S \cap I_{\mathfrak{P}}$. Therefore every $\alpha \in S$ has a representation

$$\alpha \equiv \alpha_0 + \alpha_1\pi + \cdots + \alpha_n\pi^n \pmod{\mathfrak{P}^{n+1}}$$

with $\alpha_i \in S \cap I_{\mathfrak{P}}$. This implies 4). 5) is a consequence of 4): If $g\pi \equiv \pi \pmod{\mathfrak{P}^{n+1}}$ for all $n = 1, 2, \ldots$, then $g\pi = \pi$. But $I_{\mathfrak{P}}(\pi) = N$ and therefore $g = 1$.

The monomorphisms (1.23) and (1.24) depend on the choice of a $\pi \in \mathfrak{P}$ with $\pi \notin \mathfrak{P}^2$. They are constructed as follows: Corresponding to § 3.3 let $S_{\mathfrak{P}} = \{\alpha/\beta | \alpha, \beta \in S, \beta \notin \mathfrak{P}\}$ be the localization of S at $\mathfrak{P}$. The embedding $S \hookrightarrow S_{\mathfrak{P}}$ induces an isomorphism $S/\mathfrak{P} \to S_{\mathfrak{P}}/\mathfrak{P}S_{\mathfrak{P}}$. Furthermore since any class in $S/\mathfrak{P}$ has a representative in $S \cap I_{\mathfrak{P}}$, (1.23) is induced from the homomorphism $g \to \alpha \pmod{\mathfrak{P}S_{\mathfrak{P}}}$ of V_0 into $(S_{\mathfrak{P}}/\mathfrak{P}S_{\mathfrak{P}})^{\times}$, where the class of $\alpha \in S \cap I_{\mathfrak{P}}$ is determined by $g\pi \equiv \alpha\pi \pmod{\mathfrak{P}^2 S_{\mathfrak{P}}}$, and (1.24) is induced from the homomorphism $g \to \alpha \pmod{\mathfrak{P}S_{\mathfrak{P}}}$ of V_n into $(S_{\mathfrak{P}}/\mathfrak{P}S_{\mathfrak{P}})^{+}$, where the class of $\alpha \in S \cap I_{\mathfrak{P}}$ is determined by $g\pi \equiv \pi + \alpha\pi^{n+1} \pmod{\mathfrak{P}^{n+2}S_{\mathfrak{P}}}$. □

Let $S/\mathfrak{P}$ be a finite field of characteristic p. Then (1.24) and 5) imply that V_1 is a p-group and (1.23) implies that V_0/V_1 is cyclic of order $e_0 | N(\mathfrak{P}) - 1$. Hence e_0 is the maximal divisor of e which is prime to p. Since $G_{\mathfrak{P}}/T_{\mathfrak{P}}$ also is cyclic, $G_{\mathfrak{P}}$ is a solvable group (see also § 4.6).

We mention some properties of the filtration $\{V_n | n = 0, 1, \ldots\}$ (Speiser (1919)).

Proposition 1.53. 1. *If $g \in V_0$ and $h \in V_n$, $n \geqslant 1$, then $ghg^{-1}h^{-1} \in V_{n+1}$ if and only if $g^n \in V_1$ or $h \in V_{n+1}$.*

2. *If $g \in V_m$ and $h \in V_n$, $m, n \geqslant 1$, then $ghg^{-1}h^{-1} \in V_{m+n+1}$.*

3. *Let the characteristic of $S/\mathfrak{P}$ be a prime p. If $V_m \neq V_{m+1}$ and $V_n \neq V_{n+1}$, m, $n \geqslant 1$, then $m \equiv n \pmod{p}$.* ⊠

We apply the ramification groups to the computation of the different:

Theorem 1.54. *Let $\mathfrak{P}$ be a prime ideal of S such that $(S/\mathfrak{P})/(R/\mathfrak{P} \cap R)$ is a separable extension and let v_n the order of the n-th ramification group of $\mathfrak{P}$. Then*

$$v_{\mathfrak{P}}(\mathfrak{D}_{N/K}) = \sum_{n=0}^{\infty} (v_n - 1).$$

Proof. By means of Theorem 1.44 we can assume that $G(N/K) = V_0(\mathfrak{P}, N/K)$. Then for $\alpha \in S$ we have

$$D_{N/K}(\alpha) = \prod_{\substack{g \in V_0 \\ g \neq 1}} (\alpha - g\alpha).$$

This implies

$$v_P(D_{N/K}(\alpha)) \geqslant \sum_{n=0}^{\infty} (v_n - 1)$$

and

$$v_P(D_{N/K}(\pi)) = \sum_{n=0}^{\infty} (v_n - 1)$$

if $\pi \in \mathfrak{P}$, $\pi \notin \mathfrak{P}^2$. Hence the claim of Theorem 1.54 follows from Theorem 1.44.1). □

Proposition 1.52.2) shows that the ramification groups have a nice behavior with respect to subgroups of the Galois group $G(N/K)$. The same is not true with respect to factor groups. If F is a normal intermediate field of N/K, then in general $V_n(\mathfrak{P} \cap F, F/K) \neq V_n(\mathfrak{P}, N/K)G(N/F)$ for $n > 1$. Herbrand (1931) introduced the *upper numeration* of the ramification groups which saves the situation.

The upper numeration is defined by means of the function $\varphi(x) = \varphi_{N/K}(x)$ of the real variable $x \geqslant -1$:

$$\varphi(x) = x \qquad \text{if } -1 \leqslant x \leqslant 0$$

$$\varphi(x) = (v_1 + \cdots + v_m + (x - m)v_{m+1})/v_0 \qquad \text{with } m := [x] \text{ if } x > 0.$$

$\varphi(x)$ is a continuous, strictly increasing function and thus possesses a continuous, strictly increasing inverse function $\psi(y)$. We put $V_x = V_u$ where u is the least integer $\geqslant x$. The *upper numeration* is then given by

$$V^y = V_{\psi(y)}, \qquad V^{\varphi(x)} = V_x.$$

If $\varphi(x)$ is integral, then so is x, but the converse is not true in general.

Theorem 1.55. *Let F/K be a normal subextension of N/K. Then*

$$\varphi_{N/K}(x) = \varphi_{F/K}(\varphi_{N/F}(x)) \qquad \textit{for all } x \geqslant -1$$

and

$$V^y(\mathfrak{P} \cap F, F/K) = V^y(\mathfrak{P}, N/K)G(N/F)/G(N/F) \qquad \text{for } y \geqslant -1. ⊠$$

We say that y_0 is a jump in the filtration $\{V^y\}$ if $V^{y_0} \neq V^{y_0+\varepsilon}$ for arbitrary $\varepsilon > 0$.

Theorem 1.56 (Theorem of Hasse-Arf). *Let $G = G(N/K)$ be an abelian group and y_0 a jump in the filtration $\{V^y\}$. Then y_0 is an integer.* ⊠

3.8. Ideals in Algebraic Number Fields. Let K be an algebraic number field. The maximal order $\mathfrak{O}_K$ of K is a Dedekind ring (Theorem 1.34). In the following the fundamental subject of our study are the fields K. Therefore several notations concerning $\mathfrak{O}_K$ are transfered to notations concerning K. One speaks about the *unit group of* K and means $\mathfrak{O}_K^\times$. An ideal of K is an element of $\mathfrak{M}(K) := \mathfrak{M}(\mathfrak{O}_K)$. The elements of $\mathfrak{M}(K)$ are also called fractional ideals. In this connection the ideals of $\mathfrak{O}_K$ are called integral ideals of K.

The principal ideals $(\alpha) = \alpha\mathfrak{O}_K$ form a subgroup $(K^\times)$ in $\mathfrak{M}(K)$. One of the most interesting invariants attached to K is the *(ideal) class group* $CL(K)$, i.e. the quotient group $\mathfrak{M}(K)/(K^\times)$.

Theorem 1.57. $CL(K)$ is a finite group.

Proof. This follows immediately from Theorem 1.9. □

The study of the class group of an algebraic number field is one of the main purposes of algebraic number theory. Though the literature about special cases of this question is enormous, we have very little knowledge about the structure of the class group in general. In this article results about class groups are to be find in the following sections: 2.1.2, 2.7.3–4, 3.2.6–7, 4.2.1–5, 4.3.3–4, 4.4.5–7, 5.1.2–4. The order of the class group is called the *class number*.

Lenstra had the idea to try to understand heuristically a number of experimental observations about class groups by means of a mass formula. Three of these observations are as follows:

a) The odd component of the class group of an imaginary-quadratic field seems to be quite rarely non cyclic.

b) If p is a small odd prime, the proportion of imaginary quadratic fields whose class number is divisible by p seems to be significantly greater than $1/p$ (for instance 43% for $p = 3$, 23,5% for $p = 5$).

c) It seems that a definite non zero proportion of real quadratic fields of prime discriminant (close to 76%) has class number 1, although it is not even known whether there are infinitely many.

We state the conjecture of Cohen and Lenstra in the case of imaginary quadratic fields (for other fields and more information see Cohen, Lenstra (1984) and Cohen, Martinet (1987), (1990)).

Let $CL'(K)$ be the odd component of the class group of the imaginary quadratic field K, let d_K be the discriminant of K, and let f be a non-negative function on the isomorphy classes G of finite abelian groups of odd order. We put

$$M(f) := \lim_{n\to\infty} \frac{\sum_{|d_K|\leqslant n} f(CL'(K))}{\sum_{|d_K|\leqslant n} 1}, \qquad M_0(f) := \lim_{n\to\infty} \frac{\sum_{|G|\leqslant n} f(G)/|\mathrm{Aut}\, G|}{\sum_{|G|\leqslant n} 1/|\mathrm{Aut}\, G|}.$$

Then $M(f) = M_0(f)$. □

Remark. The behavior of the 2-component of the class group of an quadratic field differs entirely from that of the odd component (2.7.3, 3.2.6). This explains why one restricts in the conjecture to the odd component of the class group. □

If P is a property of finite abelian groups of odd order and f the corresponding characteristic function, i.e. $f(G) = 1$ if G has property P and $f(G) = 0$ if G has not the property P, then we write $M_0(P) := M_0(f)$. The following values are in very good agreement with the computional results (Buell (1984), (1987)) according to the Cohen-Lenstra's conjecture:

$$M_0(G \text{ cyclic}) = 97.7575\%, \qquad M_0(3\,||G|) = 43.987\%, \qquad M_0(5\,||G|) = 23.987\%,$$
$$M_0(7\,||G|) = 16.320\%, \qquad M_0(G_3 = \mathbb{Z}/9\mathbb{Z}) = 9.335\%,$$
$$M_0(G_3 = (\mathbb{Z}/3\mathbb{Z})^2) = 1.167\%, \qquad M_0(G_3 = (\mathbb{Z}/3\mathbb{Z})^3) = 0.005\%,$$
$$M_0(G_3 = (\mathbb{Z}/3\mathbb{Z})^4 = 2.3\cdot 10^{-8}, \qquad M_0(G_5 = \mathbb{Z}/25\mathbb{Z}) = 3.802\%,$$
$$M_0(G_5 = (\mathbb{Z}/5\mathbb{Z})^2) = 0.158\%.$$

G_p denotes the p-component of G.

For any ideal $\mathfrak{a} \in \mathfrak{M}(K)$ we identify $N(\mathfrak{a}) := N_{K/\mathbb{Q}}(\mathfrak{a})$ with the positive rational number which generates $N(\mathfrak{a})$. We denote by $\mathfrak{J}(K)$ the semigroup of ideals of $\mathfrak{O}_K$ and by $\mathfrak{P}(K)$ the set of prime ideals of K.

Let $\mathfrak{S}$ be a finite set of prime ideals of K. An $\mathfrak{S}$-*unit* of K is an element α of $K^\times$ such that $\mathrm{supp}(\alpha) \subseteq \mathfrak{S}$ (§3.1).

Theorem 1.58. *Let r_1 be the number of real and $2r_2$ the number of complex isomorphisms of K. Then the group $E_{\mathfrak{S}}$ of $\mathfrak{S}$-units of K is a finitely generated group of rank $|\mathfrak{S}| + r_1 + r_2 - 1$.*

Proof. We have an exact sequence

$$\{1\} \to \mathfrak{O}_K^\times \to E_{\mathfrak{S}} \underset{()}{\to} \mathfrak{M}_{\mathfrak{S}} \to \mathfrak{M}_{\mathfrak{S}}/(E_{\mathfrak{S}}) \to \{1\}$$

where $\mathfrak{M}_{\mathfrak{S}}$ denotes the subgroup of $\mathfrak{M}(K)$ generated by $\mathfrak{S}$. Since

$$\mathfrak{M}_{\mathfrak{S}}/(E_{\mathfrak{S}}) = \mathfrak{M}_{\mathfrak{S}}/\mathfrak{M}_{\mathfrak{S}} \cap (K^\times) = \mathfrak{M}_S(K^\times)/(K^\times) \subseteq CL(K),$$

Theorem 1.13 implies $rk(E_{\mathfrak{S}}) = rk(\mathfrak{O}_K^\times) + rk(\mathfrak{M}_s) = |\mathfrak{S}| + r_1 + r_2 - 1$. □

3.9. Cyclotomic Fields. Let n be a natural number and $\zeta = \zeta_n$ a *primitive n-th root of unity*, i.e. $\zeta^n = 1$ and $\zeta^k \neq 1$ if $0 < k < n$. The field $\mathbb{Q}(\zeta_n)$ is called *n-th cyclotomic field*. In this section we study cyclotomic fields in detail as an illustration of our general theory.

Theorem 1.59. 1) *$\mathbb{Q}(\zeta_n)/\mathbb{Q}$ is a normal extension of degree $\varphi(n)$, where φ denotes the Euler function.*

2) *The conjugates of ζ_n are of the form ζ_n^a with $\bar{a} \in (\mathbb{Z}/n\mathbb{Z})^\times$ and $g_a\colon \zeta_n \to \zeta_n^a$ determines an isomorphism of $(\mathbb{Z}/n\mathbb{Z})^\times$ onto $G(\mathbb{Q}(\zeta_n)/\mathbb{Q})$.*

3) *$\mathbb{Q}(\zeta_n)/\mathbb{Q}$ is unramified for primes q with $q \nmid n$.*

Proof. Let p be a prime and m a natural number. Then ζ_{p^m} is a root of the polynomial $\phi_{p^m}(x) = (x^{p^m} - 1)/(x^{p^{m-1}} - 1)$ of degree $\varphi(p^m)$. The Eisenstein irreducibility criterion (Proposition 1.43) shows that $\phi_{p^m}(x) = (x^{p^{m-1}} - 1)^{p-1} + \cdots + p = (x-1)^{p^{m-1}(p-1)} + \cdots + p$ is irreducible and

$$(\zeta_{p^m} - 1)^{\varphi(p^m)} = (p).$$

Since

$$(\zeta_{p^m}^a - \zeta_{p^m}^b) = (\zeta_{p^m}^{a-b} - 1)|(p) \qquad \text{if } a \not\equiv b \pmod{p^m},$$

the discriminant of ζ_{p^m} is a power of p. Therefore $G(\mathbb{Q}(\zeta_{p^m})/\mathbb{Q})$ is unramified outside p.

Now let $n = \prod_{i=1}^{s} p_i^{m_i}$. Assume that Theorem 1.59 is already proved for a certain s and let $n' = np^m$, $p \nmid n$. Then in $\mathbb{Q}(\zeta_n) \cap \mathbb{Q}(\zeta_{p^m})$ the prime p is fully ramified and unramified, hence $\mathbb{Q}(\zeta_n) \cap \mathbb{Q}(\zeta_{p^m}) = \mathbb{Q}$. This implies $[\mathbb{Q}(\zeta_{n'}) : \mathbb{Q}] = \varphi(n')$. □

Theorem 1.60. *Let q be a prime with $q \nmid n$. Then for any prime divisor $\mathfrak{q}$ of q in $\mathbb{Q}(\zeta_n)$ the Frobenius automorphism F_q of $\mathfrak{q}$ maps ζ_n onto ζ_n^q.*

Proof. By definition we have $F_q\zeta_n \equiv \zeta_n^q \pmod{\mathfrak{q}}$. The equality $x^{n-1} + \cdots + 1 = \prod_{i=1}^{n-1}(x - \zeta_n^i)$ implies $(1 - \zeta_n^i)|n$ if $i \not\equiv 0 \pmod n$, hence $F_q\zeta_n = \zeta_n^q$. □

Now we compute the ramification groups for the prime divisor $\mathfrak{p} = (\zeta_{p^m} - 1)$ of p in $\mathbb{Q}(\zeta_{p^m})$: We have $G := G(\mathbb{Q}(\zeta_{p^m})/\mathbb{Q}) = V_{-1} = V_0$ and $V_i = \{g_a \in G | \zeta_{p^m}^a \equiv \zeta_{p^m} \pmod{\mathfrak{p}^{i+1}}\} = \{g_a \in G \mid \mathfrak{p}^{i+1}|(\zeta_p^{a-1} - 1)\}$, $i = 1, 2, \ldots$. We put

$$G(v) := \{g_a \in G | a \equiv 1 \pmod{p^v}\}.$$

Then

$$V_i = G(1) \quad \text{if } i = 1, \ldots, p-1, \qquad V_i = G(2) \quad \text{if } i = p, \ldots, p^2 - 1, \ldots,$$
$$V_i = \{1\} \quad \text{if } i \geqslant p^{m-1}.$$

Concerning the upper numeration, one finds that all jumps are integers (Theorem 1.56) and that $V^y = G(y)$ if $y \in \mathbb{N}$.

Now one computes the discriminant of $\mathbb{Q}(\zeta_{p^m})/\mathbb{Q}$ by means of Theorem 1.54. One finds the value $n^{\varphi(n)}/p^{n/p}$ where $n = p^m$. Furthermore it follows that $\mathbb{Z}[\zeta_{p^m}]$ is the maximal order of $\mathbb{Q}(\zeta_{p^m})$ since the discriminant of ζ_{p^m} equals $n^{\varphi(n)}/p^{n/p}$.

We go over to general $n \in \mathbb{N}$ by means of Theorem 1.46:

Theorem 1.61. *The discriminant of $\mathbb{Q}(\zeta_n)/\mathbb{Q}$ is equal to $n^{\varphi(n)}/\prod_{p|n} p^{\varphi(n)/(p-1)}$. The maximal order of $\mathbb{Q}(\zeta_n)$ is $\mathbb{Z}[\zeta_n]$.* □

In particular, Theorem 1.59 describes the splitting behavior of unramified primes q in the cyclotomic field $\mathbb{Q}(\zeta_n)$: Since all prime divisors $\mathfrak{Q}$ of q have the same degree f, it is sufficient to know f, which is the order of F_q. By Theorem 1.60, f is the order of $\bar{q}$ in the group $(\mathbb{Z}/n\mathbb{Z})^\times$. The prime q splits completely if and only if $q \equiv 1 \pmod n$.

We mentioned in the introduction to Chap. 1 that the generalization of the quadratic reciprocity law (Appendix 2) played an important role in the development of algebraic number theory. On the other hand the theory of cyclotomic fields delivers the key to the understanding of the quadratic reciprocity law, which was first proved by Gauss in his Disquisitiones arithmeticae. All together Gauss gave six proofs but these proofs gave little insight in the nature of this law. In the following we prove the quadratic reciprocity law as a consequence of Theorem 1.60.

In § 4.5, Example 30, we explain the splitting behavior of primes q in quadratic fields $\mathbb{Q}(\sqrt{D})$ with discriminant D. It is determined by the Legendre symbol (Appendix 2): q is unramified if and only if $q \nmid D$ (Theorem 1.46) and if $q \nmid 2D$, then q splits in the product of two prime ideals if and only if $\left(\frac{D}{q}\right) = 1$. Now let $p \neq q$ be an odd prime. Then $\mathbb{Q}(\zeta_p)$ contains one and only one quadratic field $L := \mathbb{Q}(\sqrt{(-1)^{(p-1)/2}p})$. By Proposition 1.52 q splits in L if and only if the restriction of F_q to L is trivial, i.e. if q is a quadratic residue mod p (Theorem 1.60). On the other hand q splits in the quadratic field L with discriminant $(-1)^{(p-1)/2}p$ if and only if $\left(\frac{(-1)^{(p-1)/2}p}{q}\right) = 1$. Hence

$$\left(\frac{q}{p}\right) = \left(\frac{(-1)^{(p-1)/2}p}{q}\right) = \left(\frac{p}{q}\right)(-1)^{(p-1)(q-1)/2}.$$

This is the quadratic reciprocity law.

From this proof we see that the quadratic reciprocity law has its root in the decomposition behavior of primes in abelian extensions of $\mathbb{Q}$. The right generalization consists therefore in the description of the decomposition behavior of the prime ideals in abelian extensions and more general in normal extensions of arbitrary number fields K by means of invariants defined in K. The solution of this problem for abelian extensions is at the heart of algebraic number theory (Chap. 2). A solution for normal extensions is now in the state of conjectures called Langlands conjectures (Chap. 5).

3.10. Application to Fermat's Last Theorem I (Main reference: Borevich, Shafarevich (1985), Chap. 3). Fermat conjectured that for $n > 2$ the equation

$$x^n + y^n = z^n \tag{1.25}$$

has no solution in natural numbers. This famous conjecture is called *Fermat's last theorem* since Fermat claimed that he had a proof of the conjecture. This proof is not known and the conjecture remained unproven until now. The only general result known at present is Faltings' theorem according to which (1.25) has only finitely many primitive solutions, i.e. solutions x, y, z with g.c.d.$(x, y, z) = 1$. This implies obviously that Fermat's last theorem is true for $n = p^m$, p a prime and m sufficiently large. But there is a vast literature dealing with special cases or with necessary conditions for the existence of a solution of (1.25). See Ribenboim (1979) for a treatment of the history of the problem and its present state.

We consider in this article only a few results with the aim to give an example of the application of algebraic number theory to the solution of a Tiophantine equation.

Since the conjecture for n implies the conjecture for any multiple of n we restrict ourselves to $n = p$ a prime number with $p > 2$. Moreover it is obviously sufficient to consider primitive solutions x, y, z. We say that the *first case of Fermat's last theorem* holds if $p \nmid xyz$ and the *second case* holds if $p|xyz$. In the following $\zeta = \zeta_p$ denotes a primitive root of unity. p is called *regular* if p does not divide the class number of $\mathbb{Q}(\zeta)$. In this section we are going to prove the first case of the Theorem of Kummer. For this purpose we need the following Lemma of Kummer.

Lemma 1.62. *Any unit ε of $\mathbb{Q}(\zeta)$ has the form $\varepsilon = \zeta^i \eta$ with $\eta \in \mathbb{R}$.*

Proof. $\bar{\varepsilon}/\varepsilon$ is a unit with $|\bar{\varepsilon}/\varepsilon| = 1$, hence $\bar{\varepsilon}/\varepsilon = \pm\zeta^j$ (1.3.b)). Furthermore $\varepsilon \equiv \bar{\varepsilon}$ (mod $\zeta - 1$), hence $\bar{\varepsilon} = \zeta^j \varepsilon$. With $j \equiv 2s$ (mod p) we have $\eta := \varepsilon\zeta^s = \bar{\eta}$. □

Now we come to the first case of the Theorem of Kummer.

Theorem 1.63. *Let p be a regular prime, $p > 2$. Then the first case of Fermat's last theorem holds.*

Proof. We assume $x^p + y^p = z^p$ for natural numbers x, y z with g.c.d$(x, y, z) = 1$ and $p \nmid xyz$ and derive a contradiction.

First let $p = 3$. The congruence $x^3 + y^3 \equiv z^3$ (mod 3) implies $x + y \equiv z$ (mod 3). We put $z = x + y + 3u$. Then $x^3 + y^3 = (x + y + 3u)^3 \equiv x^3 + y^3 + 3x^2y + 3xy^2$ (mod 9), hence $0 \equiv x^2y + xy^2 = xy(x + y) \equiv xyz$ (mod 3). This is a contradiction.

Now let $p > 3$. Using the assumption $p \nmid xyz$, one shows that $x + \zeta^i y$ is prime to $x + \zeta^j y$ for $i \not\equiv j$ (mod p). Therefore the equation

$$x^p + y^p = \prod_{i=0}^{p-1} (x + \zeta^i y) = z^p$$

implies $(x + \zeta y) = \mathfrak{a}^p$ for some ideal $\mathfrak{a}$ of $\mathbb{Z}[\zeta]$. Since the class number of $\mathbb{Q}(\zeta)$ is prime to p, $\mathfrak{a}$ is a principal ideal. Hence $x + \zeta y = \varepsilon\alpha^p$ with a unit ε of $\mathbb{Q}(\zeta)$ and $\alpha \in \mathbb{Z}[\zeta]$.

Let $\alpha = a_0 + a_1\zeta + \cdots + a_{p-2}\zeta^{p-2}$ with $a_i \in \mathbb{Z}$. Then $\alpha^p \equiv a_0 + a_1 + \cdots + a_{p-2}$ (mod p). Taking in account Lemma 1.62, we see that there is a congruence

$$\zeta^{-s}(x + \zeta y) \equiv \xi \pmod{p} \tag{1.26}$$

for some $s \in \mathbb{Z}$ and $\xi \in \mathbb{Z}[\zeta] \cap \mathbb{R}$. (1.26) together with its complex conjugate implies

$$\zeta^{-s}x + \zeta^{1-s}y \equiv \zeta^s x + \zeta^{s-1} y \pmod{p}. \tag{1.27}$$

If the exponents $-s$, $1 - s$, s, $s - 1$ are pairwise incongruent mod p we get the contradiction $x \equiv y \equiv 0$ (mod p). In general it is easy to show that (1.27) implies $x \equiv y$ (mod p). In the same way one proves $x \equiv -z$ (mod p). Hence

$$2x \equiv x + y \equiv x^p + y^p = z^p \equiv z \equiv -x \pmod{p}.$$

This implies the desired contradiction $x \equiv 0 \pmod{p}$. □

Eichler (1965) proved with a similar but more complicated argument a much stronger result. For its formulation we introduce some notations. Let $K := \mathbb{Q}(\zeta)$ and $K^+ := \mathbb{Q}(\zeta + \zeta^{-1})$ the maximal real subfield of K. The ideal norm N_{K/K^+} induces an endomorphism of the p-component CL_p of the ideal class group of K. Let CL_p^+ be the image of this endomorphism. In Chap. 2.7.3 we show that CL_p^+ can be identified with the p-component of the class group of K^+.

Theorem 1.64 (Eichler). *Let s be the p-rank of CL_p/CL_p^+. If $\sqrt{p} - 2 > s$, the first case of Fermat's last theorem holds.* ⊠

For further results on Fermat's last theorem in this article see Chap. 2.6.3, Chap. 4.2.6.

§4. Valuations

(Main references: Borevich, Shafarevich (1985), Chap. 4, Hasse (1979), Part 2)

Nowadays algebraic number fields are mostly studied by a mixture of ideal and valuation theory. In this paragraph we represent the valuation theory approach.

A valuation of a field K is the direct generalization of the absolute value of the field $\mathbb{Q}$ of rational numbers. Such as $\mathbb{R}$ is the completion of $\mathbb{Q}$ with respect to the absolute value, a valuation of K leads to a completion of K. The valuations of $\mathbb{Q}$ correspond to the prime numbers and to the absolute value. An arithmetical question about an algebraic number field K is first studied in the completions with respect to the valuations of K, called "local fields". Then one has to go over to the "global" field K by means of a "local-global principle". We shall see this procedure par excellence in §6 and in Chap. 2, §1.

4.1. Definition and First Properties of Valuations. Let K be a field. A *valuation* v of K is a homomorphism of $K^\times$ into the group of positive real numbers satisfying the triangular inequality

$$v(\alpha + \beta) \leqslant v(\alpha) + v(\beta) \qquad \text{for } \alpha, \beta \in K, \alpha + \beta \neq 0. \tag{1.28}$$

By putting $v(0) = 0$ we extend v to a function on K. Then (1.28) is fulfilled for all $\alpha, \beta \in K$. Since the trivial valuation v_0 with $v_0(\alpha) = 1$ for all $\alpha \in K^\times$ is without interest, we assume in the following that a valuation is nontrivial.

A valuation v induces a metric $d(\alpha, \beta) = v(\alpha - \beta)$ under which K becomes a topological field. Two valuations in K are called *equivalent* if they define the same topology in K. An equivalence class of valuations is called a *place*. The set of all places of a field K will be denoted by $P(K)$.

Two valuations v_1 and v_2 of K are equivalent if and only if there is a positive number c such that

$$v_2(\alpha) = v_1(\alpha)^c \qquad \text{for all } \alpha \in K.$$

In the following if we speak about valuations of a field K, we mean always pairwise inequivalent valuations of K.

A valuation v is called *archimedean* if for all α, $\beta \in K^\times$ there is a natural number n such that

$$v(n\alpha) > v(\beta).$$

All other valuations are called *non-archimedean.* An equivalence class of archimedean valuations is called an *infinite place.* An equivalence class of non-archimedean valuations is called a *finite place.*

Proposition 1.65. *Let v be a non-archimedean valuation. Then*

$$v(\alpha + \beta) \leqslant \max\{v(\alpha), v(\beta)\} \qquad \text{for all } \alpha, \beta \in K$$

and

$$v(\alpha + \beta) = \max\{v(\alpha), v(\beta)\} \qquad \text{if } v(\alpha) \neq v(\beta). \ \square$$

A valuation v is called *discrete* if the set of values $\log v(\alpha)$ for $\alpha \in K^\times$ is discrete in $\mathbb{R}$. A discrete valuation is always non-archimedean but the converse is not true in general as we will see in the following. If v is a non-archimedean valuation, then $-\log v$ is called *exponential valuation.*

Let v be a discrete valuation of the field K. The set

$$\mathfrak{O}_v = \{\alpha \in K \,|\, v(\alpha) \leqslant 1\}$$

forms a ring called the *valuation ring of v.* The ring $\mathfrak{O}_v$ has a unique prime ideal $\mathfrak{p}_v = \{\alpha \in K \,|\, v(\alpha) < 1\}$ and it is a principal ideal ring. Two elements $\alpha, \beta \in \mathfrak{O}_v$ are associated if and only if $v(\alpha) = v(\beta)$. The field $\mathfrak{O}_v/\mathfrak{p}_v$ is called the *residue class field of v.* The exponent of an element α of K with respect to $\mathfrak{p}_v$ will be denoted by $\mathrm{v}_v(\alpha)$, i.e. $\alpha\mathfrak{O}_v = \mathfrak{p}_v^{\mathrm{v}_v(\alpha)}$. We put $\mathrm{v}_v(0) = \infty$. If π is a prime element of $\mathfrak{O}_v$, then

$$v(\pi)^{\mathrm{v}_v(\alpha)} = v(\alpha) \qquad \text{for } \alpha \in K^\times.$$

Independent of valuation theory a valuation ring is defined as a domain V with quotient field K such that for all $\alpha \in K^\times$, α or α^{-1} belongs to V. The elements β of V with $\beta^{-1} \notin V$ form an ideal $\mathfrak{P}$ in V. Hence $\mathfrak{P}$ is the unique maximal ideal of V. If R is a Dedekind ring with quotient field K, then the valuation rings in K containing R are the local rings $R_\mathfrak{p}$ (§ 3.3) for prime ideals $\mathfrak{p}$ of R and R is the intersection of these valuation rings. One defines a valuation $v_\mathfrak{p}$ in K by

$$v_\mathfrak{p}(\alpha) = e^{-\mathrm{v}_\mathfrak{p}(\alpha)} \qquad \text{for } \alpha \in K^\times \text{ (§ 3.1)}. \tag{1.29}$$

Then $R_\mathfrak{p}$ is the valuation ring of $v_\mathfrak{p}$.

Valuation rings arise in a natural way in complex function theory of one variable: If $\mathfrak{p}$ is a point of a Riemann surface F, then the set of meromorphic functions on F which are holomorphic in $\mathfrak{p}$ form a valuation ring. If F is closed, one can show that the corresponding valuations $v_\mathfrak{p}$ exhaust the valuations of the field of meromorphic functions f on F with $v_\mathfrak{p}(c) = 1$ for $c \in \mathbb{C}^\times$. The fact that the number of zeroes of f is equal to the number of poles of f can be expressed in the formula

$$\prod_{\mathfrak{p} \in F} v_{\mathfrak{p}}(f) = 1. \tag{1.30}$$

The following example describes the valuations of algebraic number fields.

Example 24. Let K be an algebraic number field and $\varphi: K \to \mathbb{C}$ an isomorphism of K in $\mathbb{C}$. Then

$$v_\varphi(\alpha) := |\varphi(\alpha)| \qquad \text{for } \alpha \in K$$

is an archimedean valuation of K.

Let $\mathfrak{p}$ be a prime ideal of $\mathfrak{O}_K$. Then

$$v_{\mathfrak{p}}(\alpha) := N(\mathfrak{p})^{-\nu_{\mathfrak{p}}(\alpha)} \qquad \text{for } \alpha \in K^\times,\ v_{\mathfrak{p}}(0) = 0$$

is a non-archimedean valuation of K.

One has the relation

$$\prod_{\varphi} v_\varphi(\alpha) \prod_{\mathfrak{p}} v_{\mathfrak{p}}(\alpha) = 1 \qquad \text{for all } \alpha \in K^\times \tag{1.31}$$

where φ runs over all isomorphisms of K in $\mathbb{C}$ and $\mathfrak{p}$ runs over all prime ideals of $\mathfrak{O}_K$. (1.31) is well defined since for given $\alpha \in K$ for almost all $\mathfrak{p}$ one has $v_{\mathfrak{p}}(\alpha) = 1$. It is only a reformulation of the formula

$$|N(\alpha)| = N\left(\prod_{\mathfrak{p}} \mathfrak{p}^{\nu_{\mathfrak{p}}(\alpha)}\right) = \prod_{\mathfrak{p}} N(\mathfrak{p})^{\nu_{\mathfrak{p}}(\alpha)}. \ \square$$

Proposition 1.66. *Let K be an algebraic number field. The valuations defined in example 24 are up to equivalence all valuations of K. Two distinct valuations v_1, v_2 in example 24 are equivalent if and only if they are of the form $v_1 = v_{\varphi_1}$, $v_2 = v_{\varphi_2}$ with complex conjugated isomorphisms φ_1 and φ_2.* ⊠

The valuations defined in example 24 are called *normalized* with the exception of v_φ where φ is a complex isomorphism. In the latter case, v_φ^2 is called *normalized* though this is not a valuation since the triangular inequality is violated. But one can write (1.31) in the nice form

$$\prod_{v} v(\alpha) = 1 \qquad \text{for all } \alpha \in K^\times \tag{1.32}$$

where the product runs over all normalized valuations v of K. The *product formulas* (1.30) and (1.32) show the analogy between algebraic number fields and algebraic function fields.

In the following $|\ |_v$ denotes the normalized valuation of the places.

In §3.8 we have introduced the notion of $\mathfrak{S}$-units of an algebraic number field K, where $\mathfrak{S}$ was a set of prime ideals. In the language of valuation theory this notion can be reformulated as follows: Now $\mathfrak{S}$ denotes a finite set of places containing the set $\mathfrak{S}_\infty$ of all infinite places. $\alpha \in K^\times$ is an $\mathfrak{S}$-unit if $v(\alpha) = 1$ for all places $v \in P(K) - \mathfrak{S}$.

The group $E_{\mathfrak{S}}$ of $\mathfrak{S}$-units is a finitely generated group of rank $|\mathfrak{S}| - 1$ (Theorem 1.58). The torsion group of $E_{\mathfrak{S}}$ is the cyclic group of roots of unity in K. Let ζ be a generator of this group and let $\zeta, \varepsilon_1, \ldots, \varepsilon_{|\mathfrak{S}|-1}$ be a system of generators of $E_{\mathfrak{S}}$. Then $\varepsilon_1, \ldots, \varepsilon_{|\mathfrak{S}|-1}$ is called a fundamental system of units of $E_{\mathfrak{S}}$.

The *regulator* $R_{\mathfrak{S}}$ of $E_{\mathfrak{S}}$ is defined by

$$R_{\mathfrak{S}} := \det(\log |\varepsilon_i|_v)_{i,v}$$

where $i = 1, \ldots, |\mathfrak{S}| - 1$ and v runs through the places of $\mathfrak{S} - \{v_0\}$ excluding one arbitrary fixed place $v_0 \in \mathfrak{S}$ (compare the regulator definition in § 1.3). $R_{\mathfrak{S}}$ is independent of the choice of the fundamental system of units and of the choice of v_0. One shows $R_{\mathfrak{S}} \neq 0$ in the same way as we have shown $R \neq 0$ for the usual regulator (§ 1.3).

One has the following characterization of algebraic number fields:

Theorem 1.67. *Let K be a field of characteristic* 0. *The following conditions are equivalent:*

1) *K is an algebraic number field.*

2) *K has at least one valuation being either archimedean or discrete with finite residue class field and there are a set $\mathfrak{S}$ of valuations of K and positive numbers a_v for $v \in \mathfrak{S}$ such that for every $\alpha \in K^\times$ one has $v(\alpha) \neq 1$ only for a finite set of valuations in $\mathfrak{S}$ and*

$$\prod_{v \in \mathfrak{S}} v(\alpha)^{a_v} = 1.$$

3) *The non-archimedean valuations of K are discrete with finite residue class field. For every $\alpha \in K^\times$ there are only finitely many non-archimedean valuations such that $v(\alpha) \neq 1$.* ⊠

The chinese remainder theorem (Proposition 1.38) can be strengthened in the setting of valuation theory:

Theorem 1.68. (Strong approximation theorem). *Let v_0 be any valuation of the algebraic number field K and let $\mathfrak{S}$ be a finite set of valuations of K distinct from v_0. Furthermore let α_v, $v \in \mathfrak{S}$, be arbitrary elements of K.*

Then for every $\varepsilon > 0$ there is a $\beta \in K$ such that $v(\beta - \alpha_v) < \varepsilon$ for all $v \in \mathfrak{S}$ and $v'(\beta) \leqslant 1$ for all valuations v' of K which are not in $\mathfrak{S} \cup \{v_0\}$. (Cassels, Fröhlich (1967), Chap. 2.15). ⊠

The exceptional role of one valuation v_0 in the strong approximation theorem is natural from the point of view of the product formula (1.32) since the formula determines one valuation if all other valuations of K are given.

Example 25. Let R be a ring with divisor theory $R^* \to \mathfrak{D}$. Then any prime divisor $\mathfrak{p}$ of $\mathfrak{D}$ determines a discrete valuation $v_{\mathfrak{p}}$ of the quotient field $Q(R)$ by

$$v_{\mathfrak{p}}(\alpha) = \rho^{\nu_{\mathfrak{p}}(\alpha)} \qquad \text{for } \alpha \in Q(R)^\times$$

where ρ is an arbitrary fixed number with $0 < \rho < 1$. □

4.2. Completion of a Field with Respect to a Valuation. Let K be an arbitrary field and v a valuation of K. Exactly as in the case of the usual absolute value of the field $\mathbb{Q}$ one can use Cauchy sequences to construct the *completion of K with respect to v.*

A v-Cauchy sequence in K is a sequence $\alpha_1, \alpha_2, \ldots$ of elements of K such that for every $\varepsilon > 0$ there exists a natural number $N(\varepsilon)$ with

$$v(\alpha_n - \alpha_m) < \varepsilon \qquad \text{for } n, m > N(\varepsilon).$$

K is called *complete with respect to v* if every v-Cauchy sequence in K converges. An extension K' of K with a valuation v' which is an extension of v to K' is called completion of K with respect to v if K' is complete with respect to v' and K is dense in K'. The completion is unique in the sense that if K'', v'' is another completion, then there is a unique isomorphism $\varphi\colon K' \to K''$ of valuated fields such that the restriction of φ to K is the identity.

The set C of all v-Cauchy sequences in K form a ring with addition and multiplication defined elementwise. To C belong the sequences $\alpha_1, \alpha_2, \ldots$ with $\lim_{n\to\infty} \alpha_n = 0$ in the topology defined by v.These 0-sequences form a maximal ideal M in C. The field C/M is the completion K_v of K. The valuation v is extended to K_v in the obvious way. The extended valuation will again be denoted by v without risk of confusion.

The completions of an algebraic number field K are called *local fields.* An archimedean valuation v_φ is called real (resp. complex) if φ is a real (resp. complex) isomorphism. Then the completion of K with respect to v_φ is isomorphic to $\mathbb{R}$ (resp. $\mathbb{C}$). The completion $K_\mathfrak{p}$ of K with respect to a non-archimedian valuation $v_\mathfrak{p}$ is called $\mathfrak{p}$-adic number field. Let p be the characteristic of the residue field of $v_\mathfrak{p}$. Then $K_\mathfrak{p}$ is a finite extension of the field $\mathbb{Q}_p$ of rational p-adic numbers. The residue class field of $v_\mathfrak{p}$ is the finite field with $N(\mathfrak{p})$ elements.

In the case of archimedean valuations one has a general result about the possible complete fields due to Ostrowski.

Theorem 1.69. *Let K be a complete field with respect to an archimedean valuation. Then K is isomorphic to $\mathbb{R}$ or $\mathbb{C}$ and the valuation is in both cases the ordinary absolute value* (Hasse (1979), Chap. 13). ☒

4.3. Complete Fields with Discrete Valuation. If K is a field with discrete valuation v, then the completion of K_v has again discrete valuation and $v(K) = v(K_v)$.

Let $\mathfrak{O}$ be the valuation ring of v and π a generator of the prime ideal $\mathfrak{p}$ of $\mathfrak{O}$. Then π is called a prime element of K and the characteristic of $\mathfrak{O}/\mathfrak{p}$ is called the residue characteristic of K. We put $v := v_v$. Obviously for every sequence $\alpha_0, \alpha_1, \alpha_2, \ldots$ of elements in $\mathfrak{O}$ the series $\alpha_0 + \alpha_1\pi + \alpha_2\pi^2 + \cdots$ converges to an element in the valuation ring $\mathfrak{O}_v$ of K_v.

Proposition 1.70. *Let R be a system of representatives of the classes of $\mathfrak{O}/\mathfrak{p}$ in $\mathfrak{O}$. Then every element α of $K^\times$ can be uniquely represented in the form*

$$\alpha = \sum_{\nu=\nu_0}^{\infty} \alpha_\nu \pi^\nu \qquad \text{with } \alpha_\nu \in R, \tag{1.33}$$

$\nu_0 = \nu(\alpha)$, $\alpha_{\nu_0} \neq 0$. □

Example 26. Let $K = \mathbb{C}(z)$, u a complex number and let $v = v_u$ be the valuation of K given by

$$v_u(f) = e^{-\nu_u(f)}, \qquad f \in K^\times,$$

where $\nu_u(f)$ is the multiplicity of u in f, i.e.

$$f(z) = (z-u)^{\nu_u(f)} f_1(z), \qquad f_1(u) \in \mathbb{C}^\times.$$

(compare (1.29)). Then the valuation ring $\mathfrak{O}$ is the ring of all functions $f \in K^\times$ which have no pole in u, $\mathfrak{p}$ is the ideal of all $f \in \mathfrak{O}$ which have a zero in u and $\mathfrak{O}/\mathfrak{p} = \mathbb{C}$. Hence we can put $R := \mathbb{C}$ and $\pi = z - u$ and (1.33) is the so called formal power series

$$f(z) = \sum_{\nu=\nu_0}^{\infty} c_\nu (z-u)^\nu, \qquad c_\nu \in \mathbb{C}, \qquad \nu_0 = \nu_u(f),$$

for $f \in K$, which in this case is convergent in the usual sense of function theory. □

Remark. The origin of the valuation theoretical method in number theory was the observation made by Hensel that one can operate with series of the form (1.33) in a similar manner as with power series in function fields.

Proposition 1.71 (Hensel's lemma). *Let K be complete with respect to a discrete valuation v. Let $f(x) \in \mathfrak{O}[x]$ be a normed polynomial. Suppose that $\mathfrak{g}(x)$ and $\mathfrak{h}(x)$ are polynomials in $(\mathfrak{O}/\mathfrak{p})[x]$ which are relatively prime and such that*

$$\bar{f}(x) = \mathfrak{g}(x)\mathfrak{h}(x).$$

Then there exist polynomials $g(x), h(x) \in \mathfrak{O}[x]$ such that

$$f(x) = g(x)h(x)$$

and

$$\bar{g}(x) = \mathfrak{g}(x), \qquad \bar{h}(x) = \mathfrak{h}(x).$$

Proof. We choose polynomials $g_1(x), h_1(x) \in \mathfrak{O}[x]$ with $\bar{g}_1(x) = \mathfrak{g}(x)$, $\bar{h}_1(x) = \mathfrak{h}(x)$. Then $f(x) - g_1(x)h_1(x) = \pi w(x)$ with $w(x) \in \mathfrak{O}[x]$ and π a prime element of K. Now we consider

$$g_2(x) = g_1(x) + \pi u(x),\ h_2(x) = h_1(x) + \pi v(x)$$

with polynomials $u(x)$, $v(x)$ still to be determined. We get

$$f(x) - g_2(x)h_2(x) \equiv \pi(w(x) - g_1(x)v(x) - h_1(x)u(x)) \pmod{\pi^2}.$$

Since $g(x)$ and $h(x)$ are relatively prime we can choose $u(x)$ and $v(x)$ such that

$$w(x) - g_1(x)v(x) - h_1(x)u(x) \equiv 0 \pmod{\pi}.$$

Now assume that we have already constructed $g_n(x), h_n(x) \in \mathfrak{O}[x]$ such that $\bar{g}_n(x) = \mathfrak{g}(x), \bar{h}_n(x) = \mathfrak{h}(x)$ and $f(x) \equiv g_n(x)h_n(x) \pmod{\pi^n}$. Then we find polynomials $u_n(x), v_n(x) \in \mathfrak{O}[x]$ such that

$$g_{n+1}(x) := g_n(x) + \pi^n u_n(x), \qquad h_{n+1}(x) := h_n(x) + \pi^n v_n(x)$$

satisfy

$$f(x) \equiv g_{n+1}(x)h_{n+1}(x) \pmod{\pi^{n+1}}.$$

It follows

$$f(x) = \lim_{n\to\infty} g_n(x) \lim_{n\to\infty} h_n(x). \ \square$$

Example 27. Let K be a $\mathfrak{p}$-adic number field and $f(x) = x^{N(\mathfrak{p})} - x$. Then $\bar{f}(x)$ splits in distinct linear factors. Hensel's lemma implies that the same is true for $f(x)$ in $\mathfrak{O}[x]$. The roots of $f(x)$ are a system of representatives of the classes in $\mathfrak{O}/\mathfrak{p}$ in $\mathfrak{O}$, called the distinguished system of representatives. □

Hensel's lemma gives some knowledge about polynomials in $\mathfrak{O}[x]$ if we know something about the corresponding polynomial in $(\mathfrak{O}/\mathfrak{p})[x]$. More general one would like to compare $f \in \mathfrak{O}[x]$ with the corresponding polynomial in $(\mathfrak{O}/\mathfrak{p}^h)[x]$. In this direction one has the following more complicated version of Hensel's lemma:

Proposition 1.72. *Let $f(x), g_0(x), h_0(x)$ be polynomials in $\mathfrak{O}[x]$ such that the following conditions are fulfilled:*

1) *The highest coefficients of $f(x)$ and $g_0(x)h_0(x)$ coincide.*
2) *The exponent $r = v(R(g_0, h_0))$ of the resultant of g_0 and h_0 is finite.*
3) $f(x) \equiv g_0(x)h_0(x) \pmod{\pi^{2r+1}}$.

Then there are polynomials $g(x), h(x) \in \mathfrak{O}[x]$ such that $f(x) = g(x)h(x)$, $g(x) \equiv g_0(x)$, $h(x) \equiv h_0(x) \pmod{\pi^{r+1}}$. ⊠ (Borevich, Shafarevich (1985), Chap. 4.3).

4.4. The Multiplicative Structure of a $\mathfrak{p}$-adic Number Field. Let K be a $\mathfrak{p}$-adic number field and $\mathfrak{O}$ the valuation ring of K. The prime ideal of $\mathfrak{O}$ will be denoted by $\mathfrak{p}$ too. The residue characteristic of K will be denoted by p. In this section we want to determine the structure of $K^\times$.

Let π be a prime element of $\mathfrak{O}$, i.e. $\mathfrak{p} = \pi\mathfrak{O}$. Then $K^\times = (\pi) \times \mathfrak{O}^\times$. Furthermore let $q := N(\mathfrak{p}) = [\mathfrak{O} : \mathfrak{p}]$. By example 27 $\mathfrak{O}$ contains the group μ_{q-1} of roots of unity of order $q - 1$. Hence

$$K^\times = (\pi) \times \mu_{q-1} \times U_1$$

where $U_1 := 1 + \mathfrak{p}$ is called the group of principal units. More general we put $U_n := 1 + \mathfrak{p}^n$ for $n = 1, 2, \ldots$ and $U_0 = U := \mathfrak{O}^\times$. The groups U_n, $n = 1, 2, \ldots$, form a basis of neighbourhoods of 1 in the topology of $K^\times$ induced from the topology of K.

The group $\mathbb{Z}_p$ of p-adic integers acts on U_1 in a natural way:

$$(1 + \alpha\pi^t)^p \in 1 + \mathfrak{p}^{t+1} \qquad \text{for } \alpha \in \mathfrak{O}, t \in \mathbb{N}$$

implies

$$\eta^m \in 1 + \mathfrak{p}^{s+1} \qquad \text{for } \eta \in U_1,\ p^s | m. \tag{1.34}$$

Now let $g = \lim_{n\to\infty} g_n \in \mathbb{Z}_p$, $g_n \in \mathbb{Z}$ and $\eta \in U_1$. Then η^{g_n} converges to an element of U_1 which is independent of the choice of the sequence $g_1, g_2, \ldots$ converging to g. We put

$$\eta^g := \lim_{n\to\infty} \eta^{g_n}$$

This defines a continuous mapping $\mathbb{Z}_p \times U_1 \to U_1$ such that U_1 becomes a $\mathbb{Z}_p$-module.

Proposition 1.73. *The $\mathbb{Z}_p$-module U_1 is the direct sum of the group μ_{p^c} of roots of unity of p-power order in K and a free $\mathbb{Z}_p$-module of rank $[K:\mathbb{Q}_p]$.* □

We give a short proof for this proposition in 4.9 by means of the p-adic logarithm, but in this section we consider the finer question of the structure of U_1 as filtered group with the filtration $\{U_n | n = 1, 2, \ldots\}$.

Let e be the ramification index and f the inertia degree of $\mathfrak{O}$ with respect to $\mathbb{Z}_p$ and ζ_{p^i} a generator of μ_{p^i}, $i = 1, \ldots, c$. The map $\varphi_n \colon \mathfrak{O} \to U_n$, defined by $\varphi_n(\alpha) = 1 + \alpha\pi^n$, induces a group isomorphism of $(\mathfrak{O}/\mathfrak{p})^+$ onto U_n/U_{n+1}. Moreover

$$1 + \alpha\pi^n \to (1 + \alpha\pi^n)^p = 1 + \alpha\pi^n p + \cdots + \alpha^p\pi^{np} \qquad \text{for } \alpha \in \mathfrak{O}$$

induces a group isomorphism

$$U_n/U_{n+1} \to U_{np}/U_{np+1} \qquad \text{if } 1 \leqslant n < e/(p-1),$$

$$U_n/U_{n+1} \to U_{n+e}/U_{n+e+1} \qquad \text{if } n > e/(p-1),$$

$$U_n/U_{n+1} \to U_{n+e}/U_{n+e+1} \qquad \text{if } n = e/(p-1) \text{ and } \mu_p \not\subset U_1,$$

$$U_n/(U_{n+1}, 1+\gamma\pi^n) \to A \subset U_{n+e}/U_{n+e+1} \qquad \text{if } n = e/(p-1) \text{ and } \mu_p \subset U_1 \tag{1.35}$$

where γ is a solution of the congruence

$$1 \equiv -\gamma^{p-1}\pi^{n(p-1)}p^{-1} \pmod{\pi}$$

and A is the subgroup of index p in U_{n+e}/U_{n+e+1} of the elements $\overline{1 + \beta(1-\zeta_p)p}$ with $\beta \in \mathfrak{O}$, $\operatorname{Tr}\bar{\beta} = 0$ where Tr is the trace from $\mathfrak{O}/\mathfrak{p}$ to $\mathbb{Z}_p/p\mathbb{Z}_p$.

For the proof of (1.35) we remark that $\pi^{e/(p-1)} \sim (1-\zeta_p) =: \lambda$ and $0 = ((1-\lambda)^p - 1)/p\lambda \equiv -\lambda^{p-1}p^{-1} - 1 \pmod{\lambda}$. □

In the case $\mu_p \not\subset U_1$ we get a simple description of U_1: Let $\bar{\omega}_1, \ldots, \bar{\omega}_f$ be a set of generators of $(\mathfrak{O}/\mathfrak{p})^+$ and $\eta_{\nu\mu} := 1 + \omega_\nu\pi^\mu$. Then

$$\{\eta_{\nu\mu} | \nu = 1, \ldots, f,\ 1 \leqslant \mu \leqslant pe/(p-1),\ p \nmid \mu\} \tag{1.36}$$

is a basis of the free $\mathbb{Z}_p$-module U_1.

If $\mu_p \subset U_1$ according to (1.35) one needs one generator η_* more, which lies in $U_{ep/(p-1)}$. We can choose η_* in the form

$$\eta_* = 1 + \omega_*(1-\zeta_p)p$$

with $\operatorname{Tr}\bar{\omega}_* \neq 0$.

Between the generators (1.36) and η_* there is one generating relation corresponding to $\zeta_{p^c}^{p^c} = 1$. If

$$\mu := v_p(1 - \zeta_{p^c}) = e/(p-1)p^{c-1} \not\equiv 0 \pmod p,$$

we can take $\eta_{1\mu} = \zeta_{p^c}$. If $p|\mu$ the relation has a more complicated form. For more details see Hasse (1979), Chap. 15.

Example 28. $K = \mathbb{Q}_p$. If $p \neq 2$, then $U_1 = (1+p)$. If $p = 2$, then $U_1 = (-1, 5)$. □

If K is an algebraic number field and $\mathfrak{p}$ a prime ideal of K, then

$$\mathfrak{O}_K/\mathfrak{p}^h \cong \mathfrak{O}_\mathfrak{p}/\mathfrak{p}^h \qquad \text{for all } h = 1, 2, \ldots$$

where $\mathfrak{O}_\mathfrak{p}$ denotes the valuation ring of $K_\mathfrak{p}$. Furthermore

$$(\mathfrak{O}_\mathfrak{p}/\mathfrak{p}^h)^\times \cong \mathfrak{O}_\mathfrak{p}^\times/U_h.$$

Therefore the structure of the groups $(\mathfrak{O}_K/\mathfrak{p}^h)^\times$ is determined by the considerations above.

4.5. Extension of Valuations

Proposition 1.74. *Let v be a valuation of a field K and L a finite separable extension of K of degree n.*

1) *If K is complete with respect to v, then there is one and only one extension v' of v to L, which is given by*

$$v'(\alpha) = v(N_{L/K}(\alpha))^{1/n} \qquad \textit{for } \alpha \in L.$$

L is complete with respect to v.

2) *If K is not complete with respect to v, then the extensions of v to L correspond to the component fields $L_1, \ldots, L_s$ of the tensor product*

$$L \otimes_K K_v = L_1 \oplus \cdots \oplus L_s. \tag{1.37}$$

The fields L_ν, $\nu = 1, \ldots, s$, are complete and the corresponding valuation v_ν of L is given by the canonical injection of L into L_ν.

3) *In the case that v is a discrete valuation, $\mathfrak{O}_{v_\nu}/\mathfrak{O}_v$ is an extension of Dedekind rings with prime ideals $\mathfrak{P}_\nu|\mathfrak{p}_v$ and we have*

$$[L_\nu : K_v] = e_\nu f_\nu \tag{1.38}$$

where e_ν is the ramification index and f_ν is the inertia degree of $\mathfrak{P}_\nu$ with respect to $\mathfrak{p}_v$.

4) *If L is given by an irreducible polynomial $f(x)$ with coefficients in K, i.e. $L \cong K[x]/(f(x))$, then the complete fields L_ν, $\nu = 1, \ldots, s$, correspond to the irreducible factors $f_\nu(x)$ of $f(x)$ over K_v, i.e. $L_\nu \cong K_v[x]/(f_\nu(x))$.* ⊠

Now let K be the quotient field of a Dedekind ring R and let S be the integral closure of R in the finite separable extension L of K. Let $\mathfrak{p}$ be a prime ideal of R and $\mathfrak{P}_1, \ldots, \mathfrak{P}_s$ the prime divisors of $\mathfrak{p}$ in S. Furthermore let v resp. $v_1, \ldots, v_s$ be the valuations of K rep. L which correspond to $\mathfrak{p}$ resp. $\mathfrak{P}_1, \ldots, \mathfrak{P}_s$. Then $v_1, \ldots, v_s$ is (up to equivalence) the set of all extensions of v onto L and (1.37), (1.38) reflect the formula (1.18). We put $K_\mathfrak{p} := K_v$, $L_{\mathfrak{P}_i} := L_{v_i}$, $\mathfrak{O}_\mathfrak{p} := \mathfrak{O}_v$, $\mathfrak{O}_{\mathfrak{P}_i} := \mathfrak{O}_{v_i}$. The prime ideal of the valuation ring $\mathfrak{O}_v$ of K_v will be denoted by $\mathfrak{p}$ if there is no danger of confusion.

Theorem 1.75. $v_{\mathfrak{P}_i}(\mathfrak{D}_{S/R}) = v_{v_i}(\mathfrak{D}_{\mathfrak{O}_{\mathfrak{P}_i}/\mathfrak{O}_\mathfrak{P}})$, $i = 1, \ldots, s$,

$$v_\mathfrak{p}(\mathfrak{d}_{S/R}) = \sum_{i=1}^{s} v_v(\mathfrak{d}_{\mathfrak{O}_{\mathfrak{P}_i}/O_\mathfrak{P}}). \boxtimes$$

Theorem 1.75 shows that we can compute the different and discriminant of S/R if we know the "*local*" differents and discriminants. Roughly speaking the "*global*" different (resp. discriminant) is the product of the "*local*" differents (resp. discriminants). The computation of the local differents and discriminants is easy if we know local prime elements as we shall see in the next section.

Several notions remain unchanged and are easier computable if one goes over to the completions of L/K. E.g. let L/K be a normal extension, let $\mathfrak{P}$ be a prime ideal of S, and $\mathfrak{p} = \mathfrak{P} \cap R$. Then $L_\mathfrak{P}/K_\mathfrak{p}$ is a normal extension, and the restriction of $G(L_\mathfrak{P}/K_\mathfrak{p})$ to L defines an injection of $G(L_\mathfrak{P}/K_\mathfrak{p})$ into $G(L/K)$ whose image is the decomposition group $G_\mathfrak{P}(L/K)$. More general, the restriction of $V_n(\mathfrak{P}, L_\mathfrak{P}/K_\mathfrak{p})$ to L is $V_n(\mathfrak{P}, L/K)$, $n = -1, 0, \ldots$. In the following we identify $V_n(\mathfrak{P}, L_\mathfrak{P}/K_\mathfrak{p})$ and $V_n(\mathfrak{P}, L/K)$. If w is an archimedean valuation of L with the restriction v to K then we define the decomposition group G_w of w with respect to L/K by $G(L_w/K_v)$ considered as subgroup of $G(L/K)$.

For convenience of language we say that a *place w of L lies above the place v of K* if the valuations in w are continuations of the valuations in v. In this case we write $w|v$. Equivalence classes of archimedean (non-archimedean) valuations are called infinite (finite) places. If K is an algebraic number field, the finite places of K will be identified with the prime ideals of K. These notations have their origin in the theory of closed Riemann surfaces, where the points of the surface F correspond to the equivalence classes of valuations of the field K of meromorphic functions on F (§4.1). If L/K is a finite extension, then the Riemann surface of L is a covering of F, hence the places w of L lie over places v of K.

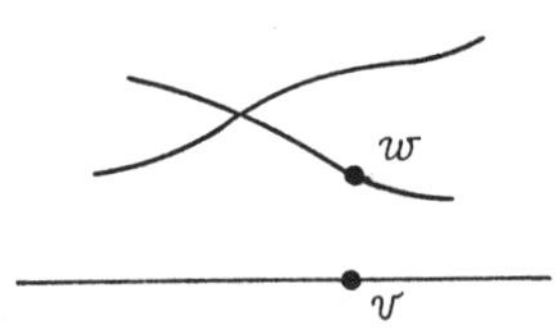

Example 29. Let $L = K[x]/(f(x))$ with an irreducible normed polynomial $f(x) \in R[x]$. Furthermore let $\mathfrak{p}$ be a prime ideal of R and

$$\bar{f}(x) = \mathfrak{f}_1(x) \ldots \mathfrak{f}_s(x)$$

the decomposition of $\bar{f}(x) = f(x) \pmod{\mathfrak{p}}$ in irreducible polynomials over $R/\mathfrak{p}$.

We assume that the discriminant of f is prime to $\mathfrak{p}$, then the polynomials $\mathfrak{f}_1(x), \ldots, \mathfrak{f}_s(x)$ are pairwise distinct. Therefore Hensel's lemma implies that $f(x) = f_1(x) \ldots f_s(x)$ with irreducible polynomials $f_\nu(x) \in R_\mathfrak{p}[x]$, $\nu = 1, \ldots, s$, with $\overline{f_\nu}(x) = \mathfrak{f}_\nu(x)$. Hence $\mathfrak{p}$ has in S the decomposition $\mathfrak{p} = \mathfrak{P}_1 \ldots \mathfrak{P}_s$ in pairwise distinct prime ideals $\mathfrak{P}_\nu$ with inertia degree $\deg f_\nu$. □

Example 30. Let d be a squarefree number in $\mathbb{Z}$, $K = \mathbb{Q}$, $L = \mathbb{Q}(\sqrt{d})$. We consider the decomposition of primes p in prime ideals in $\mathfrak{O}_L$. If $p \nmid 2d$, by definition of the Legendre symbol $\left(\frac{d}{p}\right)$ the polynomial $\bar{f}(x) = x^2 - \bar{d}$ is irreducible over $\mathbb{Z}/p\mathbb{Z}$ if and only if $\left(\frac{d}{p}\right) = -1$. For $\left(\frac{d}{p}\right) = 1$ we have $x^2 - d \equiv (x-a)(x+a) \pmod p$ with $a \in \mathbb{Z}$. Since $a \not\equiv -a \pmod p$, it follows from Hensel's lemma that $f(x)$ splits over $\mathbb{Q}_p$. Hence

$$(p) = \mathfrak{P}_1\mathfrak{P}_2 \quad \text{if } p \nmid 2d, \left(\frac{d}{p}\right) = 1, \qquad (p) = \mathfrak{P} \quad \text{if } p \nmid 2d, \left(\frac{d}{p}\right) = -1.$$

Furthermore it is easy to see, that in the first case

$$\mathfrak{P}_1 = (p, \sqrt{d} - a), \qquad \mathfrak{P}_2 = (p, \sqrt{d} + a).$$

If $p \mid d$, the polynomial $x^2 - d$ is irreducible mod p^2 and therefore irreducible over $\mathbb{Q}_p$.

$$(p) = \mathfrak{P}^2 \qquad \text{if } p \mid d \text{ with } \mathfrak{P} = (p, \sqrt{d}).$$

If $p = 2$, $d \equiv 3 \pmod 4$, the polynomial $x^2 - d$ is irreducible mod 4 and therefore irreducible over $\mathbb{Q}_2$.

$$(p) = \mathfrak{P}^2 \qquad \text{if } p = 2, d \equiv 3 \pmod 4 \text{ with } \mathfrak{P} = (2, \sqrt{d} + 1).$$

If $p = 2$, $d \equiv 1 \pmod 8$, we have

$$x^2 - d \equiv x^2 - 1 \equiv (x-1)(x+1) \pmod 8.$$

Hence $x^2 - d$ splits over $\mathbb{Q}_2$ by Proposition 1.72.

$$(p) = \mathfrak{P}_1\mathfrak{P}_2 \qquad \text{if } p = 2, p \equiv 1 \pmod 8 \text{ with } \mathfrak{P}_1 = (2, (\sqrt{d}+1)/2),$$
$$\mathfrak{P}_2 = (2, (\sqrt{d}-1)/2).$$

If $p = 2$, $d \equiv 5 \pmod 8$, the polynomial $x^2 - d$ is irreducible mod 8 and therefore irreducible over $\mathbb{Q}_2$.

$$(p) = (\mathfrak{P}) \qquad \text{if } p = 2, p \equiv 5 \pmod 8. \ \square$$

By means of the quadratic reciprocity law one shows that the decomposition behavior of a prime p in the quadratic field with discriminant D depends only on the residue class of p mod $|D|$.

4.6. Finite Extensions of $\mathfrak{p}$-adic Number Fields. Let K be a $\mathfrak{p}$-adic number field, i.e. the completion of an algebraic number field with respect to the valuation which corresponds to a prime ideal $\mathfrak{p}$ of $\mathfrak{O}_K$. Let p be the residue characteristic of $\mathfrak{p}$. Then K is a finite extension of $\mathbb{Q}_p$. On the other hand every finite extension of $\mathbb{Q}_p$ is isomorphic as a topological field to a $\mathfrak{p}$-adic number field. In this section we study more closely the finite extensions L of K. Since the valuation ring $\mathfrak{O}_L$ of L has only one prime ideal we can speak about the ramification index $e = e_{L/K}$ and the inertia degree $f = f_{L/K}$ of L/K.

Proposition 1.76. *Let π be a generator of the prime ideal $\mathfrak{P}$ of $\mathfrak{O}_L$ and let $\overline{\omega}_1, \dots, \overline{\omega}_f$ be a basis of the residue class field extension $(\mathfrak{O}_L/\mathfrak{P})/(\mathfrak{O}_K/\mathfrak{p})$. Then*

$$\{\omega_\mu \pi^\nu \mid \mu = 1, \dots, f, \nu = 0, 1, \dots, e-1\}$$

is a basis of the $\mathfrak{O}_K$-module $\mathfrak{O}_L$.

Proof. This follows easily from (1.33). □

The extension L/K is called unramified (resp. fully ramified) if $e_{L/K} = 1$ (resp. $f_{L/K} = 1$).

Proposition 1.77. 1) *Let q be the number of elements in $\mathfrak{O}_K/\mathfrak{p}$. There is one and only one intermediate field K_f of L/K such that K_f/K is unramified and L/K_f is fully ramified. Moreover $[K_f : K] = f$ and $K_f = K(\zeta)$ where ζ is a primitive root of unity of order q^{f-1}.*

2) *Let $f_\pi(x)$ be the minimal polynomial of π over K_f. Then*

$$\mathfrak{D}_{L/K} = (f'_\pi(\pi)).$$

Proof. $\mathfrak{O}_L/\mathfrak{P}$ is the field with q^f elements. Therefore the polynomial $x^{q^f-1} - 1$ has $q^f - 1$ distinct roots in $\mathfrak{O}_L/\mathfrak{P}$ and Hensel's lemma (Proposition 1.71) implies that $x^{q^f-1} - 1$ has $q^f - 1$ roots in L. Hence $K(\zeta) \subseteq L$. Since $K(\zeta)/K$ has inertia degree f and ramification index 1, we have $K(\zeta) = K_f$. (1.20), (1.21) imply $e_{L/K_f} = e$, $f_{L/K_f} = 1$. By Proposition 1.76 we have $\mathfrak{d}_{L/K_f} = (N_{L/K_f} f'_\pi(\pi))$, hence $\mathfrak{D}_{L/K} = \mathfrak{D}_{L/K_f} = (f'_\pi(\pi))$. □

Let L be contained in a field M and F a finite extension of K in M. Proposition 1.77 implies

$$e_{LF/F} \leqslant e_{L/K}. \tag{1.39}$$

Furthermore Proposition 1.77 shows that up to isomorphism there is one and only one unramified extension of K of arbitrarily given degree. This extension is normal with cyclic Galois group G and G is generated by the Frobenius automophism (Proposition 1.52.7)).

The extension L/K is called tamely ramified (resp. wildly ramified) if $e \not\equiv 0 \pmod{p}$ ($e \equiv 0 \pmod{p}$).

Proposition 1.78. *Let L/K be a tamely ramified extension with inertia degree* 1 *and let π_K be a prime element of $\mathfrak{O}_K$. Then $L = K(\sqrt[e]{\pi_K \zeta})$ where e is the degree of L/K and ζ is a $q-1$-th root of unity.*

Proof. By our assumptions, $\pi^e = \varepsilon \pi_K$ with a unit ε in $\mathfrak{O}_L$. Since L/K has inertia degree 1, we have $\varepsilon \equiv \zeta \pmod{\pi}$ with ζ a $q-1$-th root of unity. Therefore

$\varepsilon = \varepsilon_0 \zeta$ with a principal unit ε_0, i.e. $\varepsilon_0 \equiv 1 \pmod{\pi}$. Since $\mathbb{Z}_p$ operates on the group of principal units (§4.4) and $(e, p) = 1$, there is one and only one principal unit $\varepsilon_1 \in O_L$ such that $\varepsilon_1^e = \varepsilon_0$. □

It is an elementary task to classify tamely ramified extensions of K up to isomorphism. The reader is referred to Hasse (1979), Chap. 16, for this question (caution: in the three german editions of Hasse (1979) there are mistakes in Chap. 16).

The wildly ramified extensions are much more complicated. But in some sense one has also in this general situation an insight in the possible extensions. We come back to this question in Chap. 3.2.3–4. One result in this direction is the following

Proposition 1.79. *Let L/K be a fully ramified extension of degree n. Then*

$$v_{\mathfrak{P}}(\mathfrak{D}_{L/K}) \leqslant v_{\mathfrak{P}}(n) + n - 1.$$

Proof. Let π be a prime element of L and let

$$f(x) = x^n + a_1 x^{n-1} + \cdots + a_{n-1} x + a_n$$

be the minimal polynomial of π. Then the π-exponents of the summands in

$$f'(\pi) = n\pi^{n-1} + a_1(n-1)\pi^{n-2} + \cdots + a_1$$

are all distinct. Hence

$$v_{\mathfrak{P}}(\mathfrak{D}_{L/K}) = v_{\mathfrak{P}}(f'(\pi)) \leqslant v_{\mathfrak{P}}(n\pi^{n-1}) = v_{\mathfrak{P}}(n) + n - 1. \ \square$$

We use Proposition 1.79 for the proof of Theorem 1.48. Let L be an algebraic number field of degree n. By Theorem 1.12 it is sufficient to show that the discriminant d_L is bounded by a number which depends only on n and $\mathfrak{S}$. Hence it is sufficient to show that the local discriminants are bounded for $p \in \mathfrak{S}$, which follows from Proposition 1.79.

In analogy to Theorem 1.48 the number of $\mathfrak{p}$-adic number fields of given degree is finite. Moreover one has the following theorem of Krasner:

Theorem 1.80. *Let $\overline{K}$ be the algebraic closure of K and $N(n, j)$ the number of fully ramified extensions L of K in $\overline{K}$ such that $[L:K] = n$ and $\mathfrak{D}_{L/K} = \mathfrak{p}\mathfrak{P}^{j-1}$ where $\mathfrak{P}$ is the prime ideal of L. Then*

$$N(n, j) = n l_j q^h$$

where $q = [\mathfrak{O}_K : \mathfrak{p}]$, $h = E/p + E/p^2 + \cdots + E/p^\rho + [(j - E)/p^{\rho+1}]$, $l_j = 0$ if $j < sE$ or $j > mE$, $l_j = q - 1$ if $sE \leqslant j < mE$, $l_j = 1$ if $j = mE$, $E = v_{\mathfrak{P}}(p)$, $m = v_p(n)$, $s = \{\min m, v_p(j)\}$, $\rho = [j/E]$. ⊠ (Krasner (1966), Serre (1978)).

For the proof of Theorem 1.80 as in other situations the following lemma of Krasner is useful:

Proposition 1.81. *Let K be a complete field with respect to a non-archimedean valuation v and let $\alpha, \beta \in \overline{K}$ with α separable over $K(\beta)$. If*

$$v(\beta - \alpha) < v(\alpha' - \alpha)$$

for all conjugates $\alpha' \neq \alpha$ *of* α, *then* $\alpha \in K(\beta)$.

Proof. Let $L/K(\beta)$ be the normal closure of $K(\alpha, \beta)/K(\beta)$ and $g \in G(L/K(\beta))$ Then

$$v(\beta - g\alpha) = v(\beta - \alpha) < v(\alpha' - \alpha).$$

Therefore

$$v(\alpha - g\alpha) \leqslant \max(v(\alpha - \beta), v(\beta - g\alpha)) < v(\alpha' - \alpha).$$

It follows $g\alpha = \alpha$. □

Theorem 1.80 implies the following elegant *mass formula of Serre* (1978).

Theorem 1.82. *Let* Σ_n *be the set of all fully ramified extensions* L *of* K *of given degree* n *in a fixed algebraic closure of* K *and let* $j(L) := j$ *and* q *be defined as above. Then*

$$\sum_{L \in \Sigma_n} q^{-j(L)} = n. \;\square$$

4.7. Kummer Extensions. Let n be a natural number and K a field of characteristic prime to n which contains the roots of unity of order n. From basic Galois theory one knows the following facts.

Any cyclic extension L of K of degree n can be generated by a root α of an irreducible polynomial $x^n - a \in K[x]$. We associate to $\zeta \in \mu_n$ the automorphism g_ζ of L/K given by $g_\zeta \alpha = \zeta\alpha$. Then $\zeta \to g_\zeta$ is an isomorphism of μ_n onto $G(L/K)$.

A polynomial $x^n - b$, $b \in K^\times$, is irreducible if and only if $\bar{b}$ has order n in the group $K^\times/(K^\times)^n$. If β is a root of $x^n - b$ in some extension E of K, then $K(\beta)/K$ is a cyclic extension of degree n. It is called *Kummer extension* and one puts $\sqrt[n]{b} := \beta$.

Let $x^n - b_1$ and $x^n - b_2$ be irreducible polynomials in $K[x]$. Then $K(\sqrt[n]{b_1}) = K(\sqrt[n]{b_2})$ if and only if $b_2 = b_1^r c^n$ with $(r, n) = 1$, $c \in K$.

Now we assume that K is the quotient field of a Dedekind ring R.

Proposition 1.83. *Let* $x^n - a$ *be an irreducible polynomial in* $R[x]$ *and let* $L = K(\sqrt[n]{a})$.

1) *If* $\mathfrak{p}$ *is a prime ideal of* K *with* $\mathfrak{p} \nmid na$, *then* $\mathfrak{p}$ *is unramified and decomposes into* s *prime ideals in* L *where* s *is the maximal divisor of* n *such that the congruence* $x^s \equiv a \pmod{\mathfrak{p}}$ *has a solution* b *in* R.

2) *If* $\mathfrak{p}$ *is a prime ideal of* K *with* $\mathfrak{p} \nmid a$, $\mathfrak{p}^n \nmid a$, *then* $\mathfrak{p}$ *is ramified in* L/K. *If* $\mathfrak{p} \nmid n$ *and* $(v_\mathfrak{p}(a), n) := h$, *then the ramification index of* $\mathfrak{p}$ *in* L/K *is* n/h.

Proof. 1) By our assumptions $x^n = \bar{a}$ decomposes in $R/\mathfrak{p}$ into the product of pairwise distinct irreducible factors $x^{n/s} - \overline{b\zeta}$, $\zeta \in \mu_s$. Hence the claim follows from example 29. 2) is proved by an Eisenstein argument (Proposition 1.43.2)) and by 1) taking in mind that in the case $\mathfrak{p}^n | a$, $p \nmid n$, there exists a $b \in K$ such that $ab^n \in R$ and $p \nmid nab^n$. □

For the study of wild ramification in Kummer extensions we restrict ourselves to the case that $n = p$ is a prime and K is a $\mathfrak{p}$-adic number field, $\mathfrak{p}|p$.

Proposition 1.84. *Let a be an element of* $\mathfrak{p}$ *such that* $v_\mathfrak{p}(a) \not\equiv 0 \pmod{\mathfrak{p}}$, $v_\mathfrak{p}(a) < pv_\mathfrak{p}(p)/(p-1)$. *Then* $K(\sqrt[p]{1+a})/K$ *is ramified.*

Proof. We put $\beta := \sqrt[p]{1+a} - 1$. Then

$$0 = (\beta+1)^p - 1 - a \equiv \beta^p - a (\operatorname{mod} p\beta).$$

We denote the continuation of the exponential valuation $v_\mathfrak{p}$ to $K(\sqrt[p]{1+a})$ again by $v_\mathfrak{p}$. Since $v_\mathfrak{p}(\beta^p) \geqslant v_\mathfrak{p}(p\beta)$ implies $v_\mathfrak{p}(\beta^p) \geqslant pv_\mathfrak{p}(p)/(p-1)$ and $v_\mathfrak{p}(p\beta) \geqslant pv_\mathfrak{p}(p)/(p-1)$ which is impossible, we have $v_\mathfrak{p}(\beta^p) = v_\mathfrak{p}(a)$. □

According to §4.4 there is a further Kummer extension $K(\sqrt[p]{\eta})/K$ with a principal unit η, namely $\eta := \eta_* = 1 + \omega_*(1 - \zeta_p)p$. Since there exists one unramified extension L of degree p over K, it follows

Proposition 1.85. *Let* η_* *be as in* §4.4. *Then* $K(\sqrt[p]{\eta_*})/K$ *is the unramified extension of degree* p. □ (See also Chap. 2.6.1.)

4.8. Analytic Functions in Complete Non-Archimedean Valued Fields. Let K be a field which is complete with respect to the non-archimedean valuation v.

A series $\sum_{n=1}^\infty \alpha_n$ with $\alpha_n \in K$ converges if and only if $\lim_{n\to\infty} \alpha_n = 0$. If $s = \sum_{n=1}^\infty \alpha_n$ converges, then an arbitrary reordering of s converges to the same limit. It follows that the product of the convergent series $\sum_{n=1}^\infty \alpha_n$ and $\sum_{n=1}^\infty \beta_n$ is equal to $\sum_{n,m=1}^\infty \alpha_n\beta_m$.

Let $K\{\{t\}\}$ be the field of formal power series $\sum_{n=n_0}^\infty \alpha_n t^n$ in the variable t with $n_0 \in \mathbb{Z}$ and $\alpha_n \in K$. As in complex function theory we have for the series in $K\{\{t\}\}$ beside addition, subtraction, multiplication and division two other operations: Let $g(t) = \sum_{n=1}^\infty \alpha_n t^n$ and $f(t)$ an arbitrary element of $K\{\{t\}\}$. Then the composition $f(g(t))$ and differentiation $f'(t)$ are defined as for usual power series. The familiar differentiation rules carry over to formal power series.

Now consider a power series $\sum_{n=0}^\infty \alpha_n \xi^n$ with $\alpha_n, \xi \in K$. We put

$$r := \left(\overline{\lim_{n\to\infty}} \sqrt[n]{v(\alpha_n)}\right)^{-1}.$$

Then for $v(\xi) < r$ the series converges and for $v(\xi) > r$ the series diverges. r is called the convergence radius, and the set of $\xi \in K$ for which $\sum_{n=0}^\infty \alpha_n \xi^n$ converges is called the convergence region of the power series. In its convergence region a power series represents a continuous function called (*non-archimedean*) *analytic function.*

Several properties of usual analytic functions carry over to non-archimedean analytic functions. For example, if the analytic functions f_1 and f_2 have common values at a set in their convergence region C which has an accumulation point, then the functions f_1 and f_2 must be equal in the whole of C. The notions of meromorphic function, pole and residuum can be transferred to the non-archimedean situation without difficulties.

Theorem 1.86 (Mahler's theorem). *Let $f(\xi)$ be a continuous function from $\mathbb{Z}_p$ into $\mathbb{Q}_p$. Then $f(\xi)$ has a unique representation*

$$f(\xi) = \sum_{n=0}^{\infty} \alpha_n \binom{\xi}{n} \quad \text{with} \quad \binom{\xi}{n} = \frac{\xi(\xi-1)\cdots(\xi-n+1)}{n!} \tag{1.40}$$

and coefficients α_n such that $\lim_{n\to\infty} \alpha_n = 0$.

If $v(\alpha_n) \leqslant Mr^n$ for some $r < p^{-1/(p-1)}$ and some M, then $f(\xi)$ may be expressed as a power series with convergence radius at least $(rp^{1/(p-1)})^{-1}$. ⊠ (Washington (1982), § 5.1)

Some properties of the transition from formal to convergent power series are collected in the following

Proposition 1.87 (Substitution principle). *Let ξ be an element of the complete non-archimedean valued field K.*

1) *Let $f(t)$, $g(t)$ be formal power series such that $f(\xi)$, $g(\xi)$ converge and let $h(t) = f(t) \circ g(t)$ where $\circ$ stands for addition, subtraction or multiplication. Then $h(\xi)$ is convergent and $f(\xi) \circ g(\xi) = h(\xi)$.*
2) *Let $f(t) = \sum_{n=n_0}^{\infty} \alpha_n t^n$ and $g(t) = \sum_{n=1}^{\infty} \beta_n t^n$ be formal power series such that $g(\xi)$ and $f(g(\xi))$ converge and that $v(\beta_n \xi^n) \leqslant v(g(\xi))$ for all $n \geqslant 1$. Furthermore let $h(t) = f(g(t))$. Then $h(\xi)$ converges and $h(\xi) = f(g(\xi))$.*
3) *Let $f(t)$ be a formal power series such that $f(\xi)$ converges for $\xi \in K^\times$. Then also $f'(\xi)$ converges and*

$$f'(\xi) = \lim_{\eta\to 0} \frac{f(\xi+\eta) - f(\xi)}{\eta}$$

where the limit runs over all $\eta \neq 0$ in K such that $f(\xi+\eta)$ converges. ⊠

In usual analysis one has in the field $\mathbb{C}$ of complex numbers a field which is both complete and algebraically closed. We want to construct a similar field in $\mathfrak{p}$-adic analysis. Starting with $\mathbb{Q}_p$ we take the algebraic closure $\overline{\mathbb{Q}}_p$. Since any finite field extension of $\mathbb{Q}_p$ has a unique valuation extending the valuation of $\mathbb{Q}_p$, we have a unique valuation of $\overline{\mathbb{Q}}_p$ extending the valuation of $\mathbb{Q}_p$. We denote the completion of $\overline{\mathbb{Q}}_p$ with respect to this valuation by $\mathbb{C}_p$.

Proposition 1.88. *$\mathbb{C}_p$ is algebraically closed.*

Proof. Suppose α is algebraic over $\mathbb{C}_p$ and let $f(x)$ be its irreducible polynomial in $\mathbb{C}_p[x]$. Since $\overline{\mathbb{Q}}_p$ is dense in $\mathbb{C}_p$, we may choose a polynomial $g(x) \in \mathbb{Q}_p[x]$ whose coefficients are close to those of $f(x)$. Then $g(\alpha) = g(\alpha) - f(\alpha)$ is small. Hence $v(\alpha - \beta)$ must be small for some root $\beta \in \mathbb{C}_p$ of $g(x)$. We can choose $g(x)$ and β such that $v(\beta - \alpha) < v(\alpha_i - \alpha)$ for all conjugates $\alpha_i \neq \alpha$. Then Krasner's lemma (Proposition 1.81) implies $\alpha \in \mathbb{C}_p(\beta) = \mathbb{C}_p$. □

4.9. The Elementary Functions in $\mathfrak{p}$-adic Analysis. Now we apply the principles of the last section to the exponential series

$$\exp(t) = \sum_{n=0}^{\infty} \frac{t^n}{n!},$$

the logarithmic series

$$\log(1 + t) = \sum_{n=1}^{\infty} (-1)^{n-1} \frac{t^n}{n},$$

and the binomial series

$$(1 + t)^{\eta} = \sum_{n=0}^{\infty} \binom{\eta}{n} t^n, \qquad \eta \in K.$$

Since we have all natural numbers in the denominators of these series, our field K must have characteristic 0. We assume in the following that K contains the field $\mathbb{Q}_p$ of rational p-adic numbers for a certain prime p.

Let

$$v_p(\xi) := \log v(\xi)/\log v(p)$$

be the normed exponential valuation belonging to v. Then we have the following results.

Proposition 1.89. 1) $\exp(t)$ *has the convergence region*

$$C_{\exp} := \left\{\xi \in K \,|\, v_p(\xi) > \frac{1}{p-1}\right\}.$$

2) $v_p(\exp(\xi) - 1 - \xi) > v_p(\xi)$ *for* $\xi \in C_{\exp} - \{0\}$.
3) $\exp(\xi_1 + \xi_2) = \exp \xi_1 \exp \xi_2$ *for* $\xi_1, \xi_2 \in C_{\exp}$.
4) *If* $\xi_1, \xi_2 \in C_{\exp}$ *and* $\exp \xi_1 = \exp \xi_2$, *then* $\xi_1 = \xi_2$. □

Proposition 1.90. 1) $\log(1 + t)$ *has the convergence region*

$$C_{\log} := \{\xi \in K \,|\, v_p(\xi) > 0\}.$$

2) $v_p(\log(1 + \xi) - \xi) > v_p(\xi)$ *for* $v_p(\xi) > \dfrac{1}{p-1}$.

3) $\log \exp \xi = \xi$, $\exp \log(1 + \xi) = 1 + \xi$ *for* $v_p(\xi) > \dfrac{1}{p-1}$.

4) $\log(\xi_1 \xi_2) = \log \xi_1 + \log \xi_2$ *for* $\xi_1 - 1, \xi_2 - 1 \in C_{\log}$.
5) *If* $\xi_1 - 1, \xi_2 - 1 \in C_{\log}$ *and* $\log \xi_1 = \log \xi_2$, *then* $\xi_1 = \xi_2 \zeta$ *where* ζ *is a root of unity of p-power order.*

6) $\log \xi = \lim\limits_{n \to \infty} \dfrac{\xi^{p^n} - 1}{p^n}$ *for* $v_p(\xi - 1) > 0$ (Leopoldt (1961)). ⊠

Proposition 1.89 and 1.90 imply that the mapping $\xi \to \exp \xi$ defines an isomorphism from the additive group $C_{\exp}$ onto the multiplicative group $C' = \left\{\eta \in K \,|\, v_p(\eta - 1) > \dfrac{1}{p-1}\right\}$. If K is a $\mathfrak{p}$-adic number field, then $C_{\exp}$ (resp. C') is a subgroup of $\mathfrak{p}$ (resp. U_1) of finite index. This proves Proposition 1.73.

Proposition 1.91. 1) $(1+\xi)^\eta$ *converges for* $v_p(\xi) > \frac{1}{p-1}$ *and if* $\eta \in \mathbb{Z}_p$ *for* $v_p(\xi) > 0$.

2) *The function* $(1+\xi)^\eta$ *is continuous in* η.

3) $(1+\xi)^\eta = \exp(\eta \log(1+\xi))$ *for* $v_p(\xi) > \frac{1}{p-1}$, $v_p(\eta) \geqslant 0$.

4)
$$(1+\xi)^{\eta+\eta'} = (1+\xi)^\eta(1+\xi)^{\eta'},$$
$$(1+\xi)^\eta(1+\xi')^\eta = ((1+\xi)(1+\xi'))^\eta,$$
$$(1+\xi)^{\eta\eta'} = ((1+\xi)^\eta)^{\eta'}$$

for $v_p(\xi) > \frac{1}{p-1}$, $v_p(\xi') > \frac{1}{p-1}$, $v_p(\eta) \geqslant 0$, $v_p(\eta') \geqslant 0$. □

Proposition 1.92. *There is a unique extension of* log *to all of* $K^\times$ *such that* $\log p = 0$ *and* $\log(\xi\eta) = \log\xi + \log\eta$ *for* $\xi, \eta \in K^\times$. ⊠ (Washington (1982), Proposition 5.4)

4.10. Lubin-Tate Extensions (Main reference: Cassels, Fröhlich (1967), Chap. 6). Let K be a $\mathfrak{p}$-adic number field with valuation ring $\mathfrak{O}$, let q be the number of elements in the residue class field, and let $g(t)$ be an Eisenstein polynomial of degree $q-1$ with coefficients in $\mathfrak{O}$ (§ 3.4). We put $f(t) := tg(t)$. It is easy to see, that for every $\alpha \in \mathfrak{O}$ there is one and only one formal power series

$$[\alpha]_f(t) := \alpha t + \alpha_2 t^2 + \cdots$$

with coefficients in $\mathfrak{O}$ such that

$$[\alpha]_f(f(t)) = f([\alpha]_f(t)).$$

Let $\theta_n(g)$ be a root of the Eisenstein polynomial $gf^{(n-1)}(t) := g(f(\ldots f(t)))$ (f $n-1$ times iterated) in a fixed algebraic closure $\overline{K}$ of K. Then also $[\alpha]_f(\theta_n(g))$ is a root of $gf^{(n-1)}(t)$ if $\alpha \in U := \mathfrak{O}^\times$.

Theorem 1.93. 1) $K(\theta_n(g))/K$ *is an abelian extension (i.e. a normal extension with abelian Galois group), called Lubin-Tate extension. Let* σ_α *be the automorphism of this extension with*

$$\sigma_\alpha(\theta_n(g)) = [\alpha]_f(\theta_n(g)) \qquad \text{for } \alpha \in U.$$

Then $\varphi\colon \alpha \to \sigma_\alpha$ *is a homomorphism of* U *onto the Galois group* G *of* $K(\theta_n(g))/K$ *with kernel* $U_n := 1 + \mathfrak{p}^n$.

2) *Let* g_1 *and* g_2 *be Eisenstein polynomials of degree* $q-1$ *with coefficients in* $\mathfrak{O}$ *and let* L *be the unramified extension of* K *in* $\overline{K}$ *of degree* $q^{n-1}(q-1)$. *Then* $L(\theta_n(g_1)) = L(\theta_n(g_2))$. *If the absolute coefficients of* g_1 *and* g_2 *coincide, then* $K(\theta_n(g_1)) = K(\theta_n(g_2))$.

3) *For all* $u \in \mathbb{R}$, $u > 0$ *we define* $U_u = U_n$ *if* $n-1 < u \leqslant n$. *Then* φ *maps* U_u *on the* u*-th ramification group of* G *in the upper numeration* (§ 3.7). ⊠

For the proof of Theorem 1.93 one introduces the formal group law corresponding to f:

If A is an arbitrary commutative ring with 1, then a formal power series $F(x, y)$ in the indeterminates x, y with coefficients in A is called a commutative formal group law over A if $F(x, F(y, z)) = F(F(x, y), z)$, $F(y, x) = F(x, y)$, $F(0, x) = x$.

There is one and only one commutative formal group law F_f over $\mathfrak{O}$ such that $f(F_f(x, y)) = F_f(f(x), f(y))$. It is called Lubin-Tate formal group law. It follows

$$[\alpha + \beta]_f(t) = F_f([\alpha]_f(t), [\beta]_f(t)),$$
$$[\alpha\beta]_f(t) = [\alpha]_f([\beta]_f(t)) \qquad \text{for } \alpha, \beta \in \mathfrak{O}.$$

Let μ be the set of roots of the polynomials $gf^{(n-1)}$, $n = 1, 2, \ldots$, in $\bar{K}$. Then $F_f(\alpha, \beta)$ is convergent for $\alpha, \beta \in \mu$ and $F_f(\alpha, \beta) \in \mu$. It follows that F_f defines in μ a multiplication such that μ becomes an abelian group.

Example 31. Let $K = \mathbb{Q}_p$ and let $\xi_n = \zeta_{p^n}$ be a primitive p^n-th root of unity in the algebraic closure $\overline{\mathbb{Q}}_p$ of $\mathbb{Q}_p$. Then by § 3.9 $\mathbb{Q}_p(\xi_n)$ is a fully ramified extension of $\mathbb{Q}_p$ of degree $\varphi(p^n) = (p-1)p^{n-1}$ and $\Pi_n = \xi_n - 1$ is a prime element in $\mathbb{Q}_p(\xi_n)$. ξ_n is a zero of the Eisenstein polynomial

$$g_n(x) = \phi_{p^n}(x+1) = x^{p^n - p^{n-1}} + \cdots + p.$$

We put $g = g_1$. Then $gf^{(n-1)} = g_n$ and $\mathbb{Q}_p(\theta_n(g)) = \mathbb{Q}_p(\xi_n)$. The formal group law F belonging to g is the "multiplicative group"

$$F(x, y) = (x+1)(y+1) - 1.$$

Moreover $\mu' := \{\alpha - 1 \mid \alpha \in \mu\}$ is the set of roots of unity of p-power order. Hence F induces in μ' the usual multiplication of $\overline{\mathbb{Q}}_p$. □

§ 5. Harmonic Analysis on Local and Global Fields

(Main reference: Tate (1967))

In the contemporary treatment of algebraic number theory, the theory of automorphic functions associated to reductive Lie groups, and arithmetical algebraic geometry a central role plays harmonic analysis on local compact groups associated to local and global fields. In this paragraph we study harmonic analysis for the additive and multiplicative group.

General facts and notations about harmonic analysis on local compact groups are collected in Appendix 3.

There are various conventions with respect to the definition of a special additive homomorphism of a local field, the Fourier transforms, the Gaussian sums, etc. in the literature (e.g. Tate (1979), Lang (1970)). We follow here essentially the conventions of Tate (1967).

5.1. Harmonic Analysis on Local Fields, the Additive Group. Let K be a local field (§ 4.2) and v the corresponding normalized valuation. The topological space K is locally compact. More exactly we have the following

Proposition 1.94. *A subset $B \subseteq K$ is relatively compact if and only if there is a number b such that $v(\alpha) \leqslant b$ for all $\alpha \in B$.* □

Let $X(K^+)$ be the character group of the additive group of K. The structure of $X(K^+)$ is determined by the following

Proposition 1.95. *Let χ be a non-trivial character of K^+. Then for every $\alpha \in K$, the mapping $\psi_\alpha\colon \xi \to \chi(\alpha\xi)$ is also a character. The correspondence $\alpha \to \psi_\alpha$ is an isomorphism between the topological groups K^+ and $X(K^+)$.*

Proof. It is easy to see that $\alpha \to \psi_\alpha$ defines an isomorphism into $X(K^+)$. The image I is dense in $X(K^+)$ since $\psi_\alpha(\xi) = 1$ for fixed ξ and all $\alpha \in K^+$ implies $\xi = 0$. (A subgroup U of an abelian local compact group G is dense if and only if $\chi(u) = 1$ for all $\chi \in X(G)$ implies $u = 0$.) Furthermore one shows that I is closed in $X(K^+)$. □

We identify K^+ and $X(K^+)$ according to Proposition 1.95 by means of a special non-trivial character λ, which is defined as follows:

If the valuation v is archimedean, we put

$$\lambda(\alpha) = \exp(2\pi i Tr_{K/\mathbb{R}}(\alpha)).$$

If the valuation v is non-archimedean and K has residue characteristic p, we put

$$\lambda(\alpha) = \exp(-2\pi i Tr_{K/\mathbb{Q}_p}(\alpha))$$

where the function $\exp(2\pi i x)$ for $x \in \mathbb{Q}_p$ is defined by

$$\exp(2\pi i x) := \exp(2\pi i x_\nu) \qquad \text{for } x = \lim_{\nu\to\infty} x_\nu,\ x_\nu \in \mathbb{Z}\left[\frac{1}{p}\right],$$

and ν sufficiently large.

In the following examples we show directly that every character χ of $\mathbb{Q}_v^+$ for any place v of $\mathbb{Q}$ has the form $\chi(\xi) = \lambda(\alpha\xi)$ for some $\alpha \in \mathbb{Q}_v$.

Example 32. Let $K = \mathbb{R}$. Then λ defines an isomorphism between $\mathbb{R}^+/\mathbb{Z}$ and $T := \{c \in \mathbb{C}^\times \mid |c| = 1\}$. Let χ be a character of $\mathbb{R}^+$. Every homomorphism κ of $\mathbb{R}^+$ into $\mathbb{R}^+/\mathbb{Z}$ can be lifted (uniquely) to a homomorphism $\tilde\kappa\colon \mathbb{R}^+ \to \mathbb{R}^+$ (first define $\tilde\kappa$ in a small neighborhood of 0 and then extend it by means of $\tilde\kappa(n\xi) = n\tilde\kappa(\xi)$ to all of $\mathbb{R}$). Now we see that χ has the form $\chi(\xi) = \lambda(\xi\tilde\kappa(1))$, where $\kappa = \lambda^{-1}\chi$. □

Example 33. Let $K = \mathbb{Q}_p$. This case is similar to Example 32. One has to show that a homomorphism $\kappa\colon \mathbb{Q}_p^+ \to \mathbb{Q}_p^+/\mathbb{Z}_p$ can be lifted to a homomorphism $\tilde\kappa\colon \mathbb{Q}_p^+ \to \mathbb{Q}_p^+$. Let α_n be a representative in $\mathbb{Q}_p$ of the class $\kappa\left(\dfrac{1}{p^n}\right)$, $n = 0, 1, \ldots$.

Then $\alpha = \lim_{n\to\infty} p^n\alpha_n$ exists and is independent of the choice of the sequence $\{\alpha_n\}$. Furthermore $\tilde{\kappa}(\xi) = \alpha\xi$. □

Example 34. If $K = \mathbb{R}$, then obviously in our identification of $X(\mathbb{R}^+)$ with $\mathbb{R}^+$ the orthogonal complement of $\mathbb{Z} \subset \mathbb{R}$ with respect to the form

$$\langle \alpha, \beta \rangle := \lambda(\alpha\beta),\ \alpha, \beta \in \mathbb{R},$$

is $\mathbb{Z}$ itself. This explains the simple form of the (classical) Poisson sum formula for $\mathbb{R}$ and $\mathbb{Z}$ (Appendix 3.1). □

Now we consider a Haar measure μ on K^+. Let M be a measurable set in K^+. Then it is easy to prove that

$$\mu(\alpha M) = v(\alpha)\mu(M) \qquad \text{for } \alpha \in K. \tag{1.41}$$

(1.41) provides a new characterization for normalized valuations.

We fix the Haar measure μ such that for the Fourier transform

$$\hat{f}(\eta) := \int f(\xi)\lambda(\eta\xi)\, d\mu(\xi)$$

of a function $f \in L_1(K^+)$ the inversion formula

$$f(\xi) = \hat{\hat{f}}(-\xi) \qquad \text{for } \xi \in K$$

holds if $f \in S(K^+)$ where $S(K^+)$ is the space of functions $f \in L_1(K^+)$ such that f is continuous and $\hat{f} \in L_1(K^+)$.

This measure μ is uniquely determined and is the ordinary Lebesgue measure if K is real, twice the ordinary Lebesgue measure if K is complex and the measure with $\mu(\mathfrak{O}_v) = p^{-\nu/2}$ if v is non-archimedean where ν is the p-exponent of the discriminant of $\mathfrak{O}_v$ over $\mathbb{Z}_p$.

In the following we write $d\xi$ instead of $d\mu(\xi)$.

5.2. Harmonic Analysis on Local Fields, the Multiplicative Group. Concerning the multiplicative group $K^\times$ of K, we are not only interested in characters but also in *quasi-characters*, i.e. continuous homomorphisms of $K^\times$ in $\mathbb{C}^\times$. A quasi-character is called *unramified* if it is trivial on the unit group $U = U_0 := \{\alpha \in K^\times \mid v(\alpha) = 1\}$ of K.

Proposition 1.96. *An unramified quasi-character χ of $K^\times$ has the form*

$$\chi(\alpha) = v(\alpha)^s =: v^s(\alpha)$$

for some complex number s. □

For any quasi-character χ there is a uniquely determined real number σ such that $|\chi(\alpha)| = v(\alpha)^\sigma$ called the exponent of χ.

Any quasi-character χ of $\mathbb{R}^\times$ has the form

$$\chi(\alpha) = \operatorname{sgn} \alpha\, v(\alpha)^s \quad \text{or} \quad \chi(\alpha) = v(\alpha)^s \quad \text{for some } s \in \mathbb{C}.$$

v (resp. (1) is called the *conductor* of χ.

Any quasi-character χ of $\mathbb{C}^\times$ has the form

$$\chi(\alpha) = \chi_n(\alpha)v(\alpha)^s \quad \text{for some } s \in \mathbb{C},\, n \in \mathbb{Z}$$

with

$$\chi_n(\alpha) = \alpha^n v(\alpha)^{-n/2}.$$

By definition χ has always *trivial conductor.*

If K is a $\mathfrak{p}$-adic number field, then the subgroups $U_n := 1 + \mathfrak{p}^n$, $n = 1, 2, \ldots$, form a system of neighborhoods of 1 in K. Let χ be a ramified quasi-character of $K^\times$. There is a minimal n such that $\chi(U_n) = \{1\}$. The ideal $\mathfrak{p}^n$ is called the *conductor* of χ. If χ is unramified, then the ideal (1) is called the conductor of χ.

The subgroup U of $K^\times$ is compact. We select on $K^\times$ the Haar-measure $\mu^\times$ such that

$$\int_U d\mu^\times(\alpha) = 1.$$

It is easy to see, that $d\mu^\times(\alpha) = v(\alpha)^{-1}\, d\alpha$ for archimedean fields and $d\mu^\times(\alpha) = N\mathfrak{p}(N\mathfrak{p} - 1)^{-1} v(\alpha)^{-1}\, d\alpha$ for $\mathfrak{p}$-adic number fields.

5.3. Adeles. Now let K be an algebraic number field of degree n. We collect our local data for the completions of K in a global object associated to K, the *adele ring* $A(K)$. This is a subring of the direct product $\prod_v K_v$ taken over all places v of K. $\prod_v \xi_v \in \prod_v K_v$ belongs to $A(K)$ if $\xi_v \in \mathfrak{O}_v$ for almost all v. By definition $A(K)$ is a topological ring. $A(K)^+$ is the restricted product of the groups K_v^+ with respect to the compact subgroups $\mathfrak{O}_v^+$ which are defined for almost all places v. Hence $A(K)$ is locally compact (Appendix 3.1).

Dedekind's different theorem (Theorem 1.44) implies that for almost all v the measure of $\mathfrak{O}_v$, defined above, is 1. Therefore our local measures define a Haar measure $d\xi = \prod_v d\xi_v$ on $A(K)^+$. (The notations of § 5.1–3 will be used with subscripts v.)

The identifications of $X(K_v^+)$ with K_v^+ imply that $X(A(K)^+)$ is the restricted direct product of the groups K_v^+ with respect to the subgroups $\mathfrak{D}_{K_v/\mathbb{Q}_p}^{-1}$. Since $\mathfrak{D}_{K_v/\mathbb{Q}_p} = \mathfrak{O}_v$ for almost all v, we have $X(A(K)^+) = A(K)^+$. In this identification the adele $\eta = \prod_v \eta_v$ is to be identified with the character

$$\xi = \prod_v \xi_v \to \prod_v \lambda_v(\xi_v \eta_v) =: \lambda(\xi\eta).$$

We define the Fourier transform

$$\hat{f}(\eta) = \int f(\xi)\lambda(\xi\eta)\, d\xi \qquad \text{for } f \in L_1(A(K)^+).$$

Then the inversion formula

$$f(\xi) = \hat{\hat{f}}(-\xi)$$

holds for $f \in S(A(K)^+)$.

The field K is embedded in $A(K)$ by identifying $\xi \in K$ with the *principal adele* $\prod_v \xi$. The factor group $A(K)/K$ is compact, and a fundamental domain of $A(K)$ with respect to K is easily described as follows: Let S_∞ be the set of archimedean places of K. Then $A(K)_\infty := \prod_{v \in S_\infty} K_v$ is an $\mathbb{R}$-vector space of dimension $n := [K : \mathbb{Q}]$. The field K is embedded in $A(K)_\infty$ as above and $\mathfrak{O}_K$ is a lattice in $A(K)_\infty$ with fundamental domain

$$F_\infty := \{x_1\omega_1 + \cdots + x_n\omega_n | 0 \leqslant x_i \leqslant 1 \text{ for } i = 1, \ldots, n\}$$

where $\omega_1, \ldots, \omega_n$ is a basis of $\mathfrak{O}_K$ as $\mathbb{Z}$-module. Then

$$F := \prod_{v \notin S_\infty} \mathfrak{O}_v \times F_\infty$$

is a fundamental domain of $A(K)/K$. The subset K of $A(K)$ is discrete since F has an interior.

Remark. The adele class group $A(K)/K$ arises in a natural way if one considers the lattices in the n-dimensional space

$$\mathbb{R}^n = K \otimes_{\mathbb{Q}} \mathbb{R} = \prod_{v \in S_\infty} K_v$$

given by ideals $\mathfrak{A}$ of K and the corresponding factor spaces $\mathbb{R}^n/\mathfrak{A}$. The compact rings $\mathbb{R}^n/\mathfrak{A}$ with $\mathfrak{A}$ in the group $\mathfrak{M}(K)$ of fractional ideals form an inverse system in the sense of Chap. 3.1.1 with the natural mappings $\mathbb{R}^n/\mathfrak{A} \to \mathbb{R}^n/\mathfrak{B}$ for $\mathfrak{B} \supset \mathfrak{A}$ as morphisms. The inverse limit of this system is canonically isomorphic to $A(K)/K$:

For $\alpha \in K$ let $\varphi_f(\alpha)$ be the adele with component α for finite places v and component 0 for infinite places and put $\varphi_f(\mathfrak{A}) = \mathfrak{A}_f$. Then $\mathfrak{A}_f$ is compact and $\bigcap_{\mathfrak{A} \in \mathfrak{M}(K)} \mathfrak{A}_f = \{0\}$. Hence

$$A(K) = \varprojlim A(K)/\mathfrak{A}_f$$

and

$$A(K)/K = \varprojlim A(K)/(\mathfrak{A}_f + K).$$

Furthermore the map

$$\mathbb{R}^n/\mathfrak{A} \to A(K)/(\mathfrak{A}_f + K)$$

given by the natural injection $\mathbb{R}^n = \prod_{v \in S_\infty} K_v \to A(K)$ is an isomorphism. This proves the assertion. □

Our identification of $X(A(K))$ with $A(K)$ is especially nice in view of the following

Theorem 1.97. *The orthogonal complement $K^\perp$ of K with respect to the form*

$$\langle \alpha, \beta \rangle = \lambda(\alpha\beta), \qquad \alpha, \beta \in A(K),$$

is K itself (compare Example 34).

Proof. $K^\perp \supset K$ follows from $\lambda(\alpha) = 1$ for all principal adeles α. For the proof of $\lambda(\alpha) = 1$ we consider first the case $K = \mathbb{Q}$. If $a \in \mathbb{Z}$, then obviously $\lambda(a) = 1$.

Let p be any prime, then by definition

$$\lambda\left(\frac{1}{p^s}\right) = \lambda_p\left(\frac{1}{p^s}\right)\lambda_\infty\left(\frac{1}{p^s}\right) = 1.$$

Now the assertion follows from the additivity of λ. The case of arbitrary K can be reduced to the case $K = \mathbb{Q}$ since the global trace is the sum of the local traces.

$K^\perp$ can be identified with $X(A(K)/K)$. Since $A(K)/K$ is compact, $K^\perp$ is discrete. Hence $K^\perp/K$ is finite as a discrete subgroup of the compact group $A(K)/K$. Since $K^\perp/K$ is a vector space over the infinite field K, it follows $K^\perp = K$. □

As a by-product of Theorem 1.97 we see by means of Pontrjagin duality that $A(K)/K$ is the character group of the discrete group K^+.

5.4. Ideles. $J(K) := A(K)^\times$ is called the *idele group* of K. We introduce in $J(K)$ the topology which is given by the characterization of $J(K)$ as the restricted product of the groups K_v with respect to the unit groups U_v. As in the case of adeles, $K^\times$ is diagonally embedded in $J(K)$. An idele $\alpha = \prod_v \alpha_v$ has by definition the *absolute value* $|\alpha| := \prod_v v(\alpha_v)$, which is well defined since $v(\alpha_v) = 1$ for almost all v. As in the local situation, for a measurable set M in $A(K)^+$ and an idele α we have $\mu(\alpha M) = |\alpha|\mu(M)$ where μ is the measure in $A(K)^+$.

The map $\alpha \to |\alpha|$ defines a continuous homomorphism of $J(K)$ in $\mathbb{R}_+$. Its kernel will be denoted by $J^0(K)$. (1.32) implies $K^\times \subset J^0(K)$.

Theorem 1.98. *$J^0(K)/K^\times$ is compact.*

Proof. Let $\varphi: J(K) \to \mathfrak{M}(\mathfrak{O}_K)$ be the homomorphism which associates to the idele $\prod_v \alpha_v$ the ideal $\prod_{v \notin S_\infty} \mathfrak{p}_v^{\nu_v(\alpha_v)}$ where $\mathfrak{p}_v$ is the prime ideal of $\mathfrak{O}_K$ corresponding to $v \notin S_\infty$. Then we have the exact sequence

$$\{1\} \to K^\times\left(\prod_{v \notin S_\infty} U_v \times H\right)\Big/K^\times \to J^0(K)/K^\times \to CL(K) \to \{1\} \qquad (1.42)$$

where H is the group $\{\prod_{v \in S_\infty} \alpha_v \mid \prod_{v \in S_\infty} v(\alpha_v) = 1\}$.

Since

$$K^\times\left(\prod_{v \notin S_\infty} U_v \times H\right)\Big/K^\times \cong \left(\prod_{v \notin S_\infty} U_v \times H\right)\Big/\mathfrak{O}_K^\times$$

and $l(\mathfrak{O}_K^\times)$ is a lattice in $l(H)$ (§ 1.3), Theorem 1.98 follows from Dirichlet's unit theorem and the finiteness of $CL(K)$ (§ 1.2). □

We define a fundamental domain of $J^0(K)$ with respect to $K^\times$ as follows: We choose one place $v_0 \in S_\infty$. Furthermore let $\varepsilon_1, \ldots, \varepsilon_{r-1}$ be a fundamental system of units of $\mathfrak{O}_K$ (§ 1.3) and let P be the parallelotope in $l(H)$ spanned by $l(\varepsilon_1), \ldots, l(\varepsilon_{r-1})$, i.e. the set of all vectors $\sum_{i=1}^{r-1} x_i l(\varepsilon_i)$ with $0 \leqslant x_i < 1$. We define E^0 to be the set of elements α in H such that $l(\alpha) \in P$ and $0 \leqslant \arg \alpha_{v_0} < 2\pi/w$ where w denotes the order of the group of roots of unity in K. Let $\beta_1, \ldots, \beta_h$ be ideles such that $\varphi(\beta_1), \ldots, \varphi(\beta_h)$ are representatives of the ideal classes of $CL(K)$. Then it is easy to see from (1.42) that

$$E := \bigcup_{i=1}^{h} \left(\prod_{v \notin S_\infty} U_v \times E^0 \right) \beta_i$$

is a fundamental domain of $J^0(K)/K^\times$.

We associate to $t \in \mathbb{R}_+$ the idele which is equal to $t^{1/n}$ at archimedean places and to 1 at non-archimedean places ($n = [K:\mathbb{Q}]$). Then $J(K)$ is the direct product of $J^0(K)$ and $\mathbb{R}_+$. On $J(K)$ we have the product measure $\mu^\times$ in the sense of Appendix 3.2. Hence there is a uniquely determined measure $\mu_0^\times$ on $J^0(K)$ such that $\mu^\times = \mu_0^\times \times \mu_t^\times$ where $\mu_t^\times$ is the measure dt/t on $\mathbb{R}_+$.

Proposition 1.99. *Let R be the regulator of $\mathfrak{O}_K^\times$, h the ideal class number of K, w the order of the group of roots of unity in K, and r_1 (resp. r_2) the number of real (resp. complex) places of K. Then*

$$\kappa(K) := \mu_0^\times(E) = \frac{2^{r_1}(2\pi)^{r_2} hR}{w\sqrt{|d_{K/\mathbb{Q}}|}}. \boxtimes$$

5.5. Subgroups of $J(K)/K^\times$ of Finite Index and the Ray Class Groups. For any place v of K we embed $K_v^\times$ in $J(K)$ identifying $\alpha \in K_v^\times$ with the idele which is equal to α at v and to 1 at all other places.

Let A be a closed subgroup of finite index in $J(K)$. Since A contains all infinitely divisible elements of $J(K)$, it contains the subgroups $K_v^\times$ for complex v and $K_{v,+} = \{\alpha \in K_v^\times \mid \alpha > 0\}$ for real v. Since A is open, it contains U_v for almost all non-archimedean places v and it contains a subgroup $U_{n_v} \subseteq U_v$ for some $n_v \geqslant 0$ (§ 4.4) for all non-archimedean places v.

A *defining module*[1] $\mathfrak{m}$ in K is the formal product of an ideal $\mathfrak{m}_0$ of $\mathfrak{O}_K$ and some real places $v_1, \ldots, v_s$, $\mathfrak{m} := \mathfrak{m}_0 v_1 \ldots v_s$. The support supp $\mathfrak{m}$ of $\mathfrak{m}$ is defined by $\operatorname{supp} \mathfrak{m} := \operatorname{supp} \mathfrak{m}_0 \cup \{v_1, \ldots, v_s\}$. We say that $\mathfrak{m}$ is smaller or equal to $\mathfrak{m}' = \mathfrak{m}_0' v_1' \ldots v_t'$ if $\mathfrak{m}_0 \mid \mathfrak{m}_0'$ and $\{v_1, \ldots, v_s\} \subseteq \{v_1', \ldots, v_t'\}$. Let $U_\mathfrak{m} \subset J(K)$ be the direct product of the groups $U_{\mathfrak{m},v}$ for all places v where $U_{\mathfrak{m},v} := U_{n_v} \subseteq U_v$ with $n_v := v_\mathfrak{p}(\mathfrak{m}_0)$ if v is the non-archimedean valuation associated to the prime ideal $\mathfrak{p}$,

$$U_{\mathfrak{m},v} := K_{v,+} \text{ if } v \in \{v_1, \ldots, v_s\},$$

$$U_{\mathfrak{m},v} := K_v^\times \text{ if } v \text{ archimedean, } v \notin \{v_1, \ldots, v_s\}.$$

It is easy to see that $U_\mathfrak{m} K^\times$ is a closed subgroup of finite index in $J(K)$. The defining module $\mathfrak{f}$ such that $U_\mathfrak{f}$ is maximal with $U_\mathfrak{f} \subseteq A$, i.e. the smallest defining module, is called the *conductor* of A.

The factor group $J(K)/U_\mathfrak{m} K^\times$ can be interpreted as an ideal class group, called *ray class group*, as follows. Let $\mathfrak{A}_\mathfrak{m} = \mathfrak{A}_S$ be the group of all ideals $\mathfrak{a}$ in $\mathfrak{M}(\mathfrak{O}_K)$ which are prime to $S := \operatorname{supp} \mathfrak{m}$, i.e. $\operatorname{supp} \mathfrak{a} \cap \operatorname{supp} \mathfrak{m} = \varnothing$. The ray group $R_\mathfrak{m}$ is the subgroup of principal ideals (α) in $\mathfrak{A}_S$ such that $\alpha \in U_{\mathfrak{m},v}$ for $v \in \operatorname{supp} \mathfrak{m}$.

---

[1] This is the translation of the german "Erklärungsmodul". Other translations are "interpretation modulus" (Hasse (1967)), "cycle of declaration" (Neukirch (1986), Chap. 4).

Example 35. Let $K = \mathbb{Q}$ and let m be a natural number. We put $\mathfrak{m} := (m)v_\infty$, where v_∞ is the real place of $\mathbb{Q}$. In every class $\bar{a}$ of $(\mathbb{Z}/m\mathbb{Z})^\times$ there is a positive number a. We associate to $\bar{a}$ the class $\overline{(a)}$ in $\mathfrak{A}_\mathfrak{m}$ and we get an isomorphism of $(\mathbb{Z}/m\mathbb{Z})^\times$ onto $\mathfrak{A}_\mathfrak{m}/R_\mathfrak{m}$. □

Proposition 1.100. *There is a canonical isomorphism of* $J(K)/U_\mathfrak{m}K^\times$ *onto* $\mathfrak{A}_S/R_\mathfrak{m}$.

Proof. We put

$$J_\mathfrak{m} = \left\{\prod_v \alpha_v \in J(K) | \alpha_v = 1 \text{ if } v \in \operatorname{supp} \mathfrak{m}\right\}.$$

Then $J_\mathfrak{m} U_\mathfrak{m} K^\times = J(K)$ by Proposition 1.68. Hence

$$J(K)/U_\mathfrak{m}K^\times \cong J_\mathfrak{m}/J_\mathfrak{m} \cap U_\mathfrak{m}K^\times \cong A_S/R_\mathfrak{m}. \ \square$$

$\mathfrak{A}_\mathfrak{m}/R_\mathfrak{m}$ is called *ray class group mod* $\mathfrak{m}$.

§6. Hecke L-Series and the Distribution of Prime Ideals

(Main references: Lang (1970), Chap. 6–8, Narkiewicz (1974), Chap. 6, 7)

The distribution of prime numbers in the sequence of integers is one of the most exciting questions of number theory. The most celebrated result about this question is the prime number theorem: If $\pi(x)$ denotes the number of primes which are smaller or equal the real number x, then

$$\lim_{x \to \infty} \pi(x) \Big/ \left(\frac{x}{\log x}\right) = 1 \tag{1.43}$$

The classical approach to the proof of (1.43) consists in a detailed study of *Riemann's zeta function*

$$\zeta(s) = \sum_{n=1}^{\infty} n^{-s} \qquad \text{for } s \in \mathbb{C}, \operatorname{Re} s > 1.$$

The connection of $\zeta(s)$ with prime numbers is given by *Euler's product formula*

$$\zeta(s) = \prod_p (1 - p^{-s})^{-1} \qquad \text{for } \operatorname{Re} s > 1,$$

where the product runs over all primes p.

$\zeta(s)$ can be extended to a meromorphic function on the whole complex plan, which is holomorphic beside $s = 1$ where it has a simple pole. Furthermore the function

$$\psi(s) = \pi^{-s/2}\Gamma(s/2)\zeta(s) \tag{1.44}$$

satisfies the functional equation

$$\psi(1 - s) = \psi(s) \tag{1.45}$$

For the proof of (1.43) one needs the inequality $\zeta(1 + it) \neq 0$ for $t \in \mathbb{R}$. For more information about the distribution of primes one has to know more about the distribution of zeroes of $\zeta(s)$ in the "*critical stripe*" $0 \leqslant \operatorname{Re} s \leqslant 1$. The famous *Riemann conjecture* asserts that all zeroes in this stripe lie on the straight line $\operatorname{Re} s = \frac{1}{2}$. It is unproved until now.

Riemann (1859) gave two proofs of the functional equation (1.45). His second proof uses the functional equation

$$\vartheta\left(\frac{1}{x}\right) = \sqrt{x}\,\vartheta(x) \qquad \text{for } x > 0 \tag{1.46}$$

of Jacobi's theta function

$$\vartheta(x) = \sum_{n=-\infty}^{\infty} e^{-n^2\pi x}$$

and proceeds as follows: Put

$$\omega(x) := \sum_{n=1}^{\infty} e^{-n^2\pi x} = (\vartheta(x) - 1)/2.$$

Then

$$\psi(s) = \pi^{-s/2}\Gamma(s/2)\zeta(s) = \int_0^\infty x^{s/2-1}\omega(x)\,dx$$

$$= \int_1^\infty x^{s/2-1}\omega(x)\,dx + \int_1^\infty x^{-s/2-1}\omega(1/x)\,dx,$$

$$\times \int_1^\infty x^{-s/2-1}\omega(1/x) = \int_1^\infty x^{-s/2-1}\left(-\frac{1}{2} + \frac{1}{2}\sqrt{x} + \sqrt{x}\,\omega(x)\right)dx$$

$$= -\frac{1}{s} + \frac{1}{s-1} + \int_1^\infty x^{-(s+1)/2}\omega(x)\,dx$$

Hence

$$\psi(s) = \frac{1}{s(s-1)} + \int_1^\infty (x^{s/2-1} + x^{(1-s)/2-1})\omega(x)\,dx. \tag{1.47}$$

The integral on the right side of (1.47) defines a holomorphic function in the whole complex plane, which is symmetric with respect to the straight line $\operatorname{Re} s = \frac{1}{2}$. This proves (1.45).

(1.46) follows easily from Poisson's summation formula. In this paragraph we will explain the proof of a generalization of (1.45) based directly on Poisson's summation formula.

For an even integer $s = k \geqslant 2$ one has the well known formula of Euler

$$\zeta(k) = (-1)^{k/2+1}\frac{(2\pi)^k}{k!2}B_k,$$

where B_k denotes the k-th Bernoulli number. By means of (1.45) one gets

$$\zeta(1-k) = -B_k/k, \tag{1.48}$$

a rational number. The consequences and generalizations of (1.48) will be studied in Chap. 4.

In connection with the distribution of primes in arithmetic progressions Dirichlet studied more generally series

$$L(s, \chi) := \sum_{n=1}^{\infty} \chi(n) n^{-s} = \prod_p (1 - \chi(p) p^{-s})^{-1} \qquad \text{for } \operatorname{Re} s > 1,$$

where χ is a character mod m for some natural number m, i.e. χ is a homomorphism of $(\mathbb{Z}/m\mathbb{Z})^\times$ into $\mathbb{C}^\times$, $\chi(n) = \chi(\bar{n})$ if $(n, m) = 1$, $\chi(n) = 0$ if $(n, m) \neq 1$. χ is called *Dirichlet character*. Let m' be a multiple of m. Then χ induces a character χ' mod m' defined as $\chi'(n) = \chi(n)$ if $(n, m') = 1$ and $\chi'(n) = 0$ if $(n, m') \neq 1$. A character mod f is called *primitive* and f its *conductor* if it is not induced from a character mod m with $m | f$. For every character χ' mod m' there exists a uniquely determined primitive character χ such that χ' is induced from χ. The conductor of χ' is defined as the conductor of χ. The characters mod m form a group. The unit element of this group will be denoted by χ_0. It is called the *trivial* (or *unit*) *character*.

$L(s, \chi)$ is called *Dirichlet L-function*. Again $L(s, \chi)$ can be extended to a meromorphic function in the whole complex plane being holomorphic if χ is distinct from the trivial character χ_0 and satisfying a functional equation analogous to (1.45).

Dirichlet showed that the existence of primes in arithmetic progressions can be rather easily proved if one knows that $L(1, \chi) \neq 0$:

Let l be a natural number with $(l, m) = 1$. We want to show that there are infinitely many primes p with $p \equiv l \pmod{m}$. Let a be a natural number with $al \equiv 1 \pmod{m}$ and let $\sigma > 1$. Then

$$\begin{aligned}\sum_\chi \chi(a) \log L(\sigma, \chi) &= -\sum_\chi \chi(a) \prod_p \log(1 - \chi(p) p^{-\sigma}) \\ &= \sum_\chi \chi(a) \prod_p \chi(p) p^{-\sigma} + f(\sigma) = \sum_p \left(\sum_\chi \chi(ap) \right) p^{-\sigma} + f(\sigma) \\ &= \sum_{p \equiv l (\mathrm{mod}\, m)} p^{-\sigma} + f(\sigma),\end{aligned}$$

where $f(\sigma)$ is a continuous function for $\sigma > 0$. If $L(1, \chi) \neq 0$ for $\chi \neq \chi_0$, then $\sum_\chi \chi(a) \log L(\sigma, \chi)$ is divergent for $\sigma \to 1$ since $\lim_{\sigma \to 1} L(\sigma, \chi_0) = \infty$. Hence $\lim_{\sigma \to 1} \sum_{p \equiv l (\mathrm{mod}\, m)} p^{-\sigma} = \infty$ and this proves the assertion.

$L(1, \chi) \neq 0$ can be proved by various methods but the conceptional proof uses our knowledge about the cyclotomic field $\mathbb{Q}(\zeta_m)$ (§ 3.9): We introduce the following notations: h is the ideal class number, w the number of roots of unity, R the regulator, D the discriminant, r_1 the number of real and r_2 half the number of complex conjugates of $\mathbb{Q}(\zeta_m)$. Furthermore $F(s)$ denotes the elementary function

$$F(s) := \prod_{\mathfrak{p}|m} (1 - N(\mathfrak{p})^{-s})^{-1} \prod_{p|m} (1 - p^{-s})$$

where the first product runs over all prime divisors $\mathfrak{p}$ of m in $\mathbb{Q}(\zeta_m)$. Then one has the following analytic class number formula

$$h = \frac{w\sqrt{|D|}}{2^{r_1+r_2}\pi^{r_2}R} F(1) \prod_{\chi \neq \chi_0} L(1, \chi), \tag{1.49}$$

where the product runs over all non trivial characters χ of $(\mathbb{Z}/m\mathbb{Z})^\times$. (1.49) shows immediately $L(1, \chi) \neq 0$, since $h \neq 0$ by definition.

The proof of (1.49) is intimately related to the study of the Dedekind zeta function of $\mathbb{Q}(\zeta_m)$ and the decompositions of primes in $\mathbb{Q}(\zeta_m)$. We will prove a much more general formula in Chap. 2.1.13.

The *Dedekind zeta function*

$$\zeta_K(s) := \sum_{\mathfrak{a}} N(\mathfrak{a})^{-s} = \prod_{\mathfrak{p}} (1 - N(\mathfrak{p})^{-s})^{-1} \qquad \text{for Re } s > 1$$

of the algebraic number field K, where the sum runs over all integral ideals $\mathfrak{a}$ of K and the product runs over all prime ideals $\mathfrak{p}$ of K, is a first generalization of Riemann's zeta function $\zeta(s)$. More general one associates a zeta function $\zeta_X(s)$ to any scheme X of finite type over Spec $\mathbb{Z}$: $\zeta_X(s)$ is defined by the *Euler product*

$$\zeta_X(s) := \prod_x (1 - N(x)^{-s})^{-1},$$

where the product runs over all closed points x of X and $N(x)$ denotes the cardinality of the residue field of x. The product converges absolutely if Re $s > \dim X$. In particular we get Riemann's zeta function for $X = \text{Spec } \mathbb{Z}$ and $\zeta_K(s) = \zeta_X(s)$ for $X = \text{Spec } \mathfrak{O}_K$. Let $\mathbb{F}_q$ be the finite field with q elements. Then

$$\zeta_X(s) = (1 - q^{-s})^{-1} \qquad \text{for } X = \text{Spec } \mathbb{F}_q$$

and if z is a variable

$$\zeta_X(s) = \prod_f (1 - q^{-s \deg f})^{-1} = (q^{s-1} - 1)^{-1} \qquad \text{for } X = \text{Spec } \mathbb{F}_q[z],$$

where the product runs over all irreducible polynomials $f \in \mathbb{F}_q[z]$.

If X is an arbitrary algebraic variety over Spec $\mathbb{F}_q$, then by a famous theorem of Deligne (see e.g. Katz (1976)) one has the following results:

1. $\zeta_X(s)$ is a rational function $Z_X(y)$ of $y := q^s$.
2. $\zeta_X(s)$ satisfies a functional equation similar to (1.45).
3. The zeroes and poles of $\zeta_X(s)$ lie on the straight lines Re $s = \frac{j}{2} \log q$, $j = 0, 1, \ldots, 2 \dim X$.

3. is the analogue of the Riemann conjecture for $\zeta(s)$ mentioned above.

Dirichlet L-functions can be generalized to L-functions for certain sheaves on schemes of finite type and to L-functions for representations of reductive groups and Galois groups of local and global fields. The interplay between these L-functions is the main subject of the Langlands philosophy (Chap. 5).

As a resumé of our considerations we have four main problems in the study of zeta functions and L-functions:

1. Meromorphic continuation of the functions to the whole plane and functional equation.
2. Situation of poles, computation of their residues.
3. Values in integral points.
4. Distribution of zeroes, generalization of Riemann's conjecture and proof of the conjecture.

In this paragraph we study mainly the first two problems for L-functions connected with number fields and we apply our knowledge about L-functions to the distribution of ideals of number fields. The third problem will be studied in special situations in Chap. 4.

Following the thesis of J. Tate (1967) we consider L-series in the setting of harmonic analysis on adele and idele groups (§ 5) and we study at first rather general functions. The L-functions we are really interested in, then will appear in a natural way. We keep the notation of § 5.

6.1. The Local Zeta Functions. Let K be a local field and v the corresponding normalized valuation. A complex valued function f on K^+ is called admissible if $f \in S(K^+)$ and $f(\alpha)v(\alpha)^\sigma$, $\hat{f}(\alpha)v(\alpha)^\sigma$ are in $L_1(K^\times)$ for all positive numbers σ.

For any quasi-character χ with positive exponent and any admissible function f we define

$$\zeta(f, \chi) := \int f(\alpha)\chi(\alpha)\, d\mu^\times(\alpha).$$

$\zeta(f, \chi)$ is called a *zeta function of* K.

Let $\Gamma(s)$ be the Gamma function. For any quasi-character χ we define the corresponding L-function or *Euler factor* $L(\chi) \in \mathbb{C}^\times \cup \{\infty\}$ as follows:

a) $K = \mathbb{R}$, then

$$L(\chi) := \pi^{-s/2}\Gamma(s/2) \qquad \text{if } \chi = v^s,$$

$$L(\chi) := \pi^{-(s+1)/2}\Gamma((s+1)/2) \qquad \text{if } \chi = \operatorname{sgn} v^s.$$

b) $K = \mathbb{C}$, then

$$L(\chi) := 2(2\pi)^{-s-|n|/2}\Gamma(s + |n|/2) \qquad \text{if } \chi = \chi_n v^s.$$

c) $K = \mathfrak{p}$-adic number field, $\mathfrak{p} = (\pi)$, then

$$L(\chi) = \begin{cases} (1 - \chi(\pi))^{-1} & \text{if } \chi \text{ is unramified} \\ 1 & \text{if } \chi \text{ is ramified.} \end{cases}$$

In every case $L(\chi v^s)$ is a meromorphic function of s and L has no zeroes.

Theorem 1.101. *Let f be an admissible function and χ a quasi-character with exponent* 0. *Then $\zeta(f, \chi v^s)$ is a holomorphic function for* $\operatorname{Re} s > 0$. *There is a meromorphic function $\varepsilon(\chi v^s)$ of $s \in \mathbb{C}$, called local ε-factor, which is independent of f, such that*

$$\frac{\zeta(\hat{f}, \chi^{-1}v^{1-s})}{L(\chi^{-1}v^{1-s})} = \varepsilon(\chi v^s)\frac{\zeta(f, \chi v^s)}{L(\chi v^s)} \qquad \textit{for } 0 < \operatorname{Re} s < 1. \tag{1.50}$$

(1.50) defines the analytic continuation of $\zeta(f, \chi v^s)$ *to a meromorphic function of* $s \in \mathbb{C}$.

Proof. Let f and g be arbitrary admissible functions. By means of Fubini's theorem one shows that

$$\zeta(f, \chi v^s)\zeta(\hat{g}, \chi^{-1}v^{1-s}) = \zeta(\hat{f}, \chi^{-1}v^{1-s})\zeta(g, \chi v^s) \qquad \text{for } 0 < \operatorname{Re} s < 1.$$

Therefore it is sufficient to prove Theorem 1.101 for a special admissible function f.

a) $K = \mathbb{R}$. If $\chi = 1$, we put $f(\xi) = \exp(-\pi\xi^2)$ and find $\hat{f}(\xi) = f(\xi)$, $\zeta(f, v^s) = \pi^{-s/2}\Gamma(s/2)$, $\varepsilon(v^s) = 1$. If $\chi = \operatorname{sgn}$, we put $f(\xi) = \xi\exp(-\pi\xi^2)$ and find $\hat{f}(\xi) = if(\xi)$, $\zeta(\operatorname{sgn} v^s) = \pi^{-(s+1)/2}\Gamma((s+1)/2)$, $\varepsilon(\operatorname{sgn} v^s) = i$.

b) $K = \mathbb{C}$. We put

$$f_n(\xi) = \begin{cases} \bar{\xi}^{|n|}\exp(-2\pi v(\xi)) & \text{if } n \geqslant 0, \\ \xi^{|n|}\exp(-2\pi v(\xi)) & \text{if } n \leqslant 0. \end{cases}$$

Then $\hat{f}_n(\xi) = i^{|n|}f_{-n}(\xi)$, $\zeta(f_n, \chi_n v^s) = (2\pi)^{1-s+|n|/2}\Gamma(s + |n|/2)$, $\varepsilon(\chi_n v^s) = i^{|n|}$.

c) $K = \mathfrak{p}$-adic number field with residue characteristic p. Without loss of generality let χ be a character of $K^\times$ with conductor $\mathfrak{p}^n$ and $\chi(\pi) = 1$. We put $\mathfrak{D}_{K/\mathbb{Q}_p} =: \mathfrak{p}^d$,

$$f(\xi) = \begin{cases} \lambda^{-1}(\xi) & \text{if } \xi \in \mathfrak{p}^{-d-n}, \\ 0 & \text{if } \xi \notin \mathfrak{p}^{-d-n}. \end{cases}$$

Then

$$\hat{f}(\xi) = \begin{cases} p^{d/2+n} & \text{if } \xi \in U_n, \\ 0 & \text{if } \xi \notin U_n. \end{cases}$$

In the unramified case, $n = 0$, one has

$$\zeta(f, \chi v^s) = (N\mathfrak{p}^d)^{s-1/2}(1 - (N\mathfrak{p})^{-s})^{-1}, \zeta(f, \chi^{-1}v^{1-s}) = (1 - (N\mathfrak{p})^{s-1})^{-1},$$
$$\varepsilon(\chi v^s) = (N\mathfrak{p}^d)^{1/2-s},$$

and in the ramified case, $n > 0$, one has

$$\zeta(f, \chi v^s) = (N\mathfrak{p}^{d+n})^s\left(\sum_\beta \chi(\beta)\lambda^{-1}(\beta/\pi^{d+n})\right)\int_{U_n} d\mu^\times(\alpha)$$

where β runs through a set of representatives of the elements of U/U_n, and

$$\zeta(f, \chi^{-1}v^{1-s}) = (N\mathfrak{p})^{d/2+n}\int_{U_n} d\mu^\times(\alpha), \varepsilon(\chi v^s) = (N\mathfrak{p}^{d+n})^{1/2-s}\varepsilon_0(\chi)$$

where

$$\varepsilon_0(\chi) := (N\mathfrak{p}^n)^{-1/2} \sum_\beta \chi^{-1}(\beta)\lambda(\beta/\pi^{d+n})$$

is the so called *root number* of χ and has absolute value 1. □

For any non trivial character ψ of K^+ and any character χ of $K^\times$ with conductor $\mathfrak{p}^n$ the sum

$$\tau(\chi, \psi) := \sum_\beta \chi^{-1}(\beta/\pi^{d+n})\psi(\beta/\pi^{d+n})$$

is called a *local Gaussian sum*. Now d denotes the largest integer such that $\psi(\mathfrak{p}^{-d}) = \{1\}$. If $\psi = \lambda$, then $\mathfrak{p}^d = \mathfrak{D}_{K/\mathbb{Q}_p}$.

6.2. The Global Functional Equation. Now let K be an algebraic number field. Let χ be a quasi-character of the idele class group $J(K)/K^\times$. The induced quasi character on $J(K)$ will be denoted by χ, too. Since $J^0(K)/K^\times$ is compact, the restriction of χ to $J^0(K)$ is a character. It follows that χ has the form

$$\chi(\alpha) = \chi_0(\alpha)|\alpha|^s \qquad \text{with } s \in \mathbb{C}\ (\chi = \chi_0|\ |^s)$$

where χ_0 is a character of $J(K)/K^\times$ which is trivial on $\mathbb{R}_+$. We call χ_0 a *normalized character*. As in the local situation, we call Re s the exponent of χ.

We want to define global zeta functions. An admissible function f on $A(K)$ is a complex-valued function on $A(K)$ with the following properties:

1. f and $\hat{f}$ are continuous and in $L_1(A(K))$.

2. $\sum_{\xi\in K} f(\alpha(\beta+\xi))$ and $\sum_{\xi\in K} \hat{f}(\alpha(\beta+\xi))$ are both convergent for each $\alpha \in J(K)$, $\beta \in A(K)$. The convergence is uniform for α, β ranging over compact subsets of $J(K)$ resp. $A(K)$.

3. $f(\alpha)|\alpha|^\sigma$ and $\hat{f}(\alpha)|\alpha|^\sigma \in L_1(J(K))$ for $\sigma > 1$.

Then

$$\zeta(f, \chi) := \int f(\alpha)\chi(\alpha)\, d\mu^\times(\alpha)$$

is defined for any quasi-character χ with exponent $\sigma > 1$ and $\zeta(f, \chi|\ |^s)$ is a holomorphic function of s for Re $s > 1 - \sigma$.

The following theorem contains the solution of the main problems 1) and 2) from the introduction to § 6 for $\zeta(f, \chi|\ |^s)$ considered as function of s.

Theorem 1.102. *Let χ be a normalized character and f an admissible function on $A(K)$. Then $\zeta(f, \chi|\ |^s)$ can be analytically continued to a meromorphic function on $s \in \mathbb{C}$. If χ is non trivial, then $\zeta(f, \chi|\ |^s)$ is holomorphic on $\mathbb{C}$. If χ is trivial, then $(f, \chi|\ |^s)$ is holomorphic except at $s = 0$ and $s = 1$ where it has simple poles with residues $-\kappa f(0)$ and $\kappa\hat{f}(0)$ respectively with $\kappa := \mu_0^\times(E)$* (Proposition 1.99). *$\zeta(f, \chi|\ |^s)$ satisfies the functional equation*

$$\zeta(f, \chi|\ |^s) = \zeta(\hat{f}, \chi^{-1}|\ |^{1-s}). \tag{1.51}$$

Proof. We write $\zeta(f, \chi|\ |^s)$ in the form

$$\zeta(f, \chi|\ |^s) = \int_{|\alpha|\leqslant 1} f(\alpha)\chi(\alpha)|\alpha|^s \, d\mu^\times(\alpha) + \int_{|\alpha|>1} f(\alpha)\chi(\alpha)|\alpha|^s \, d\mu^\times(\alpha).$$

The second integral is convergent for all $s \in \mathbb{C}$. With the first integral we proceed as follows: We put $\alpha = \beta t$ with $\beta \in J^0(K)$, $t \in \mathbb{R}_+$. Then

$$\int_{|\alpha|\leqslant 1} f(\alpha)\chi(\alpha)|\alpha|^s \, d\mu^\times(\alpha) = \int_0^1 \int_{J^0(K)} f(\beta t)\chi(\beta)\, d\mu_0^\times(\beta) t^{s-1}\, dt,$$

$$\begin{aligned}\int_{J^0(K)} f(\beta t)\chi(\beta)\, d\mu_0^\times(\beta) &= \sum_{\alpha\in K^\times} \int_E f(\alpha\beta t)\chi(\beta)\, d\mu_0^\times(\beta)\\ &= \sum_{\alpha\in K} \int_E f(\alpha\beta t)\chi(\beta)\, d\mu_0^\times(\beta) - f(0)\int_E \chi(\beta)\, d\mu_0^\times(\beta)\\ &= \int_E \left(\sum_{\alpha\in K} f(\alpha\beta t)\right)\chi(\beta)\, d\mu_0^\times(\beta) - f(0)\delta_\chi \mu_0^\times(E)\end{aligned}$$

$$\text{where } \delta_\chi := \begin{cases} 1 & \text{if } \chi \text{ is trivial} \\ 0 & \text{if } \chi \text{ is non-trivial.}\end{cases}$$

The essential step in the proof is now the application of the Poisson sum formula (Appendix 3) for the function $g(\xi) := f(\xi\beta t)$, its dual $\hat{g}(\xi) = \frac{1}{|\beta|t}\hat{f}\left(\frac{\xi}{\beta t}\right)$, $\xi \in A(K)$, and for the discrete subgroup K of $A(K)$ (Theorem 1.97):

$$\sum_{\alpha\in K} f(\alpha\beta t) = \frac{1}{|\beta| t}\sum_{\alpha\in K}\hat{f}\left(\frac{\alpha}{\beta t}\right).$$

We get

$$\begin{aligned}\int_E \left(\sum_{\alpha\in K} f(\alpha\beta t)\right)\chi(\beta)\, d\mu_0^\times(\beta) &= \int_E \left(\sum_{\alpha\in K} \hat{f}\left(\frac{\alpha}{\beta t}\right)\right)\frac{\chi(\beta)}{t}\, d\mu_0^\times(\beta)\\ &= \int_{J^0(K)} \hat{f}(\beta/t)(\chi(\beta)t)^{-1}\, d\mu_0^\times(\beta) + \hat{f}(0)\mu_0^\times(E)\delta_\chi t^{-1}.\end{aligned}$$

Putting our computations together we find

$$\begin{aligned}\zeta(f, \chi|\ |^s) = {} & \int_{|\alpha|\geqslant 1} f(\alpha)\chi(\alpha)|\alpha|^s\, d\mu^\times(\alpha) + \int_{|\alpha|\geqslant 1} \hat{f}(\alpha)\chi^{-1}(\alpha)|\alpha|^{1-s}\, d\mu^\times(\alpha)\\ & + \delta_\chi\mu_0^\times(E)\left(\frac{\hat{f}(0)}{s-1} - \frac{f(0)}{s}\right). \end{aligned} \tag{1.52}$$

The expression on the right side is a meromorphic function with the properties of Theorem 1.102. □

6.3. Hecke Characters. In the following section we will choose special global functions f which lead to a generalization of the functional equation for the

Riemann zeta function. In this section we compare the characters of the idele class group with Hecke characters.

Let $\mathfrak{m}$ be a defining module in K (§ 5.5). A character λ of $\mathfrak{A}_{\mathfrak{m}}$ is called *Hecke character* (or *groessen character*) mod $\mathfrak{m}$ if there exists a character λ_∞ of $\prod_{v \in S_\infty} K_v^\times$ such that

$$\lambda((\alpha)) = \lambda_\infty(\alpha) \qquad \text{for all } \alpha \in K^\times \text{ with } (\alpha) \in R_{\mathfrak{m}_0},$$

where $K^\times$ is diagonally embedded in $\prod_{v \in S_\infty} K_v^\times$. The character λ_∞ is uniquely determined by λ (Theorem 1.68) and is called the *infinite component* of λ. A character λ_∞ of $\prod_{v \in S_\infty} K_v^\times$ appears as infinite component of a Hecke character mod $\mathfrak{m}$ if and only if $\lambda_\infty(\varepsilon) = 1$ for all units ε of K with $\varepsilon \in U_{\mathfrak{m},v}$ for $v \in \operatorname{supp} \mathfrak{m}_0$.

Let $\mathfrak{m}'$ be a defining module in K and λ' a Hecke character mod $\mathfrak{m}'$ such that $\lambda_\infty = \lambda'_\infty$ and

$$\lambda(a) = \lambda'(a) \qquad \text{for all } \mathfrak{a} \in \mathfrak{A}_{\mathfrak{m}} \cap \mathfrak{A}_{\mathfrak{m}'}.$$

Then $\mathfrak{m}'$ is called a defining module of λ. The smallest defining module of λ is called the *conductor* of λ and is denoted by $\mathfrak{f}_\lambda$. A Hecke character mod $\mathfrak{m}$ is called *primitive* if $\mathfrak{m} = \mathfrak{f}_\lambda$.

Now let χ be a character of the idele class group $J(K)/K^\times$ and let χ_v be the restriction of χ to $K_v^\times$. We denote the conductor of χ_v by $\mathfrak{f}_v$ (§ 5.2). By the definition of the topology of $J(K)$, $\mathfrak{f}_v = (1)$ for almost all places. We call $\prod_v \mathfrak{f}_v$ the conductor of χ.

To any character χ of $J(K)/K^\times$ with conductor $\mathfrak{f}$ we associate the primitive Hecke character λ_χ defined by $\lambda_\chi(\mathfrak{p}) = \chi(\pi)$ for all prime ideals $\mathfrak{p}$ in $\mathfrak{A}_{\mathfrak{f}}$, where π is a prime element in $K_{\mathfrak{p}}$. It is easy to see that for every primitive Hecke character λ there corresponds one and only one character χ of $J(K)/K^\times$ such that $\lambda = \lambda_\chi$ (compare § 5.5). If there is no danger of confusion we identity χ with λ.

In the following S denotes a finite set of places of K containing all archimedean places and we consider characters χ of $J(K)/K^\times$ which are unramified outside S, i.e. $\mathfrak{f}_v = (1)$ for $v \notin S$. Such a character χ is the product of local characters χ_v,

$$\chi\left(\prod_v \alpha_v\right) = \prod_v \chi_v(\alpha_v),$$

satisfying the conditions:

1) χ_v is unramified if $v \notin S$.
2) $\prod_v \chi_v(\alpha) = 1$ for $\alpha \in K^\times$.

To construct a Hecke character with conductor $\mathfrak{f}$ we first look at S-units of K. Let ε_0 be a generator of the group of roots of unity in K and $\varepsilon_1, \ldots, \varepsilon_m$ a fundamental system of S-units in K (§ 3.8). We have to choose the local characters χ_v for $v \in S$ with conductor $\mathfrak{f}_v$ such that

$$\prod_{v \in S} \chi_v(\varepsilon_i) = 1 \quad \text{for } i = 0, 1, \ldots, m. \tag{1.53}$$

It remains to choose the Hecke character χ such that

$$\chi(\varphi_S(\alpha)) = \prod_{v \in S} \chi_v(\alpha)^{-1} \quad \text{for } \alpha \in K^\times,$$

where

$$\varphi_S(\alpha) := \prod_{\mathfrak{p} \notin S} \mathfrak{p}^{\nu_\mathfrak{p}(\alpha)}.$$

This means that χ is given on the subgroup $\varphi_S(K^\times)$ of finite index in $\mathfrak{A}_S$ and one has to extend χ arbitrarily to $\mathfrak{A}_S$.

Example 36. Let $H_\mathfrak{m}$ be the ray class group mod $\mathfrak{m}$ (§ 5.5). An arbitrary character of $H_\mathfrak{m}$ is a Hecke character mod $\mathfrak{m}$. Taking into account the isomorphism of Example 35 we see that Dirichlet's characters χ mod m considered in the introduction to § 6 are special Hecke characters. □

Example 37. $K = \mathbb{Q}(\sqrt{-1})$, $S = \{\infty\}$, the archimedean place of K. A character of $K_\infty^\times = \mathbb{C}^\times$ has the form $\chi_\infty(\xi) = |\xi|^s \xi^n$ with $s \in \mathbb{C}$, $n \in \mathbb{Z}$, $\operatorname{Re} s = -n$. (1.53) means $4|n$. Hence the most general Hecke character unramified outside S has the form

$$\chi(\alpha) = |\alpha|^s \alpha^{4n} \qquad \text{for } \alpha \in K^\times \text{ with } s \in \mathbb{C}, n \in \mathbb{Z}, \operatorname{Re} s = -4n$$

(here $|\alpha|$ denotes the usual absolute value). χ is normalized if and only if $s = -4n$. □

6.4. The Functional Equation for Hecke L-Series. Let χ be a normalized character of $J(K)/K^\times$, unramified outside S. We choose the function f on $A(K)$ as product of the local functions f_v defined in § 6.1:

$$f\left(\prod_v \xi_v\right) = \prod_v f_v(\xi_v).$$

Proposition 1.103. *f is admissible,*

$$\hat{f}\left(\prod_v \xi_v\right) = \prod_v \hat{f}_v(\xi_v),$$

$$\zeta(f, \chi|\ |^s) = \prod_v \zeta(f_v, \chi_v v^s). \boxtimes$$

An essential step in the proof of Proposition 1.86 is the absolute convergence of

$$\prod_v \zeta(f_v, \chi_v v^s) = \prod_v \frac{1}{1 - \chi(\mathfrak{p})N(\mathfrak{p})^{-s}} \qquad \text{for } \operatorname{Re} s > 1 \tag{1.54}$$

where the products runs over all $\mathfrak{p} = v \notin S$ for which $K_v/\mathbb{Q}_p$ is unramified. (1.54) follows from the convergence of

$$\prod_\mathfrak{p} \frac{1}{1 - N(\mathfrak{p})^{-s}} \qquad \text{for } \sigma > 1$$

where the product runs over all prime ideals $\mathfrak{p}$ of K.

We extend the Hecke character χ to a function on $\mathfrak{A}_\varnothing$ putting $\chi(\mathfrak{a}) = 0$ if $\mathfrak{a} \in \mathfrak{A}_\varnothing - \mathfrak{A}_S$.

$$L(s, \chi) := \prod_{v \notin S} \frac{1}{1 - \chi(\mathfrak{p})N(\mathfrak{p})^{-s}} = \sum_{\mathfrak{a} \in \mathfrak{A}_\varnothing} \chi(\mathfrak{a})N(\mathfrak{a})^{-s}$$

is called the *Hecke L-series* attached to the Hecke character χ. If χ_0 is the trivial character,

$$\zeta_K(s) := L(s, \chi_0) = \sum_{\mathfrak{a} \in \mathfrak{A}_\varnothing} N(\mathfrak{a})^{-s}$$

is the Dedekind zeta function of K mentioned in the introduction to § 6.

If $K = \mathbb{Q}$ and χ a character of $(\mathbb{Z}/m\mathbb{Z})^\times$ (Example 31), then χ can be considered as a Hecke character by means of the isomorphism $(\mathbb{Z}/m\mathbb{Z})^\times \to \mathfrak{A}_\mathfrak{m}/R_\mathfrak{m}$ in Example 31. The function

$$L(s, \chi) = \prod_p \frac{1}{1 - p^{-s}} = \sum_{a=1}^{\infty} \chi(a)a^{-s}$$

is the Dirichlet L-series mod m mentioned in the introduction to § 6.

Combining Theorem 1.102 and Proposition 1.103 we come to the following result.

Theorem 1.104. *Let χ be a normalized primitive Hecke character. If χ is non-trivial, then $L(s, \chi)$ is holomorphic in the whole complex plan. $\zeta_K(s)$ is holomorphic in the complex plan beside $s = 1$ where it has a simple pole with residuum $\kappa(K)$* (Proposition 1.99).

We put

$$L_\infty(s, \chi) := \prod_{v \in S_\infty} L(\chi_v v^s), \; \Lambda(s, \chi) := L(s, \chi)L_\infty(s, \chi).$$

Then $\Lambda(s, \chi)$ satisfies the functional equation

$$\Lambda(s, \chi) = \varepsilon(s, \chi)\Lambda(1 - s, \chi^{-1})$$

where

$$\varepsilon(s, \chi) := \prod_v \varepsilon(\chi_v v^s)$$

is the global ε-factor. □

$\Lambda(s, \chi)$ is called the *enlarged L-function.*

6.5. Gaussian Sums. Let K be an algebraic number field, $\mathfrak{D}$ the different of $K/\mathbb{Q}$, $\mathfrak{f}$ an ideal of $\mathfrak{O}_K$, χ a character mod $\mathfrak{f}$, i.e. a character of $(\mathfrak{O}_K/\mathfrak{f})^\times$, and $\alpha \in \mathfrak{D}^{-1}\mathfrak{f}^{-1}$. Then

$$\tau_\alpha(\chi) := \sum_{\bar{\xi} \in (\mathfrak{O}_K/\mathfrak{f})^\times} \chi(\xi) \exp(2\pi i \operatorname{Tr}_{K/\mathbb{Q}}(\alpha\xi))$$

is called the *Gaussian sum* attached to χ and α. The expression $\operatorname{Tr}_{K/\mathbb{Q}}(\alpha\xi)$ is well defined since $\xi \equiv \xi'$ (mod $\mathfrak{f}$) implies $\operatorname{Tr}_{K/\mathbb{Q}}(\alpha\xi) \equiv \operatorname{Tr}_{K/\mathbb{Q}}(\alpha\xi')$ (mod 1).

Example 38. For $K = \mathbb{Q}$ and $\mathfrak{f} = (f)$

$$\tau_{1/f}(\chi) = \sum_{\substack{1 \leqslant x < f \\ (x, f) = 1}} \chi(x) \exp(2\pi i x/f)$$

is the sum studied by Gauss. □

Proposition 1.105. *Let $\beta \in \mathfrak{O}_K$, $(\beta, \mathfrak{f}) = 1$. Then*

$$\tau_{\alpha\beta}(\chi) = \overline{\chi(\beta)}\tau_\alpha(\chi).$$

Proof. $\tau_{\alpha\beta}(\chi) = \sum_\xi \chi(\xi) \exp(2\pi i \operatorname{Tr}_{K/\mathbb{Q}}(\alpha\beta\xi))$

$$= \sum_\xi \chi(\xi\beta^{-1}) \exp(2\pi i \operatorname{Tr}_{K/\mathbb{Q}}(\alpha\xi)) = \overline{\chi(\beta)}\tau_\alpha(\chi). \ \square$$

Proposition 1.106. *Let $\mathfrak{f}_1$ and $\mathfrak{f}_2$ be ideals of $\mathfrak{O}_K$ which are relatively prime, let χ_k be a character* mod $\mathfrak{f}_k$, $k = 1, 2$, *and χ the corresponding character* mod $\mathfrak{f} := \mathfrak{f}_1\mathfrak{f}_2$ *according to the canonical isomorphism*

$$(\mathfrak{O}_K/\mathfrak{f})^\times \cong (\mathfrak{O}_K/\mathfrak{f}_1)^\times \times (\mathfrak{O}_K/\mathfrak{f}_2)^\times \ (\S\, 3.2).$$

Furthermore let α, α_1, $\alpha_2 \in \mathfrak{O}_K$ be such that $\alpha \in \mathfrak{D}^{-1}\mathfrak{f}^{-1}$, $\alpha_k \in \mathfrak{D}^{-1}\mathfrak{f}_k^{-1}$, $\alpha - \alpha_k \in \mathfrak{D}^{-1}\mathfrak{f}^{-1}\mathfrak{f}_k$, $k = 1, 2$. Then

$$\tau_\alpha(\chi) = \tau_{\alpha_1}(\chi_1)\tau_{\alpha_2}(\chi_2). \ \square$$

The connection with the local Gaussian sums defined in §6.1 is given by the following

Proposition 1.107. *Let $\mathfrak{f} = \mathfrak{p}^m$, $m > 0$, be a power of the prime ideal $\mathfrak{p}$ of K and χ a character of $K_\mathfrak{p}^\times$ with conductor $\mathfrak{f}$. The character of $(\mathfrak{O}_K/\mathfrak{f})^\times$ induced by χ will be denoted by χ, too. Let $\alpha \in \mathfrak{D}^{-1}\mathfrak{f}^{-1}$ be such that $\alpha \notin \mathfrak{D}^{-1}\mathfrak{f}^{-1}\mathfrak{p}$. Then*

$$\chi(\alpha)\tau_\alpha(\chi) = \overline{\tau(\chi, \lambda)}, \text{ where } \lambda \text{ is the function defined in } \S\, 5.1.$$

Proof. Let p be the residue characteristic of $K_\mathfrak{p}$. Then

$$\chi(\alpha)\tau_\alpha(\chi) = \sum_{\bar\xi} \chi(\alpha\xi) \exp\left(2\pi i\right) \sum_{\mathfrak{p}|p} \operatorname{Tr}_{K_\mathfrak{q}/\mathbb{Q}_p}(\alpha\xi) = \sum_{\bar\xi} \chi(\alpha\xi)\lambda^{-1}(\alpha\xi)$$

$$= \overline{\sum_{\bar\xi} \chi(\alpha\xi)^{-1}\lambda(\alpha\xi)}. \ \square$$

Now let χ be a character of $J(K)/K^\times$ and $\mathfrak{f}_\chi$ the finite part of the conductor of χ. The Gaussian sum $\tau(\chi)$ attached to χ is defined by

$$\tau(\chi) := \prod_{\mathfrak{p}|\mathfrak{f}_\chi} \tau(\chi_\mathfrak{p}, \lambda_\mathfrak{p}).$$

The character of $(\mathfrak{O}_K/\mathfrak{f}_\chi)^\times$ induced by χ according to the isomorphism

$$(\mathfrak{O}_K/\mathfrak{f}_\chi)^\times \cong \prod_{\mathfrak{p}|\mathfrak{f}_\chi} U_\mathfrak{p}/U_{\mathfrak{p}, v_\mathfrak{p}(\mathfrak{f}_\chi)}$$

will be denoted by χ, too. The preceding propositions imply

Proposition 1.108. *Let $\alpha \in \mathfrak{D}^{-1}\mathfrak{f}^{-1}$ such that $\alpha \notin \mathfrak{D}^{-1}\mathfrak{f}^{-1}\mathfrak{p}$ for $\mathfrak{p} \in \mathfrak{f}_\chi$. Then*

$$|\tau_\alpha(\chi)| = |\tau(\chi)|. \square$$

Example 39. For $K = \mathbb{Q}$ and $\mathfrak{f} = (f)$ we have

$$\tau_{1/f}(\chi) = \overline{\tau(\chi)}. \square$$

In § 6.1 we have defined local root numbers for ramified characters. For unramified characters of $K_\mathfrak{p}^\times$ the local root number is 1 and for infinite places it is equal to the ε-factor, which in this case is a constant. The *global root number* $\varepsilon_0(\chi)$ is defined as the product of the local root numbers. Hence

$$\varepsilon_0(\chi) = \tau(\chi) i^r (N\mathfrak{f}_\chi)^{-1/2} \tag{1.55}$$

where r is the number of real places in the conductor of χ. Since $\varepsilon_0(\chi)$ has absolute value 1, we have

$$|\tau(\chi)| = (N\mathfrak{f}_\chi)^{1/2}. \tag{1.56}$$

6.6. Asymptotical Distribution of Ideals and Prime Ideals (Main reference: Narkiewicz (1974), Chap. 7). Now we apply our results about Hecke L-series to the distribution of ideals and prime ideals. In fact we need one more theorem about Hecke L-series:

Theorem 1.109. *Let χ be a Hecke character. Then $L(s, \chi) \neq 0$ for* Re $s \geqslant 1$. ⊠

Without loss of generality let χ be normalized. The proof of $L(1 + it, \chi) \neq 0$ in the case $t \neq 0$ or $\chi^2 \neq \chi_0$ is not much more difficult as the proof of $\zeta(1 + it) \neq 0$ for Riemann's zeta function (see Narkiewicz (1974), Theorem 7.5). $L(1, \chi) \neq 0$ for a character χ of finite order can be proved as explained for Dirichlet characters in the introduction to § 6.

Let A be a set of prime ideals of the algebraic number field K. The set A is called *regular with Dirichlet density* $a \geqslant 0$ if

$$\prod_{\mathfrak{p} \in A} N(\mathfrak{p})^{-s} - a \log(s - 1)$$

can be continued to a holomorphic function in Re $s \geqslant 1$.

If A has Dirichlet density $a > 0$, then of course A is infinite. In fact in many interesting situations the usual method to prove that A is not empty consists in proving that A has Dirichlet density $a > 0$.

It is easy to see that a is uniquely determined by A, i.e. our definition makes sense. In the following $g_i(s)$, $i = 1, 2, \ldots$, denote holomorphic functions in Re $s \geqslant 1$.

Theorem 1.110. *Let χ be a normalized Hecke character unramified outside S and let S' be the subset of S consisting of the non-archimedean places for which χ is unramified. Then*

$$L(s, \chi) = \alpha_\chi/(s - 1) + g_1(s) \tag{1.57}$$

and

$$\sum_{\mathfrak{p}} \chi(\mathfrak{p}) N(\mathfrak{p})^{-s} = \delta_\chi \log \frac{1}{s-1} + g_2(s) \tag{1.58}$$

where the sum runs over all prime ideals $\mathfrak{p}$ *of* K *and*

$$\alpha_\chi = \delta_\chi = 0 \qquad \text{if } \chi \text{ is non-trivial,}$$

$\alpha_\chi = \kappa(K) \prod_{\mathfrak{p} \in S'} (1 - N(\mathfrak{p})^{-1})$, $\delta_\chi = 1$ *if* χ *is trivial.*

Proof. (1.57) follows immediately from Theorem 1.104. (1.58) follows from Theorem 1.109 and (1.57) taking in account

$$\log L(s, \chi) = \sum_{\mathfrak{p}} \chi(\mathfrak{p}) N(\mathfrak{p})^{-s} + g_3(s). \square$$

If we take for χ the trivial character, (1.58) shows that the set of all prime ideals of K is regular with Dirichlet density 1. A finite set of prime ideals has Dirichlet density 0.

Theorem 1.111. *Let* $\mathfrak{m}$ *be a defining module and* $\mathfrak{C}$ *a class in the ray class group* $H_\mathfrak{m}$. *Then* $\mathfrak{C}$ *is regular with Dirichlet density* $1/|H_\mathfrak{m}|$.

Proof. By the orthogonality properties of characters χ of finite abelian groups one gets

$$\sum_{\mathfrak{p} \in \mathfrak{C}} N(\mathfrak{p})^{-s} = |H_\mathfrak{m}|^{-1} \sum_{\chi} \chi(\mathfrak{C}^{-1}) \sum_{\mathfrak{p}} \chi(\mathfrak{p}) N(\mathfrak{p})^{-s} = |H_\mathfrak{m}|^{-1} \log \frac{1}{s-1} + g_4(s). \square$$

According to Example 35 Theorem 1.111 implies as a special case that for given natural numbers k and l with $(k, l) = 1$ there are infinitely many prime numbers p with $p \equiv l (\mathrm{mod}\ k)$ (*Dirichlet's theorem on primes in arithmetic progressions*).

Theorem 1.112. *The set* A *of all prime ideals* $\mathfrak{p}$ *of* K *with* $f_{K/\mathbb{Q}}(\mathfrak{p}) = 1$ *is regular with Dirichlet density* 1.

Proof. $\sum_{\mathfrak{p} \notin A} N(\mathfrak{p})^{-s}$ converges uniformly for $\mathrm{Re}\, s \geqslant 3/4$. $\square$

The following theorem, which generalizes the prime number theorem mentioned in the introduction to this paragraph, can be proved by means of the Tauberian theorem of Delange-Ikehara (Delange (1954)).

Theorem 1.113. *Let* A *be a regular set of prime ideals with Dirichlet density* a *and put*

$$\prod_A (x) := |\{\mathfrak{p} \in A \mid N(\mathfrak{p}) \leqslant x\}|.$$

Then

$$\lim_{x \to \infty} \frac{\prod_A (x)}{x/\log x} = a. \boxtimes$$

With the same method and (1.57) one gets quantitative results about the distribution of ideals. We state here only a theorem of Dedekind (1894):

Theorem 1.114. *Let $M(x)$ be the number of ideals $\mathfrak{a}$ of $\mathfrak{O}_K$ with $N(\mathfrak{a}) \leqslant x$. Then*

$$\lim_{x\to\infty} \frac{M(x)}{x} = \kappa(K). \boxtimes$$

The next theorem gives the general scheme for uniform distribution of ideals and prime ideals.

The sequence $g_1, g_2, \ldots$ of elements of a compact abelian group G with Haar measure μ is *uniformly distributed* in G if for any complex valued continuous function f on G one has

$$\lim_{n\to\infty} \frac{1}{n} \sum_{i=1}^{n} f(g_i) = \frac{1}{\mu(G)} \int_G f(x)\,dx. \tag{1.59}$$

$g_1, g_2, \ldots$ is uniformly distributed in G if and only if (1.59) is fulfilled for all characters $f = \chi$ of G.

Theorem 1.115. *Let G be a compact abelian group and let f be a continuous homomorphism from $J^0(K)/K^\times$ onto G. We extend f to $J(K) = J^0(K) \times \mathbb{R}_+$ putting $f(\mathbb{R}_+) = 1$. Furthermore we put $g(\mathfrak{p}) := f(\pi)$ for prime ideals $\mathfrak{p}$ of K where π is some fixed prime element of $K_\mathfrak{p}$ and we extend g by multiplicativity to the semigroup I of all ideals of $\mathfrak{O}_K$.*

Then the sequence $\{g(\mathfrak{a}) | \mathfrak{a} \in I\}$ (resp. $\{g(\mathfrak{p}) | \mathfrak{p} \in P(K)\}$) ordered with respect to the ideal norm is uniformly distributed in G. $\square$

Example 40. Theorem 1.111 in the quantitative form given by Theorem 1.113 is a special case of Theorem 1.115. One takes $G = H_\mathfrak{m}$. $\square$

Example 41. Let K be an imaginary quadratic number field, h its class number, and w the number of roots of unity contained in K. Furthermore let $a, b \in \mathbb{R}$ and

$$N(x) = |\{\alpha \in \mathfrak{O}_K | N\alpha \leqslant x,\ a \leqslant \arg\alpha \leqslant b,\ (\alpha) \text{ prime ideal}\}|.$$

Then

$$\lim_{x\to\infty} \frac{N(x)}{x/\log x} = \frac{(b-a)w}{2\pi h}. \tag{1.60}$$

For the proof of (1.60) one puts $G = T \times CL(K)$, where T is the unit circle and

$$f(\alpha) = \exp(i \arg(\varphi(\alpha)/\alpha_\infty)^{hw}) \times \overline{\varphi(\alpha)} \qquad \text{for } \alpha \in J(K).$$

$\varphi(\alpha)^{hw}$ means the uniquely determined number β^w in K such that $\varphi(\alpha)^h = (\beta)$. $\square$

6.7. Chebotarev's Density Theorem (Main reference: Narkiewicz (1974), Chap. 7). Let L be a finite normal extension of the algebraic number field K and $\mathfrak{P}$ a prime ideal of L which is unramified in L/K. We put $G := G(L/K)$. It is easy to prove that

$$\left(\frac{L/K}{g\mathfrak{P}}\right) = g\left(\frac{L/K}{\mathfrak{P}}\right)g^{-1} \qquad \text{for } g \in G \tag{1.61}$$

(Proposition 1.52.5). Let $\mathfrak{p} := \mathfrak{P} \cap K$. We put

$$\left[\frac{L/K}{\mathfrak{p}}\right] := \left\{ g \left(\frac{L/K}{\mathfrak{P}}\right) g^{-1} | g \in G \right\}.$$

By (1.61) and Proposition 1.51 one has

$$\left[\frac{L/K}{\mathfrak{p}}\right] = \left\{ \left(\frac{L/K}{\mathfrak{Q}}\right) \middle| \mathfrak{Q} | \mathfrak{p} \right\}. \tag{1.62}$$

Given a conjugacy class C in G one may ask whether there is a prime ideal $\mathfrak{p}$ of K unramified in L/K such that $C = \left[\frac{L/K}{\mathfrak{p}}\right]$. The positive answer to this question is the weakest form of *Chebotarev's density theorem*, which we state now in its strong form:

Theorem 1.116. *The set of all prime ideals $\mathfrak{p}$ of K with $\left[\frac{L/K}{\mathfrak{p}}\right] = C$ is regular with Dirichlet density $|C|/|G|$.* ⊠

Artin's reciprocity theorem (Theorem 2.12) shows that Theorem 1.116 is a generalization of Theorem 1.111. In the case $\mathbb{Q}(\zeta_m)/\mathbb{Q}$ of cyclotomic fields corresponding to Dirichlet's theorem on primes in arithmetic progressions Artin's reciprocity theorem is already proved in § 3.9.

For the proof of the strong version of Chebotarev's density theorem one needs Artin's reciprocity theorem. On the other hand Chebotarev proved his density theorem in its weak form without using class field theory (which will be developed in Chap. 2). On the contrary his method of proof was the tool for Artin's proof of the reciprocity theorem (Artin (1927)): One proves the theorem first for cyclotomic extensions (Chap. 2.4.6) and then by the method of *Durchkreuzung* one goes over to the general case. The idea of this proof is also the basis of the existence of canonical classes in Chap. 2.4 and will be explained there.

6.8. Kronecker Densities and the Theorem of Bauer (Main reference: Hasse (1970), II, § 25). Let L be an arbitrary finite extension of the algebraic number field K and h a non negative integer. We denote by A_h the set of all prime ideals of K which are unramified in L/K and have exactly h prime divisors of inertia degree 1 in L. Let N/K be the normal closure of L/K. We consider $G(N/K)$ as a subgroup of S_n, $n := [L:K]$. Then $\mathfrak{p} \in A_h$ if and only if the permutations in $\left[\frac{N/K}{\mathfrak{p}}\right]$ let h letters invariant. Hence A_h has a Dirichlet density d_h by Theorem 1.116, called *Kronecker density*. By the method of § 6.6 one shows

$$\sum_h h d_h = 1. \tag{1.63}$$

We put $P(L/K) := \bigcup_{h \geqslant 1} A_h$ and $\quad d(L/K) := \sum_{h \geqslant 1} d_h$.

Theorem 1.117. *Let n be the degree of L/K, then $d(L/K) \geqslant 1/n$. The extension L/K is normal if and only if $d(L/K) = 1/n$.*

Proof. This follows from (1.63) and the considerations above. □

One goal of the theory of algebraic number fields is the description of extensions L/K by means of invariants attached to the ground field K. The following theorem of Bauer shows that the set $P(L/K)$ is a candidate for such an invariant if L/K is normal. It remains then to characterize the set of prime ideals of K which have the form $P(L/K)$ for some normal extension L/K. This is possible if one restricts to abelian extensions L/K (Chap. 2.1.7).

Theorem 1.118 (Theorem of Bauer). *Let L be an arbitrary finite extension of the algebraic number field K and let M/K be a finite normal extension. Then $P(M/K) \supseteq P(L/K)$ if and only if $M \subseteq L$. In particular a finite normal extension M of K is characterized by the set of prime ideals of K which split completely in M/K.*

Proof. If L/K is normal, then Theorem 1.118 follows from the elementary fact that

$$P(LM/K) = P(L/K) \cap P(M/K)$$

(Proposition 1.52). In general one uses Chebotarev's density theorem and some arguments concerning Frobenius automorphisms. ⊠

A finite extension F/K is called a *Bauerian extension* (Schinzel (1966)) if for every finite extension L/K the inclusion $P(L/K) \subseteq P(F/K)$ implies that L contains a subfield F' isomorphic to F over K. By Theorem 1.118 any normal extension F/K is a Bauerian extension. Schinzel (1966) shows that any extension of degree $\leqslant 4$ of $\mathbb{Q}$ is a Bauerian extension but the extensions of degree 5 with symmetric Galois group are non-Bauerian extensions.

The following theorem is a consequence of the theorem of Bauer:

Theorem 1.119. *Let L be a finite extension of the algebraic number field K. Then L/K is normal if and only if all prime ideals in $P(L/K)$ split completely in L.* □

Since Theorems 1.117–119 are based on Dirichlet densities, they are also true if one neglects a finite set of prime ideals of K.

As an application of Theorem 1.117 we have the following

Theorem 1.120. *Let $f(x)$ be a normed polynomial with coefficients in $\mathbb{Z}$ and let $f(x)$ be the corresponding polynomial with coefficients in $\mathbb{Z}/p\mathbb{Z}$. Then $f(x)$ splits in linear factors for almost all primes p if and only if $f(x)$ splits in linear factors.*

Proof. Without loss of generality let $f(x)$ be irreducible. Let N be the splitting field of $f(x)$. Then by the lemma of Hensel (Proposition 1.71) $f(x)$ splits in linear factors for almost all primes $p \Leftrightarrow N \subset \mathbb{Q}_p$ for almost all primes $p \Leftrightarrow d(N/\mathbb{Q}) = 1 \Leftrightarrow [N : \mathbb{Q}] = 1$. □

Example 42. Let M be the quadratic field with discriminant D and L the cyclotomic field of $|D|$-th roots of unity. Since the decomposition behavior of a prime p in M depends only on the residue class of $p \pmod{|D|}$ (Example 34), and the primes p which split completely in L are the primes $p \equiv 1 \pmod{|D|}$ (Theorem 1.59), one has $P(M/\mathbb{Q}) \supset P(L/\mathbb{Q})$ hence $M \subset K$. Of course this can also be shown by other methods. □

6.9. The Prime Ideal Theorem with Remainder Term (Main reference: Narkiewicz (1974), Chap. 7.2). Theorem 1.113 is not the best possible approximation for the function π_A by means of elementary functions. With methods of complex function theory one proves the following estimation with remainder term for π_A:

Theorem 1.121. *Let A be the set of prime ideals in a class of the ray class group $H_{\mathfrak{m}}$. Then there are positive constants c_1, c_2 such that for all $x \geqslant 2$*

$$\left| \pi_A(x) - |H_{\mathfrak{m}}|^{-1} \int_2^x \frac{dt}{\log t} \right| < c_1 x \exp(-c_2 \sqrt{\log x}). \boxtimes$$

A fundamental role in the proof of Theorem 1.121 plays the investigation of the zeroes of the functions $L(s, \chi)$ for characters χ of $H_{\mathfrak{m}}$ in the critical strip $0 < \operatorname{Re} s < 1$. The *generalized Riemann hypothesis* asserts that all these zeroes lie on the line $\operatorname{Re} s = \frac{1}{2}$. In the case of the Riemann zeta function $\zeta(s) := \zeta_{\mathbb{Q}}(s)$ this conjecture goes back to Riemann (1859).

Theorem 1.122. *Let A be the set of prime ideals in a class of the ray class group $H_{\mathfrak{m}}$. If the generalized Riemann hypothesis is fulfilled for all characters of $H_{\mathfrak{m}}$, then for all $\varepsilon > 0$ there is a positive constant c such that for all $x \geqslant 2$*

$$\left| \pi_A(x) - |H_{\mathfrak{m}}|^{-1} \int_2^x \frac{dt}{\log t} \right| < c x^{1/2+\varepsilon}. \boxtimes$$

6.10. Explicit Formulas (Main reference: Lang (1970), Chap. 10). For natural n we define $\Lambda(n) := \log p$ if n is a power of a prime p and $\Lambda(n) = 0$ otherwise. Then

$$-\zeta'(s)/\zeta(s) = \sum_{n=1}^{\infty} \frac{\Lambda(n)}{n^s}.$$

$$\psi(x) := \sum_{n \leqslant x} \Lambda(n)$$

is called *Chebyshev function*. It plays an important role in the proof of Theorem 1.121 in the case that A is the set of all primes.

A classical formula connects $\psi(x)$ with the zeroes of $\zeta(s)$: We put $\psi_0(x) := \psi(x) - \frac{1}{2}\Lambda(x)$ if x is a natural number and $\psi_0(x) := \psi(x)$ otherwise. Then

$$\psi_0(x) = x - \sum_{\rho} \frac{x^{\rho}}{\rho} - \frac{\zeta'(0)}{\zeta(0)} - \frac{1}{2} \log(1 - x^{-2}), \tag{1.64}$$

where the sum runs over all zeroes ρ of $\zeta(s)$ with $0 < \operatorname{Re} \rho < 1$. More exactly

$$\sum_\rho \frac{x^\rho}{\rho} := \lim_{T\to\infty} \sum_{\operatorname{Im}\rho<T} \frac{x^\rho}{\rho}.$$

Weil (1952), (1972) generalized (1.64) in the context of Hecke L-functions for a broad class of functions (Lang 1970, Chap. 10). We formulate here this generalization only for the zeroes of Dedekind zeta functions $\zeta_K(s)$ and for functions which will be of interest in the next section.

Let $F(x)$, $x \in \mathbb{R}$, be a real even function which is differentiable and satisfies

$$|F(x)| < c \exp(-(\tfrac{1}{2}+\varepsilon)|x|), |F'(x)| < c \exp(-(\tfrac{1}{2}+\varepsilon)|x|), F(0) = 1$$

for some constants $c > 0$, $\varepsilon > 0$. We put

$$\phi(s) := \int_{-\infty}^{\infty} F(x) \exp(s - \tfrac{1}{2})x\, dx,\ \varphi(t) := \phi(\tfrac{1}{2} + it).$$

Theorem 1.123. *Let K be an algebraic number field with r_1 real and r_2 complex places. We put*

$$g(s) := (\pi^{-s/2}\Gamma(s/2))^{r_1}((2\pi)^{-s}\Gamma(s))^{r_2}.\ \textit{Then}$$

$$\log|d_K| + \frac{1}{2n}\int_{-\infty}^{\infty} \varphi(t) 2 \operatorname{Re}\frac{g'}{g}(\tfrac{1}{2}+it)\, dt$$

$$= -\phi(0) - \phi(1) + \sum_\rho \phi(\rho) + \sum_{\mathfrak{p},m} 2\frac{\log N\mathfrak{p}}{N\mathfrak{p}^{m/2}} F(\log N\mathfrak{p}^m), \tag{1.65}$$

where the first sum runs over all zeroes ρ of $\zeta_K(s)$ with $0 < \operatorname{Re}\rho < 1$ and the second over all prime ideals $\mathfrak{p}$ of K and all natural numbers m. ⊠

6.11. Discriminant Estimation (Main reference: Poitou (1976)). Let K be an algebraic number field of degree n. If one applies Stirling's formula to Minkowski's discriminant estimation (Theorem 1.10) one gets

$$|d_K| \geqslant \left(\frac{\pi}{4}e^2\right)^n (2\pi n)^{-1} \exp(-1/6n), \tag{1.66}$$

called the Minkowski bound. This estimation was improved by several authors by a refinement of Minkowski's method of proof (see Narkiewicz (1974), Chap. 2.4.9) More recently Odlyzko related d_K to the zeroes of the zeta function of K and got results which are much better than the previously known. Developing an idea of Serre, Poitou (1976) showed that one gets *Odlyzko's bounds* by means of the explicit formula (1.65) for certain choices of the function F:

Theorem 1.124. *Let $M(n, r_2)$ be the set of algebraic number fields with r_1 real and r_2 complex places, $n = r_1 + 2r_2$. Then*

$$|d_K|^{1/n} \geqslant (60,8)^{r_1/n}(22,3)^{2r_2/n} \qquad \textit{if } K \in M(n, r_2)$$

for n sufficiently large.

Moreover

$$|d_K|^{1/n} \geqslant (215,3)^{r_1/n}(44,7)^{2r_2/n} \qquad \textit{if } K \in M(n, r_2)$$

for n sufficiently large if the generalized Riemann hypothesis is fulfilled for $\zeta_K(s)$. □

For small n Theorem 1.123 leads to estimations of $|d_K|^{1/n}$ which, for $2 \leqslant n \leqslant 8$, we state in the following table. At the place of the table with coordinates n, r_2 stand three numbers $h_1(n, r_2)$, $h_2(n, r_2)$, $h_3(n, r_2)$. $h_1(n, r_2) := |d_{K_0}|^{1/n}$ for the field $K_0 \in M(n, r_2)$ which has the smallest absolute value of the discriminant among all known fields in $M(n, r_2)$. It is proved that these are the fields in $M(n, r_2)$ with the smallest absolute value of the discriminant if $n \leqslant 6$, if $n = 7$, $r_2 = 0, 2, 3$, or if $n = 8$, $r_2 = 4$. There is always only one field of this kind (Martinet (1979)), $h_3(n, r_2)$ (resp. $h_2(n, r_2)$) is the lower bound for $|d_K|^{1/n}$, $K \in M(n, r_2)$ derived from Theorem 1.123 (under the assumption of the generalized Riemann hypothesis).

n \ r_2	0	1	2	3	4
2	2,236 2,225 2,228	1,732 1,730 1,730			
3	3,659 3,639 3,613	2,844 2,820 2,819			
4	5,189 5,124 5,067	4,072 4,036 4,014	3,289 3,263 3,258		
5	6,809 6,640 6,524	5,381 5,322 5,264	4,378 4,345 4,318		
6	8,182 8,143 7,942	6,728 6,638 6,524	5,512 5,484 5,419	4,622 4,592 4,558	
7	11,051 9,611 9,303	8,110 7,960 7,766	6,710 6,653 6,536	5,654 5,619 5,549	
8	11,385 11,036 10,597	9,544 9,266 8,975	7,905 7,834 7,645	6,779 6,675 6,554	5,787 5,734 5,659

Chapter 2
Class Field Theory

(Main references: Cassels, Fröhlich (1967), Artin, Tate (1968), Neukirch (1986))

There are two main problems in the theory of algebraic number fields: On the one hand the description of the arithmetical properties of a given number field and on the other hand the description of number fields with given arithmetical properties.

The theory of complex function fields in one variable delivers us an example of the resolution of the second problem in the case of function fields: The finite extensions L of $\mathbb{C}(z)$ which are unramified outside a set S of points of the Riemann surface $F = \mathbb{C} \cup \{\infty\}$ of $\mathbb{C}(z)$ correspond to the subgroups U of finite index of the fundamental group of the topological space $F - S$. Thereby the covering space of F given by U corresponds to the Riemann surface of L (see also Chap. 3, Example 13). Unfortunately up to now we have no result in the theory of number fields of this generality. We can express this saying that up to now we have not sufficiently good understood the notion of the fundamental group in the case of number fields.

Before we come to a more detailed discussion of the two main problems we are going to reformulate them for the purposes of this chapter. With respect to arithmetical properties we ask more concretely for the decomposition behavior of prime ideals in finite extensions and since every finite extension is contained in a normal extension we restrict to normal extensions and describe the arithmetical properties of the extensions (and its subextensions) by means of the decomposition and ramification groups (Chap. 1.3.7).

In Chap. 1.3.9 we have seen that we have a very detailed description of the arithmetical properties of the cyclotomic field $\mathbb{Q}(\zeta_{p^i})$: The Galois group of $\mathbb{Q}(\zeta_{p^i})$ over $\mathbb{Q}$ is canonically isomorphic to the group $(\mathbb{Z}/p^i\mathbb{Z})^\times$. For a prime $q \neq p$ the Frobenius automorphism F_q is given by $F_q(\zeta_{p^i}) = \zeta_{p^i}^q$ and the j-th ramification group of p in the upper numeration corresponds by this isomorphism to the subgroup $\{\bar{a} \in (\mathbb{Z}/p^i\mathbb{Z})^\times \mid a \equiv 1 \pmod{p^j}\}$.

It turns out that such a beautiful description of the arithmetical properties is possible for any abelian extension L/K of number fields (an abelian extension L/K is a normal extension L of K with abelian Galois group). The corresponding theory is called *class field theory*. It is the subject of this chapter.

According to the local-global principle (Chap. 1.4), class field theory begins with the study of local fields K. The abelian extensions of K correspond to the closed subgroups of finite index of the multiplicative group $K^\times$ of K (§ 1.3). For global fields K the idele class group (Chap. 1.5.5) plays the role of the multiplicative group in the local case: The abelian extensions of the global field K correspond to the closed subgroups of finite index of the idele class group of K (§ 1.5). By the theorem of Bauer (Theorem 1.118) a finite normal extension L of K is characterized by the set of prime ideals which split completely in L. For abelian extensions L/K class field theory delivers a beautiful description of such sets of prime ideals in terms of the subgroup of the idele class group corresponding to L (§ 1.6).

Beside the two main problems the most interesting problems in class field theory are the explicit construction of class fields and of reciprocity laws. Both problems played a decisive role in the development of the theory. Explicit constructions of abelian extensions are known for local fields (Lubin-Tate extensions (Chap. 1.4.10)), for $\mathbb{Q}$ (§ 1.1), and for imaginary-quadratic fields (§ 2). For other global fields one has only partial results. Explicit reciprocity laws are the direct generalization of the quadratic reciprocity law. They allow (in principle) to decide whether an integral number in a field K is an n-th power residue with respect to a prime ideal. For a smooth formulation of the results it is necessary to assume that K contains the n-th roots of unity. Thus Gauss studied the field $\mathbb{Q}(\sqrt{-1})$ for his formulation of the explicit biquadratic reciprocity law.

The notion of class field was already present in the mind of Kronecker who knew that every abelian extension of $\mathbb{Q}$ is contained in a cyclotomic field and it was his "liebster Jugendtraum" that analogously every abelian extension of an imaginary-quadratic number field can be obtained by "singular moduli" connected with modular functions and elliptic functions (§ 2). The general concepts of class field theory were developed at the end of the last century by Hilbert and Weber. Hilbert wrote in the introduction to his famous Zahlkörperbericht (Hilbert (1895), p. VII):

Die Theorie der Zahlkörper ist wie ein Bauwerk von wunderbarer Schönheit und Harmonie; als der am reichsten ausgestattete Teil dieses Bauwerks erscheint mir die Theorie der Abel'schen und relativ-Abel'-schen Körper, die uns Kummer durch seine Arbeiten über die höheren Reciprocitätsgesetze und Kronecker durch seine Untersuchungen über die complexe Multiplication der elliptischen Functionen erschlossen haben. Die tiefen Einblicke, welche die Arbeiten dieser beiden Mathematiker in die genannte Theorie gewähren, zeigen uns zugleich, dass in diesem Wissensgebiet eine Fülle der kostbarsten Schätze noch verborgen liegt, winkend als reicher Lohn dem Forscher, der den Wert solcher Schätze kennt und die Kunst, sie zu gewinnen, mit Liebe betreibt.

In the definition of Weber (1891), (1898) the class field L of an algebraic number field K with respect to a subgroup $H/R_{\mathfrak{m}}$ of the ray class group $\mathfrak{A}_{\mathfrak{m}}/R_{\mathfrak{m}}$ (Chap. 1.5.5) is the (uniquely determined (Chap. 1.6.8)) normal extension of K such that a prime ideal $\mathfrak{p}$ of K with $\mathfrak{p} \nmid \mathfrak{m}$ splits completely in L/K if and only if it belongs to H. The Galois group of L/K is isomorphic to $\mathfrak{A}_{\mathfrak{m}}/H$. The existence of the class field for a given group H was proved by Takagi (1920) and Artin (1927) showed that there is a canonical isomorphism of $\mathfrak{A}_{\mathfrak{m}}/H$ onto $G(L/K)$ given by the Frobenius symbol (Chap. 1.3.7). This isomorphism can be considered as a far reaching generalization of the quadratic reciprocity law. Artin's result completed the foundation of class field theory.

At that time the proofs of the main theorems were extremely complicated (see Hasse (1970)) and the following period is characterized by a simplification and rebuilding of the whole theory. In particular Hasse (1930) developed the local class field theory, i.e. the theory of abelian extensions of local fields, and Chevalley (1940) introduced the notion of idele which allows a formulation of class field theory by means of a local global principle (Chap. 1.4). In connection with the computation of the Brauer group of local and global fields it became clear that technically the best formulation of class field theory is by means of cohomology of groups. In particular Tate (1952) generalized Artin's reciprocity law to a theorem about cohomology groups of arbitrary normal extensions of local and global fields. (See Hasse (1967) for more details about the history of class field theory).

The content of Chapter 2 is as follows. In § 1 the main theorems of class field theory are formulated without proofs and without regard to cohomology groups. § 2 contains the description of the class field theory of imaginary-quadratic fields by means of complex multiplication. In § 3 we develop the cohomology of groups as preparation for the survey of the proofs of the main theorems of class field theory in § 4 and as preparation for the application of cohomological methods to non-abelian Galois theory of algebraic number fields in Chapter 3. § 5 is devoted to the theory of the Brauer group of local and global fields. In § 6 we consider explicite reciprocity laws and in § 7 we represent some further results of class field theory.

There are several books which serve as an introduction to class field theory. The standard cohomological approach is presented in Chevalley (1954), Cassels, Fröhlich (1967), Neukirch (1969) and Iyanaga (1975). Goldstein (1971) and Lang (1970) use analytical methods and Weil's treatment (1967) is based on the theory of simple algebras. Neukirch (1986) uses only the cohomology groups of dimension $-1, 0, 1$ and gives a new proof for Artin's reciprocity law. There is no book with a full treatment of class field theory, but Artin, Tate (1968), sometimes called the bible of class field theory, contains much of the material which should be included in such a book.

§ 1. The Main Theorems of Class Field Theory

We begin with the description of the main theorems of class field theory.

1.1. Class Field Theory for Abelian Extensions of $\mathbb{Q}$. We gave already a detailed description of the abelian extensions $\mathbb{Q}(\zeta_n)/\mathbb{Q}$ where ζ_n is a primitive root of unity of order n (Chap. 1.3.9). On the other hand one has the following theorem of Kronecker and Weber:

Theorem 2.1. *For every finite abelian extension $L/\mathbb{Q}$ there exists a natural number n such that $L \subseteq \mathbb{Q}(\zeta_n)$.* ⊠

Disregarding the proof of Weber (1887) (see also Neumann (1981)) the proofs of Theorem 2.1 can be considered as variations of the proof in Hilbert (1894/95), Capitel XXIII (Speiser (1919), Shafarevich (1951), further references in Narkiewicz (1974), Chap. VI, § 3). Hilberts proof consists in the reduction of the problem to fields with a smaller number of ramified primes using Chap. 1, Proposition 1.4. One ends up with an unramified field, which coincides with $\mathbb{Q}$ by Minkowski's discriminant theorem (Chap. 1.1.2).

The smallest n with $L \subseteq \mathbb{Q}(\zeta_n)$ is called the *conductor* of L.

Since $G(\mathbb{Q}(\zeta_n)/\mathbb{Q})$ is canonical isomorphic to $(\mathbb{Z}/n\mathbb{Z})^\times$, the field L is given by the subgroup U of $(\mathbb{Z}/n\mathbb{Z})^\times$ which corresponds to $G(\mathbb{Q}(\zeta_n)/L)$. Hence the pair n, U determines L. The Galois group of $L/\mathbb{Q}$ is canonical isomorphic to $(\mathbb{Z}/n\mathbb{Z})^\times/U$, a prime is ramified in $L/\mathbb{Q}$ if and only if it divides the conductor of L, and the Frobenius automorphism of an unramified prime p is given by $\bar{p}U$.

This description of the abelian extensions of $\mathbb{Q}$ has the advantage of being of a very elementary nature, but it has the disadvantage of being not smooth enough for the formulation of further properties of abelian extensions concerning higher ramification groups for instance. We know already that every finite factor group of the idele class group $J(\mathbb{Q})/\mathbb{Q}^\times$ factors through $(\mathbb{Z}/n\mathbb{Z})^\times$ for a certain n (Chap. 1.5.5). It turns out that the idele class group is the best setting for the description of abelian extensions in general.

1.2. The Hilbert Class Field. Let K be an algebraic number field and L a finite extension of K. An infinite place v of K is called *ramified* in L/K if there exists a place w of L above v with $L_w \neq K_v$ (hence complex places can not be ramified).

If v is an arbitrary place of K which is unramified in the extensions L_1 and L_2 of K, then v is unramified in $L_1 L_2$ (1.20). The compositum of all abelian extensions of K which are unramified at all places of K is called the *Hilbert class field*. It will be denoted by $H(K)$.

Theorem 2.2. *$H(K)$ is a finite extension of K. The Galois group of $H(K)/K$ is canonically isomorphic to the ideal class group $CL(K)$. A prime ideal $\mathfrak{p}$ of K splits in $[H(K):K]$ distinct prime divisors in $H(K)$ if and only if $\mathfrak{p}$ is a principal ideal.* ☒

Example 1. $K = \mathbb{Q}(\sqrt{6})$ has class number one. Hence there is no non-trivial unramified abelian extension. $Q(\sqrt{-2}, \sqrt{-3})/K$ is unramified at finite places but ramified at the infinite places. ☒

Example 2. $K = \mathbb{Q}(\sqrt{-5})$ has class number 2, $H(K) = Q(\sqrt{-1}, \sqrt{5})$. ☒

Example 3. $K = \mathbb{Q}(\sqrt{-23})$ has class number 3. $H(K) = \mathbb{Q}(\sqrt{-23}, \theta)$ with $\theta^3 - \theta - 1 = 0$. The extension $K(\theta)/K$ is unramified: $x^3 - x - 1$ has discriminant -23, hence only $(\sqrt{-23})$ could be ramified in $K(\theta)$, but then the inertia group of a prime divisor of $(\sqrt{-23})$ with respect to $\mathbb{Q}$ must be cyclic of order 6 (Proposition 1.52), which is impossible because the Galois group of $K(\theta)/\mathbb{Q}$ is not cyclic. ☒

Example 4. The class number of $\mathbb{Q}(\sqrt{-47})$ is 5 (Example 11, Chap. 1.1.7). It needs already much more effort to determine the Hilbert class field of $\mathbb{Q}(\sqrt{-47})$ (Hasse (1964)). ☒

1.3. Local Class Field Theory. In this section K denotes a $\mathfrak{p}$-adic number field. We keep the notations of Chap. 1.4.10. K_{nr} denotes the maximal unramified extension of K.

A central role in local class field theory plays the *norm symbol* also called *norm residue symbol* or *Hasse symbol*. For an abelian extension L/K the norm symbol $(\alpha, L/K)$, $\alpha \in K^\times$, is a homomorphism of $K^\times$ onto $G(L/K)$. It is characterized by the following properties.

Let L/K be any finite abelian extension and let M be an intermediate field of L/K. Then

(i) $(\alpha, M/K) = \varphi(\alpha, L/K)$ for $\alpha \in K^\times$, where φ is the natural projection $G(L/K) \to G(M/K)$.

(ii) $(\beta, L/M) = (N_{M/K}\beta, L/K)$ for $\beta \in M^\times$.

(iii) If L/M is unramified and π_M a prime element of M, then $(\pi_M, L/M)$ is the Frobenius automorphism of L/M (Proposition 1.52.7)).

The name *norm symbol* is justified by the following theorem.

Theorem 2.3 (Main theorem of local class field theory). *There is a one to one correspondence ϕ between the finite abelian extensions L of K and the closed subgroups of $K^\times$ of finite index. This correspondence has the following properties:*

a) $\phi(L) = N_{L/K}L^\times = \mathrm{Ker}(\ , L/K)$.

b) *The norm symbol induces an isomorphism of $K^\times/\phi(L)$ and $G(L/K)$ which maps $U_n\phi(L)/\phi(L)$ onto the n-th ramification group of L/K in the upper numeration.*

c) *Let L and L' be finite abelian extensions of K. Then $L \subseteq L'$ if and only if $\phi(L) \supseteq \phi(L')$. If $\phi(L) \supseteq \phi(L')$, then L is characterized as the fixed field of $(\phi(L), L'/K)$.* ⊠

The norm symbol can be defined by means of Lubin-Tate extensions (Chap. 1.4.10): First of all we have the following analogue to the theorem of Kronecker and Weber:

Theorem 2.4. *Let L/K be a finite abelian extension and let g be an Eisenstein polynomial of degree $q - 1$ with coefficients in $\mathfrak{O}$. Then there is a natural number n such that $L \subset K_{nr}(\theta_n(g))$.*

Let n be the smallest natural number such that $L \subset K_{nr}(\theta_n(g))$ if L/K is ramified and let $n = 0$ if L/K is unramified. Then n is independent of the choice of g and $\mathfrak{p}^n$ is called the conductor of L/K. ⊠

Let π be the absolute coefficient of g and $\alpha = \pi^\nu\varepsilon$, $\varepsilon \in U$. Then $(\alpha, L/K)$ is the restriction to L of the automorphism of $K_{nr}(\theta_n(g))/K$ which induces on K_{nr}/K the ν-th power of the Frobenius automorphism F and on $K(\theta_n(g))/K$ the automorphism given by $\theta_n(g) \to [\varepsilon^{-1}]_g(\theta_n(g))$. (The field K_{nr} is generated over K by the roots of unity ζ of order prime to q and F is determined by $F\zeta = \zeta^q$ (Chap. 1.4.6)).

Another construction of the norm symbol will be given in §4.

A closed subgroup V of $K^\times$ of finite index is also open and contains therefore U_m for a certain m. Let n be the smallest number with $U_n \subset V$. Then $\mathfrak{p}^n$ is called the conductor of V. The conductor of L is equal to the conductor of $\phi(L)$.

In Chap. 1.5.2 we have defined the conductor $\mathfrak{f}_\chi$ of a character χ of $K^\times$. The following theorem allows to determine the discriminant of L/K if $\phi(L)$ is given.

Theorem 2.5 (Führerdiskriminantenproduktformel). $\mathfrak{d}_{L/K} = \prod_\chi \mathfrak{f}_\chi$ *where the product is taken over all characters* χ *of* $K^\times$ *which are trivial on* $\phi(L)$. ☒

In Chap. 1.4.4 we have seen that we have an excellent insight into the structure of $K^\times$. Therefore Theorem 2.3 shows that we have also an excellent insight into the possible abelian extensions of K.

Example 5. Let $K = \mathbb{Q}_p$. The number of cyclic extensions of K of degree p is equal to the number of closed subgroups of $K^\times$ of index p. For $p \neq 2$ we have $[\mathbb{Q}_p^\times : \mathbb{Q}_p^{\times p}] = p^2$. Hence there are $p + 1$ such extensions of $\mathbb{Q}_p$. If $p = 2$, then $[\mathbb{Q}_2^\times : \mathbb{Q}_2^{\times 2}] = 2^3$ and we have seven quadratic extensions of $\mathbb{Q}_2$. The last result follows of course also by Kummer theory of cyclic fields (Chap. 1.4.7). We recomment the reader to compute the subgroups of $\mathbb{Q}_2^\times$ belonging to the seven fields $\mathbb{Q}_2(\sqrt{2^{i_1}(-1)^{i_2}5^{i_3}})$, $i_1, i_2, i_3 \in \{0, 1\}$, $i_1 + i_2 + i_3 \neq 0$. □

In the following we need the symbol $(\alpha, L/K)$ also if K is equal to $\mathbb{R}$ or $\mathbb{C}$. We put

$$(\alpha, L/K) = 1 \qquad \text{if } L = K,$$

$$(\alpha, \mathbb{C}/\mathbb{R}) = \begin{cases} 1 & \text{if } \alpha > 0, \\ c & \text{if } \alpha < 0, \end{cases}$$

where $G(\mathbb{C}/\mathbb{R}) = \{1, c\}$.

1.4. The Idele Class Group of a Normal Extension. In the next section we formulate the main theorems of global class field theory. In this section we make some preparations.

Let L be an arbitrary finite extension of an algebraic number field K. Let w be a place of L lying above the place v of K. Then we denote the natural injection of K_v in L_w by ι_w. We define an injection ι of the idele group $J(K)$ in the idele group $J(L)$ by

$$\iota\left(\prod_{v \in P_K} a_v\right) := \prod_{w \in P_L} \iota_w a_v \quad \text{(Chap. 1.5.4).}$$

ι maps principal ideles onto principal ideles. There is also an injection τ of $J(K)/K^\times$ into $J(L)/L^\times$. In the following we identify $J(K)$ (resp. $J(K)/K^\times$) with its image in $J(L)$ (resp. $J(L)/L^\times$). Furthermore we put $\mathfrak{C}(K) := J(K)/K^\times$.

Let $\alpha \in L^\times$ and let v be a place of K. By Proposition 1.74.2) we have

$$N_{L/K}\alpha = \prod_{w|v} N_{L_w/K_v}\alpha. \tag{2.1}$$

We define the norm map of $J(L)$ into $J(K)$ by

$$N_{L/K}(\pi\alpha_w) = \prod_v \left(\prod_{w|v} \alpha_w\right).$$

Obviously this is a continuous homomorphism. (2.1) shows that $N_{L/K}$ maps principal ideles onto principal ideles. Hence it induces a homomorphism of $\mathfrak{C}(L)$ into $\mathfrak{C}(K)$, again denoted by $N_{L/K}$.

Let $g: L \to L'$ be a field isomorphism. To every valuation w of L there corresponds a valuation gw of L' given by

$$gw(\alpha) = w(g^{-1}\alpha) \qquad \text{for } \alpha \in L'.$$

It is easy to see that g induces an isomorphism of L_w onto L_{gw}, of $J(L)$ onto $J(L')$, and of $\mathfrak{C}(L)$ onto $\mathfrak{C}(L')$, again denoted by g.

Now let L/K be a normal extension with Galois group G. Then under the action of $g \in G$, just explained, $J(L)$ and $\mathfrak{C}(L)$ become G-modules. The set $J(L)^G$ of ideles which are fixed by the automorphisms of G is equal to $J(K)$ in the sense of our identification of $J(K)$ with its image in $J(L)$. Furthermore it is clear that $\mathfrak{C}(K) \subseteq \mathfrak{C}(L)^G$. We want to show $\mathfrak{C}(K) = \mathfrak{C}(L)^G$. This means that for each idele α of L such that $g\alpha = f(g)\alpha$ with some $f(g) \in L^\times$ for all $g \in G$, there is an $a \in J(K)$ such that $\alpha = \beta a$ for some $\beta \in L^\times$. f is a function on G with values in L which satisfies the equation

$$f(gh) = gf(h)f(g) \qquad \text{for all } g, h \in G.$$

Such a function is called a (one dimensional) factor system of G with values in L. We have to show the following proposition, which is E. Noether's generalization of Hilbert's Satz 90 in the Zahlkörperbericht (Hilbert (1894/95)).

Proposition 2.6. *Let K be an arbitrary field and L a finite separable normal extension of K with Galois group G. Furthermore let f be a factor system of G with values in $L^\times$. Then there is a $\beta \in L^\times$ such that*

$$f(g) = g\beta/\beta \qquad \textit{for } g \in G.$$

Proof. The linear independence of the automorphisms in G implies that there exists a $\gamma \in L^\times$ with

$$\delta := \sum_{h \in G} f(h)h\gamma \neq 0.$$

Hence

$$g\delta = \sum_{h \in G} gf(h)gh\gamma = \sum_{h \in G} f(gh)gh\gamma/f(g) = \delta/f(g). \; \square$$

It is easy to see that $N_{L/K}\alpha = \prod_{g \in G} g\alpha$ for $\alpha \in J(L)$.

1.5. Global Class Field Theory. Let L be a finite normal extension of the algebraic number field K with Galois group G and let w be a place of L above the place v of K. Then $G(L_w/K_v)$ will be identified with the decomposition group $G_w = \{g \in G \mid gw = w\}$ of w with respect to L/K (Chap. 1.4.5). Since

$$G_{gw} = \{h \in G \mid hgw = gw\} = gG_wg^{-1} \qquad \text{for } g \in G,$$

the decomposition group of w is determined up to conjugacy by v.

Hence if L/K is abelian, G_w depends only on v and we can speak about the decomposition group of v with respect to L/K. For any idele $\prod \alpha_v \in J(K)$ almost all $(\alpha_v, L_w/K_v) = 1$, since $(\alpha_v, L_w/K_v) = 1$ if α_v is a unit in K_v and L_w/K_v is unramified. $(\alpha_v, L_w/K_v)$ is independent of the choice of w above v. For every v we choose one place w of L above v. Then

$$\left(\prod \alpha_v, L/K\right) := \prod_v (\alpha_v, L_w/K_v) \tag{2.2}$$

is a well defined homomorphism of $J(K)$ into $G(L/K)$. It is called the (global) norm symbol.

The following theorem is sometimes called *Artin's reciprocity law*. As we will see later on (§ 1.8), it can be considered as a far reaching generalization of Gauss' quadratic reciprocity law.

Theorem 2.7. $(\alpha, L/K) = 1$ *for* $\alpha \in K^\times$. ⊠

Theorem 2.7 shows that the norm symbol $(\ , L/K)$ can be considered as a homomorphism of $\mathfrak{C}(K)$ into $G(L/K)$. For the formulation of the following main theorem of global class field theory we remember that we have identified $K_v^\times$ with a subgroup of $J(K)$ (Chap. 1.5.5).

Theorem 2.8. *There is a one to one correspondence ϕ between the finite abelian extensions L of K and the closed subgroups of $\mathfrak{C}(K)$ of finite index. This correspondence has the following properties*:

a) $\phi(L) = N_{L/K}\mathfrak{C}(L) = \mathrm{Ker}(\ , L/K)$.

b) The norm symbol induces an isomorphism of $\mathfrak{C}(K)/\phi(L)$ onto $G(L/K)$ and maps $K_v^\times$ onto the decomposition group of v for all places v of K.

c) Let L and L' be finite abelian extensions of K. Then $L \subseteq L'$ if and only if $\phi(L) \supseteq \phi(L')$. If $\phi(L) \supseteq \phi(L')$, then L is characterized as the fixed field of $(\phi(L), L'/K)$. ⊠

Remark. The statement about $K_v^\times$ follows immediately from the definition of the norm symbol and from local class field theory, the n-th group $U_{v,n}$ of principal units in K_v is mapped onto the n-th ramification group of v in the upper numeration.

If U is a closed subgroup of finite index in $\mathfrak{C}(K)$, then $\phi^{-1}(U)$ is called the class field of U (or of $\mathfrak{C}(K)/U$).

For the following we keep the notations of Chap. 1.5.5.

Example 6. Let m be a natural number. The class field of $J(\mathbb{Q})/\mathbb{Q}^\times U_{m\infty} \cong (\mathbb{Z}/m\mathbb{Z})^\times$ is the field $\mathbb{Q}(\zeta_m)$. (§ 1.1). ⊠

Example 7. Let $U_1 = \prod_v U_v$, i.e. $U_v := K_v^\times$ for v infinite. Then the class field of $J(K)/K^\times U_1 \cong CL(K)$ is the Hilbert class field (§ 1.2). ⊠

Let $\mathfrak{m}$ be an arbitrary defining module. Then the class field of $J(K)/K^\times U_{\mathfrak{m}} \cong H_{\mathfrak{m}}$ is called the ray class field mod $\mathfrak{m}$.

1.6. The Functorial Behavior of the Norm Symbol. In this section we consider the functorial properties of the norm symbol in the local and global case. Since we want to study both cases simultaneously, we put $A_K := K^\times$ if K is a local field and $A_K := \mathfrak{C}(K)$ if K is a global field. We call A_K the class module of K. If

$g: K \to K'$ is an isomorphism of local or global fields, the corresponding isomorphism of A_K onto $A_{K'}$ will be denoted by g, too.

It will be conveniant to consider finite normal extensions L/K. Let L^{ab} be the maximal abelian subextension of L/K. We put

$$G(L/K)^{ab} := G(L^{ab}/K), (\alpha, L/K) := (\alpha, L^{ab}/K) \qquad \text{for } \alpha \in A_K.$$

Theorem 2.9. *Let L_1/K and L_2/K be finite abelian extensions with $L_1 \subseteq L_2$ and let $\pi: G(L_2/K) \to G(L_1/K)$ be the projection. Then*

$$(\alpha, L_1/K) = \pi(\alpha, L_2/K) \qquad \text{for } \alpha \in A_K. \boxtimes$$

For the next theorem we need the notion of *group transfer* (*Verlagerung*). Let G be a finite group and H a subgroup of G. We denote the commutator group of G by $[G, G]$ and put $G^{ab} := G/[G, G]$. The transfer Ver from G into H is a homomorphism of G^{ab} into H^{ab} which is defined as follows: Let R be a system of representatives of G/H in G and denote by $\tilde{g}$ the representative of g. Then

$$\mathrm{Ver}(g[G, G]) := \prod_{r \in R} h(g, r)[H, H]$$

where $h(g, r) \in H$ is given by

$$gr = \widetilde{gr}h(g, r).$$

In § 3.5 we shall see that the transfer map is a special case of a cohomological operation called restriction. There it will become clear that Ver is a well defined homomorphism.

Theorem 2.10. *Let L/K be a finite normal extension, K'/K an arbitrary subextension, and* Ver *the transfer from $G(L/K)$ into $G(L/K')$. Then*

$$\mathrm{Ver}(\alpha, L/K) = (\alpha, L/K') \qquad \text{for } \alpha \in A_K. \tag{2.3}$$

Let κ be the homomorphism from $G(L/K')^{ab}$ into $G(L/K)^{ab}$ induced by the inclusion $G(L/K') \subseteq G(L/K)$. Then

$$\kappa(\alpha, L/K') = (N_{K'/K}\alpha, L/K) \qquad \text{for } \alpha \in A_{K'}. \boxtimes \tag{2.4}$$

Theorem 2.11. *Let L/K be a finite abelian extension and g an isomorphism of L onto gL. Then*

$$(g\alpha, gL/gK) = g(\alpha, L/K)g^{-1} \qquad \text{for } \alpha \in A_K. \boxtimes$$

The property c) in the main theorems of local and global class field theory follows from (2.4): If $L \subseteq L'$, then $N_{L/K}A_L \supseteq N_{L/K}(N_{K'/L}A_{L'}) = N_{L'/K}A_{L'}$. If $\phi(L) \supseteq \phi(L')$, then

$$(\phi(L), LL'/K) = (N_{L/K}A_L, LL'/K) = (A_L, LL'/L) \subseteq G(LL'/L)$$

and $(\phi(L'), LL'/K) \subseteq G(LL'/L')$ hence $(\phi(L'), LL'/K) = \{1\}$. This implies $\phi(L') \subseteq \phi(LL')$, $[L' : K] = [LL' : K]$ and $L \subseteq L'$.

1.7. Artin's General Reciprocity Law. We keep the notations of Chap. 1.5.5. Let L be a finite normal extension of the algebraic number field K and let $\mathfrak{P}$ be a prime ideal of L which is unramified in L/K. Then the Frobenius automorphism $\left(\frac{\mathfrak{P}}{L/K}\right)$ of $\mathfrak{P}$ (Chap. 1. Proposition 1.52.3)) is an element of $G(L/K)$. We have

$$\left(\frac{g\mathfrak{P}}{L/K}\right) = g\left(\frac{\mathfrak{P}}{L/K}\right)g^{-1} \qquad \text{for } g \in G(L/K). \tag{2.5}$$

If L/K is abelian, then $\left(\frac{\mathfrak{P}}{L/K}\right)$ depends only on $\mathfrak{p} := \mathfrak{P} \cap O_K$ and we put $\left(\frac{\mathfrak{p}}{L/K}\right) := \left(\frac{\mathfrak{P}}{L/K}\right)$. Let S be the set of prime ideals of K which are ramified in the abelian extension L/K. We put

$$\left(\frac{\mathfrak{a}}{L/K}\right) := \prod_{\mathfrak{p} \notin S} \left(\frac{\mathfrak{p}}{L/K}\right)^{v_{\mathfrak{p}}(\mathfrak{a})} \qquad \text{for } \mathfrak{a} \in \mathfrak{A}_S.$$

$\left(\frac{\mathfrak{a}}{L/K}\right)$ is called the Artin symbol of $\mathfrak{a}$. By definition $\left(\frac{\quad}{L/K}\right)$ is a homomorphism of $\mathfrak{A}_S$ into $G(L/K)$.

Theorem 2.12 (Artin's general reciprocity law). *Let L/K be a finite abelian extension unramified outside S. Then $\left(\frac{\quad}{L/K}\right)$ is a homomorphism of $\mathfrak{A}_S$ onto $G(L/K)$ with kernel containing the ray group $R_{\mathfrak{m}}$ for some defining module $\mathfrak{m}$.*

By Theorem 2.12 the decomposition of a prime ideal $\mathfrak{p}$ of K in an abelian extension L/K depends only on the class of $\mathfrak{p}$ modulo the defining module $\mathfrak{m}$. This is the direct generalization of the law for the decomposition of primes in cyclotomic extensions (Chap. 1.3.9), which in turn stands behind the quadratic reciprocity law.

Proof of Theorem 2.12. By definition the kernel of the global norm symbol $(\ , L/K)\colon J(K) \to G(L/K)$ contains the subgroup $U_{\mathfrak{m}}$ for some defining module $\mathfrak{m}$ with supp $\mathfrak{m}_0 = S$. Let π be a prime element in $K_{\mathfrak{p}}$. Then we have

$$\left(\frac{\mathfrak{p}}{L/K}\right) = (\pi, L/K) \qquad \text{for } \mathfrak{p} \notin S$$

by definition. The claim follows now easily from the Theorems 2.7, 2.8. □

The smallest defining module is called the *conductor* of L/K. It is the product of the local conductors. $\mathfrak{A}_{\mathfrak{m}}/\mathrm{Ker}\left(\frac{\quad}{L/K}\right)$ is called the class group of L/K with respect to $\mathfrak{m}$.

1.8. The Power Residue Symbol. In this and the next two sections we consider the connection between Artin's general reciprocity law, Gauss' quadratic reciprocity law, its direct generalization to the reciprocity law for the m-th power residue symbol, and the Hilbert symbol (Main reference: Hasse (1970), Teil II).

As already mentioned (Chap. 1, Introduction) Gauss considered the numbers in the ring $\mathbb{Z}[\sqrt{-1}]$ and formulated the biquadratic reciprocity law. Jacobi and Eisenstein gave the first proofs for the biquadratic and cubic reciprocity law. On the basis of his theory of ideal numbers Kummer formulated and proved the reciprocity law for the p-th power residue symbol in cyclotomic fields $\mathbb{Q}(\zeta_p)$ (Chap. 1.3.9) if p is a regular prime number, i.e. if the class number of $\mathbb{Q}(\zeta_p)$ is prime to p. Takagi settled the case of an arbitrary prime number p. Hilbert asked in his ninth problem for the "most general reciprocity law in an arbitrary algebraic number field". In some sense the answer was given by the general reciprocity law of Artin (§ 1.7), but in a more direct sense the problem was solved by Shafarevich (§ 6.1) and more explicitly independently by Brückner and Vostokov (§ 6.2).

Let m be a natural number and K an algebraic number field containing the m-th roots of unity. For $\alpha \in K^\times$ we denote by $\mathfrak{f}_\alpha$ (resp. $\mathfrak{d}_\alpha$) the conductor (§ 1.7) (resp. the discriminant) of the abelian extension $K(\sqrt[m]{\alpha})/K$. Then for any fractional ideal $\mathfrak{b}$ of K which is prime to $\mathfrak{f}_\alpha$ the *m-th power residue symbol* $\left(\frac{\alpha}{\mathfrak{b}}\right)_{m,k} = \left(\frac{\alpha}{\mathfrak{b}}\right) \in \mu_m$ is defined by

$$\left(\frac{\mathfrak{b}}{K(\sqrt[m]{\alpha})/K}\right)\sqrt[m]{\alpha} = \left(\frac{\alpha}{\mathfrak{b}}\right)\sqrt[m]{\alpha}.$$

The symbol $\left(\frac{\alpha}{\mathfrak{b}}\right)$ has the following properties which follow from the properties of the Artin symbol (§ 1.6–7).

Theorem 2.13.

1. $$\left(\frac{\alpha_1}{\mathfrak{b}}\right)\left(\frac{\alpha_2}{\mathfrak{b}}\right) = \left(\frac{\alpha_1\alpha_2}{\mathfrak{b}}\right) \quad \textit{if } \operatorname{supp} \mathfrak{f}_{\alpha_1}\mathfrak{f}_{\alpha_2} \cap \operatorname{supp} \mathfrak{b} = \varnothing. \tag{2.6}$$

2. $$\left(\frac{\alpha}{\mathfrak{b}_1}\right)\left(\frac{\alpha}{\mathfrak{b}_2}\right) = \left(\frac{\alpha}{\mathfrak{b}_1\mathfrak{b}_2}\right) \quad \textit{if } \operatorname{supp} \mathfrak{f}_{\alpha} \cap (\operatorname{supp} \mathfrak{b}_1 \cup \operatorname{supp} \mathfrak{b}_2) = \varnothing. \tag{2.7}$$

3. *Let $g\colon K \to gK$ be an isomorphism, then* (2.8)

$$\left(\frac{g\alpha}{g\mathfrak{b}}\right) = g\left(\frac{\alpha}{\mathfrak{b}}\right). \tag{2.9}$$

4. *Let d be a divisor of m. Then*

$$\left(\frac{\alpha}{\mathfrak{b}}\right)_m^d = \left(\frac{\alpha}{\mathfrak{b}}\right)_{m/d}. \tag{2.10}$$

5. *Let L/K be a finite extension, $\alpha \in K^\times$, and $\mathfrak{B}$ an ideal of L which is prime to*

$\mathfrak{f}_\alpha$. *Then*

$$\left(\frac{\alpha}{\mathfrak{B}}\right)_L = \left(\frac{\alpha}{N_{L/K}(\mathfrak{B})}\right)_K. \tag{2.11}$$

6. *Let $\mathfrak{p}$ be a prime ideal of K which is prime to m. Then $\chi\colon \bar{\alpha} \to \left(\frac{\alpha}{\mathfrak{p}}\right)$ is a character of $(\mathfrak{O}_K/\mathfrak{p})^\times$ with $\chi^m = 1$ and χ generates the group of characters of $(\mathfrak{O}_K/\mathfrak{p})^\times$ with this property.*

7. (Generalized Euler criterion) *If $\alpha \in \mathfrak{O}_K$, $(\alpha m, \mathfrak{p}) = 1$, then*

$$\left(\frac{\alpha}{\mathfrak{p}}\right) \equiv \alpha^{(N(\mathfrak{p})-1)/m} \pmod{\mathfrak{p}},$$

in particular $\left(\frac{\alpha}{\mathfrak{p}}\right) = 1$ if and only if $\alpha \equiv \beta^m \pmod{\mathfrak{p}}$ for some $\beta \in \mathfrak{O}_K$.

8. *Let $H_\alpha \subseteq \mathfrak{A}_{\mathfrak{f}_\alpha}$ be the ideal group which belongs to $K(\sqrt[m]{\alpha})/K$ in the sense of class field theory. Then*

$$\left(\frac{\alpha}{\mathfrak{b}_1}\right) = \left(\frac{\alpha}{\mathfrak{b}_2}\right) \qquad \textit{for } \mathfrak{b}_1, \mathfrak{b}_2 \in \mathfrak{A}_{\mathfrak{f}_\alpha}$$

if and only if $\mathfrak{b}_1\mathfrak{b}_2^{-1} \in H_\alpha$. In particular $\left(\frac{\alpha}{\mathfrak{b}}\right)$ depends only on the ray class mod $\mathfrak{f}_\alpha$ *of* $\mathfrak{b}$. □

Let $K = \mathbb{Q}$. If $a \in \mathbb{Z}$ and $p \neq 2$ is a prime number with $p \nmid a$, then $\left(\frac{a}{(p)}\right) = \left(\frac{a}{p}\right)$ is the Legendre symbol. If $b \in \mathbb{Z}$ and $(a, 2b) = 1$, then $\left(\frac{a}{(b)}\right) = \left(\frac{a}{b}\right)$ is the Jacobi symbol.

1.9. The Hilbert Norm Symbol. In this section K denotes a local field. Let m be a natural number such that K contains the m-th roots of unity. For $\alpha, \beta \in K^\times$ the *Hilbert norm symbol* $(\beta, \alpha)_{m,K} = (\beta, \alpha) \in \mu_m$ is defined by

$$(\beta, K(\sqrt[m]{\alpha})/K)\sqrt[m]{\alpha} = (\beta, \alpha)\sqrt[m]{\alpha}.$$

It has the following properties which follow from local class field theory (§ 1.3, § 1.6):

Theorem 2.14. 1. *$(\beta, \alpha) = 1$ if and only if $\beta \in N_{K(\sqrt[m]{\alpha})/K}(K(\sqrt[m]{\alpha})^\times)$.*

2. $(\beta_1\beta_2, \alpha) = (\beta_1, \alpha)(\beta_2, \alpha)$, $\quad(\beta, \alpha_1\alpha_2) = (\beta, \alpha_1)(\beta, \alpha_2)$. (2.12)
3. $(\alpha, \beta) = (\beta, \alpha)^{-1}$, $\quad(-\alpha, \alpha) = 1$. (2.13)
4. $(1 - \alpha, \alpha) = 1$ if $\alpha \neq 1$.
5. *The map $\beta, \alpha \to (\beta, \alpha)$ is continuous.*
6. *Let $g\colon K \to gK$ be an isomorphism, then*

$$(g\beta, g\alpha)_{gK} = g(\beta, \alpha)_K \qquad \text{for } \alpha, \beta \in K. \tag{2.14}$$

7. *Let d be a divisor of m, then*

$$(\beta, \alpha)_m^d = (\beta, \alpha)_{m/d}. \tag{2.15}$$

8. *Let L/K be a finite extension, then for $\beta \in L^\times$, $\alpha \in K^\times$*

$$(\beta, \alpha)_L = (N_{L/K}\beta, \alpha)_K. \tag{2.16}$$

9. *Let K/K_0 be a normal extension with Galois group G. Then $(\ ,\ )$ induces a non-degenerate pairing* (§ 3.7) *of the G-modules $K^\times/K^{\times m}$, $K^\times/K^{\times m}$ onto the G-module μ_m* (non-degenerate means here that for every $\bar{\alpha} \in K^\times/K^{\times m}$ of order d there exists a $\bar{\beta} \in K^\times/K^{\times m}$ such that (β, α) has order d).

Perhaps the most striking property of the Hilbert norm symbol is 4. since it involves the additive structure of K. It is the starting point of the theory of symbols (§ 6.6) and of the K-theory of Milnor (1970). 4. is proved as follows: Let d be the largest divisor of m such that $\alpha = \gamma^d$ for some $\gamma \in K$. We put $n := m/d$. Then $K(\sqrt[m]{\alpha}) = K(\sqrt[n]{\gamma})$ has degree n over K (Chap. 1.4.7). Hence

$$\prod_{i=0}^{d-1} (1 - \zeta_m^i \sqrt[m]{\alpha}),$$

where ζ_m is a primitive m-th root of unity, has the norm $1 - \alpha$. Therefore 4. follows from 1. □

It is easy to see that if $K = \mathbb{R}$, then

$$(\beta, \alpha) = (-1)^{(\operatorname{sgn}\beta - 1)(\operatorname{sgn}\alpha - 1)/4}. \tag{2.17}$$

1.10. The Reciprocity Law for the Power Residue Symbol. In this section K denotes an algebraic number field containing the m-th root of unity. For any place v of K and $\alpha, \beta \in K^\times$ we write

$$\left(\frac{\beta, \alpha}{v}\right) := (\beta, \alpha)$$

where $(\ ,\ )$ is the Hilbert norm symbol of K_v.

Let $\mathfrak{p}$ be a prime ideal of K with $\mathfrak{p} \nmid \mathfrak{f}_\alpha$. The connection between the power residue symbol and the Hilbert norm symbol is given by the following formula

$$\left(\frac{\beta, \alpha}{\mathfrak{p}}\right) = \left(\frac{\alpha}{\mathfrak{p}}\right)^{-v_\mathfrak{p}(\beta)}. \tag{2.18}$$

Theorem 2.7 implies the product formula

$$\prod_v \left(\frac{\beta, \alpha}{v}\right) = 1 \tag{2.19}$$

where v runs over all places of K. From (2.13), (2.18), (2.19) one derives

Theorem 2.15 (Reciprocity law). *Let $\alpha, \beta \in K^\times$ and*

$$\mathfrak{a} = \prod_{\mathfrak{p} \nmid \mathfrak{f}_\beta} \mathfrak{p}^{v_\mathfrak{p}(\alpha)}, \qquad \mathfrak{b} = \prod_{\mathfrak{p} \nmid \mathfrak{f}_\alpha} \mathfrak{p}^{v_\mathfrak{p}(\beta)}.$$

Then

$$\left(\frac{\alpha}{\mathfrak{b}}\right)\left(\frac{\beta}{\mathfrak{a}}\right)^{-1} = \prod_{v \in V}\left(\frac{\beta, \alpha}{v}\right)$$

where $V = \operatorname{supp} \mathfrak{f}_\alpha \cap \operatorname{supp} \mathfrak{f}_\beta$. □

The following special case of Theorem 2.15 can be considered as the direct generalization of Jacobi's quadratic reciprocity law.

Theorem 2.16. *Let* α, β *be numbers of* K *which are prime to* m *and to each other. Then*

$$\left(\frac{\alpha}{\beta}\right)\left(\frac{\beta}{\alpha}\right)^{-1} = \prod_{v | m\infty}\left(\frac{\beta, \alpha}{v}\right) \tag{2.20}$$

where $\left(\frac{\alpha}{\beta}\right) := \left(\frac{\alpha}{(\beta)}\right)$. □

We show that Theorem 2.16 implies the reciprocity law for the Jacobi symbol if $m = 2$, $K = \mathbb{Q}$:

$$\left(\frac{a}{b}\right)\left(\frac{b}{a}\right) = (-1)^{(a-1)(b-1)/4} \qquad \text{for } a, b \in N, 2 \nmid ab.$$

One has to show

$$\left(\frac{a, b}{2}\right) = (-1)^{(a-1)(b-1)/4}. \tag{2.21}$$

Since

$$(a_1 a_2 - 1)/2 \equiv (a_1 - 1)/2 + (a_2 - 1)/2 \pmod 2$$

it is sufficient to prove (2.21) for a basis of $\mathbb{Z}_2^\times$, e.g. for $-1, 5$. Since $5 = N(2 + \sqrt{-1})$ and $2 = N(1 + \sqrt{-1})$, we have $(5, -1) = 1$, $(2, -1) = 1$ and therefore $(-1, -1) = -1$. (Theorem 2.14.9). Furthermore $(5, 5) = (5, -5) = 1$. □

As complement to Theorem 2.16 one has

Theorem 2.17 (Complementary theorem for the reciprocity law). *Let* λ, $\alpha \in K^\times$ *with* $\operatorname{supp} \lambda \subseteq \operatorname{supp} m\infty$ *and* $\operatorname{supp} \alpha \cap \operatorname{supp} m = \phi$. *Then*

$$\left(\frac{\lambda}{\alpha}\right)^{-1} = \prod_{v | m\infty}\left(\frac{\lambda, \alpha}{v}\right). \ \square$$

1.11. The Principal Ideal Theorem. Hilbert conjectured the following *principal ideal theorem*:

Theorem 2.18. *Let* K *be an algebraic number field and* H *the Hilbert class field of* K. *Then every ideal of* K *becomes principal considered as ideal of* H.

By means of Theorem 2.10 we can reformulate Theorem 2.18 as a purely group theoretical statement: Let H' be the Hilbert class field of H. Then H'/K is normal and H/K is the maximal abelian subextension of H'/K. Therefore Theorem 2.18 is equivalent to the statement

$$\left(\frac{\mathfrak{a}}{H'/H}\right) = 1 \qquad \text{for } \mathfrak{a} \in \mathfrak{M}(K). \tag{2.22}$$

By Theorem 2.10 (2.22) is equivalent to

$$\operatorname{Ver}\left(\frac{\mathfrak{a}}{H'/K}\right) = 1 \qquad \text{for } \mathfrak{a} \in \mathfrak{M}(K).$$

Hence Theorem 2.18 follows from the group theoretical ideal theorem:

Theorem 2.19. *Let G be a finite group with abelian commutator group $H := [G, G]$. Then the transfer map from G to H is trivial.* ⊠ (Artin, Tate (1968), Chap. 13).

Beginning with Taussky (1932) several authors investigated the process of "*capitulation*", i.e. becoming a principal ideal, of the ideals of K in the subfields of the Hilbert class field. For a more general point of view and as reference source see Schmithals (1985).

1.12. Local-Global Relations (Main reference: Artin-Tate (1968), Chap. 10). Let L be an abelian extension of the algebraic number field K, let $N = N_{L/K}\mathfrak{C}(L)$ be the corresponding subgroup of the idele class group $\mathfrak{C}(K)$, and let w be a place of L above the place v of K. Then Theorem 2.8 implies that the norm group of L_w/K_v is equal to $N \cap K_v^\times$. On the other hand assume that for some finite set S of places of K we have local abelian extensions L_w/K_v for $v \in S$. The natural question arises whether there exists a global abelian extension L of K such that the localization of L/K at places v in S are isomorphic to L_w/K_v. Class field theory translates this question in a question about subgroups of $\mathfrak{C}(K)$ and the positive answer is given by the following theorem:

Theorem 2.20. *Let P be the group $\prod_{v \in S} K_v^\times$ with the topology given by the natural injection*

$$\prod_{v \in S} K_v^\times \to \mathfrak{C}(K)$$

and let $\tilde{P}$ be the group $\prod_{v \in S} K_v^\times$ with the product topology.

Then a subgroup V of finite index is closed in P if and only if it is closed in $\tilde{P}$. For any such subgroup V there exists a closed subgroup N of $\mathfrak{C}(K)$ such that $P \cap N = V$. ⊠

One may ask whether it is possible to find a global extension L/K with the local properties described above and the extra condition that $[L:K]$ is the least common multiple of the local degrees $[L_w:K_v]$ for $v \in S$. Grunwald (1933) claimed that this is true for cyclic extensions but Wang (1950) found a mistake in the proof of Grunwald and gave the right formulation of the theorem (Theorem 2.22 below), now called Grunwald-Wang theorem.

There are "exceptional" cases in which Grunwald's claim is wrong. These cases are connected with the following theorem on m-powers in K:

Theorem 2.21. *Let m be a natural number and S a finite set of places of K. Moreover let s be the greatest natural number such that $K(\eta_s) = K$, where $\eta_s = \zeta_{2^s} + \zeta_{2^s}^{-1}$ and ζ_{2^s} is a primitive root of unity of order 2^s. Then the group*

$$P(m, S) := \{\alpha \in K^\times \,|\, \alpha \in K_v^{\times m} \text{ for } v \notin S\}$$

is equal to $K^{\times m}$ except under the following conditions which will be called the special case:

1. $-1, 2 + \eta_s$, *and* $-(2 + \eta_s)$ *are non-squares in* K.
2. $m = 2^t m'$, *where* m' *is odd, and* $t > s$.
3. $S_0 \subseteq S$, *where* S_0 *is the set of prime divisors* $\mathfrak{p}$ *of* 2 *in* K *for which* $-1, 2 + \eta_s$, *and* $-(2 + \eta_s)$ *are non-squares in* $K_\mathfrak{p}$.

In the special case

$$P(m, S) = K^{\times m} \cup \alpha_0 K^{\times m},$$

where $\alpha_0 = (2 + \eta_s)^{m/2}$. ⊠

Example 8. Let $K = \mathbb{Q}$. Then $s = 2$. The special case occurs if $8|m$ and if $2 \in S$. Since $\eta_s = 0$, we have $\alpha_0 = 2^{m/2}$. Especially 16 is an 8-th power at ∞ and at all odd primes. On the other hand 16 is not an 8-th power in $\mathbb{Q}$ or in $\mathbb{Q}_2$. □

Now we come to the formulation of the *Grunwald-Wang theorem*:

Theorem 2.22. *Let S be a finite set of places of K, let χ_v be local characters of period n_v for each $v \in S$, and let m be the least common multiple of the n_v. Then there exists a global character χ on $\mathfrak{C}(K)$ whose local restrictions are the given χ_v. There exists such a χ with period m if in the special case the condition*

$$\prod_{p \in S_0} \chi_p(\alpha_0) = 1$$

is satisfied. If this condition is not satisfied, one can only achieve the period $2m$. ⊠

If locally only the degrees of the extensions L_w/K_v are prescribed, one can avoid the special case. The resulting theorem below has an interesting application in the theory of simple algebras over number fields (Theorem 2.89).

Theorem 2.23. *Let S be a finite set of places of K and n_v positive integers associated with each $v \in S$. If v is archimedean, n_v should be a possible degree for an extension of K_v. Then there exists a cyclic extension L/K whose degree is the least common multiple of the n_v, and such that the completions L_w/K_v have degree n_v for all $v \in S$.* ⊠

1.13. The Zeta Function of an Abelian Extension. Let L be a finite abelian extension of an algebraic number field K,

$$\zeta_L(s) = \prod_{\mathfrak{P}} \frac{1}{1 - N_{L/\mathbb{Q}}(\mathfrak{P})^{-s}}$$

the Dedekind zeta function, and let $\Lambda_L(s)$ be the corresponding enlarged zeta function (Chap. 1.6.3).

Theorem 2.24. *Let $U = N_{L/K}J(L)$ be the subgroup of $J(K)$ corresponding to L and let X be the group of characters χ of $J(K)$ which are trivial on U. Then*

$$\zeta_L(s) = \prod_{\chi \in X} L(s, \chi), \tag{2.23}$$

$$\Lambda_L(s) = \prod_{\chi \in X} \Lambda(s, \chi). \tag{2.24}$$

Proof. For the proof of (2.23) one has to show

$$\prod_{\mathfrak{P}|\mathfrak{p}} (1 - N_{L/\mathbb{Q}}(\mathfrak{P})^{-s}) = (1 - N_{K/\mathbb{Q}}(\mathfrak{p})^{-fs})^g = \prod_{\chi \in X} (1 - \chi(\mathfrak{p}) N_{K/\mathbb{Q}}(\mathfrak{p})^{-s}), \tag{2.25}$$

where g denotes the number of prime divisors $\mathfrak{P}$ in L of the prime ideal $\mathfrak{p}$ and f is the inertia degree of $\mathfrak{P}$ in L/K. The equation (2.25) is equivalent to the polynomial equation,

$$(1 - x^f)^g = \prod_{\chi \in X} (1 - \chi(\mathfrak{p})x). \tag{2.26}$$

By Theorem 2.8 the characters of $J(K)/U$ correspond to the characters of $G(L/K)$. The product in (2.26) is to be taken over all characters which are unramified, i.e. characters which are trivial on the inertia group $T_\mathfrak{p}$. Since f is the order of $G_\mathfrak{P}/T_\mathfrak{P}$ and $G_\mathfrak{P}/T_\mathfrak{P}$ is generated by $(\pi, L/K)$ with π a prime element of $K_\mathfrak{p}$, we see that there are exactly g such characters with $\chi(\mathfrak{p}) = \chi(\pi)$ a given root of unity in μ_f. This proves (2.23).

For the proof of (2.24) one uses Legendre's formula

$$\Gamma(s) = 2^{s-1}\pi^{-1/2}\Gamma(s/2)\Gamma((s+1)/2). \ \square$$

(2.23) together with Theorem 1.104 implies

Theorem 2.25 (Analytical class number formula).

$$\kappa(L) = \kappa(K) \prod_{\chi \in X - \{\chi_0\}} L(1, \chi). \ \square$$

Theorem 2.25 shows immediately that $L(1, \chi) \neq 0$ for $\chi \neq \chi_0$. This is sufficient to prove the existence of infinitely many prime ideals in the classes of the ray class group $H_\mathfrak{m}$ (compare Theorem 1.111).

Furthermore the functional equation for $\Lambda(s, \chi)$ (Theorem 1.104) together with (2.24) shows

Theorem 2.26 (Product formula for the ε-factors)

$$|d_{L/Q}|^{s-1/2} = \prod_{\chi \in X} \varepsilon(s, \chi), \tag{2.27}$$

$$\prod_{\chi \in X} \varepsilon_0(\chi) = 1. \ \square \tag{2.28}$$

For the computation of $\varepsilon_0(\chi)$ we can without loss of generality assume that L/K is the cyclic extension corresponding to the kernel of χ. Let $\chi \neq \chi_0$ be a real character, i.e. $\chi^2 = \chi_0$. Then $|X| = 2$ and since $\varepsilon_0(\chi_0) = 1$, (2.28) implies $\varepsilon_0(\chi) = 1$. Hence (1.55) determines the Gaussian sum $\tau(\chi)$:

Theorem 2.27. *Let χ be a real character. Then*

$$\tau(\chi) = i^{-r}(N_{K/\mathbb{Q}}\mathfrak{f}_\chi)^{1/2}$$

where r is the number of real places in the conductor of χ and $\mathfrak{f}_\chi$ is the finite part of the conductor of χ. □

Example 9. Let $K = \mathbb{Q}$ and let χ be a real character with conductor fv^r, $r = 0, 1$. Then Theorem 2.27 together with Proposition 1.106 imply Gauss' formula

$$\tau_{1/f}(\chi) = \begin{cases} f^{1/2} & \text{if} \quad \chi(-1) = 1, \\ if^{1/2} & \text{if} \quad \chi(-1) = -1. \end{cases} \square$$

§ 2. Complex Multiplication

(Main reference: Deuring (1958))

Every abelian extension of $\mathbb{Q}$ is contained in a cyclotomic field, i.e. in a field which is generated by a value of the function $\exp(2\pi i x)$ at a division point x_1 of the lattice $\mathbb{Z}$ in $\mathbb{Q}$ (§ 1.1).

Something similar one has for an imaginary quadratic field K as base field: The ring $\mathfrak{O}_K = \mathbb{Z}\omega_1 + \mathbb{Z}\omega_2$, $\operatorname{Im} \omega_1/\omega_2 > 0$, of integers in K form a lattice in $K_\infty = \mathbb{C}$ with *complex multiplication*, i.e. $\alpha\mathfrak{O}_K \subseteq \mathfrak{O}_K$ for $\alpha \in \mathfrak{O}_K$. Let $j(z)$ be the elliptic modular function. Then $K(j(\omega_1/\omega_2))$ is the Hilbert class field of K (§ 1.2) and every abelian extension of K is contained in a field $K(j(\omega_1/\omega_2)), \tau(z_1))$ where $\tau(z)$ is Weber's τ-function which up to a constant factor is equal to Weierstrass' $\wp$-function, and z_1 is a division point of the lattice $\mathfrak{O}_K$.

It is possible to develop the class field theory of imaginary quadratic number fields on the basis of the functions $j(z)$ and $\tau(z)$ (Deuring (1958)). We confine ourselves here to the study of the number theoretical properties of special values of these functions on the basis of the main theorems of class field theory.

2.1. The Main Polynomial. A matrix $A = \begin{pmatrix} a & b \\ c & d \end{pmatrix}$ with $a, b, c, d \in \mathbb{Z}$ and $\det A > 0$ is called primitive if g.c.d. $(a, b, c, d) = 1$. Let A_s be the set of such matrices with determinant s. The orbits of A_s with respect to left multiplication by matrices from $SL_2(\mathbb{Z})$ are represented by the triangular matrices $\begin{pmatrix} a & b \\ 0 & d \end{pmatrix}$ with $a > 0$, $ad = s$, $(a, b, d) = 1$ and $0 \leqslant b < d$. Let $\mathfrak{S}$ be the set of this matrices.

Proposition 2.28. *The polynomial*

$$J_s(x, j(z)) := \prod_{S \in \mathfrak{S}} (x - j(S(z))) \quad \textit{with} \quad S(z) = \frac{az + b}{cz + d} \quad \textit{for} \quad S = \begin{pmatrix} a & b \\ c & d \end{pmatrix}$$

has coefficients in $\mathbb{Z}[j(z)]$. *The highest coefficient of the polynomial* $J_s(x, x)$ *is equal to* ± 1. ⊠

The main point of the proof is to use that the Fourier expansion

$$j(z) = q^{-1} + c_0 + c_1 q + \cdots, q = e^{2\pi i z},$$

has integer coefficients and that $j(z)$ generates the field of modular functions on $SL_2(\mathbb{Z})$.

Proposition 2.29. *Let* $s = p$ *be a prime. Then*

$$J_p(x, j(z)) \equiv (x^p - j(z))(x - j(z)^p) \pmod{p}. \boxtimes$$

2.2. The First Main Theorem. Let K be an imaginary quadratic field and $\mathfrak{O}_f$ the order in K with conductor f (Chap. 1.1.1). The complete modules $\mathfrak{a}$ with order $\mathfrak{O}_f$ such that the ideal $\mathfrak{a}\mathfrak{O}_K$ of K is prime to f form a group $\mathfrak{M}_f$ (Chap. 1.1.6). The correspondence $\varphi\colon \mathfrak{a} \to \mathfrak{a}\mathfrak{O}_K$ defines an isomorphism of $\mathfrak{M}_f$ onto $\mathfrak{A}_f$ (Chap. 1.5.5). We put $\varphi^{-1}(\mathfrak{b}) = \mathfrak{b}_f$ for $\mathfrak{b} \in \mathfrak{A}_f$.

Since there are modules of $\mathfrak{M}_f$ in every class of $CL(\mathfrak{O}_f) = \mathfrak{M}(\mathfrak{O}_f)/\mathfrak{H}(\mathfrak{O}_f)$, we have $CL(\mathfrak{O}_f) = \mathfrak{M}_f/(\mathfrak{M}_f \cap \mathfrak{H}(\mathfrak{O}_f))$.

Let $\mathfrak{G}_f$ be the group of principal ideals $\alpha\mathfrak{O}_K$ such that $\alpha\mathfrak{O}_K \in \mathfrak{A}_f$ and $\alpha \equiv r \pmod{f}$ for some $r \in \mathbb{Q}$ (Chap. 1.5.5) ($\alpha \equiv r \pmod{f}$ means $v_p(\alpha - r) \geqslant v_p(f)$ for $p | f$). Then $\mathfrak{a} \to \mathfrak{a}\mathfrak{O}_K$ induces an isomorphism of $CL(\mathfrak{O}_f)$ onto $\mathfrak{A}_f/\mathfrak{G}_f$. We understand $CL(\mathfrak{O}_f)$ as a class group in the sense of class field theory (§ 1.7) by means of this isomorphism. It is called *ring class group* mod f.

Let $\mathfrak{a} = \alpha_1\mathbb{Z} + \alpha_2\mathbb{Z}$, $\operatorname{Im}(\alpha_1/\alpha_2) > 0$, be any complete module with order $\mathfrak{O}_f$. Then $j(\mathfrak{a}) := j(\alpha_1/\alpha_2)$ depends only on the class $\bar{\mathfrak{a}} \in CL(\mathfrak{O}_f)$ of $\mathfrak{a}$ and determines this class. We put $j(\bar{\mathfrak{a}}) := j(\mathfrak{a})$.

Theorem 2.30. $\{j(\bar{\mathfrak{a}}) | \bar{\mathfrak{a}} \in CL(\mathfrak{O}_f)\}$ *is a full set of conjugated algebraic integers with respect to* K. *The field* $K_f := K(j(\mathfrak{a}))$ *is an abelian extension, the class field of the ring class group* $CL(\mathfrak{O}_f)$.

Proof. By means of general class field theory (§ 1) it is sufficient to show 1. that $j(\bar{\mathfrak{a}})$ for $\bar{\mathfrak{a}} \in CL(\mathfrak{O}_f)$ is an algebraic integer and 2. that for almost all prime ideals $\mathfrak{p}$ of K of degree 1, $\mathfrak{P}_f$ is a principal module of $\mathfrak{O}_f$ if and only if $\mathfrak{p}$ splits completely in K_f (Theorem 1.118) (Weber's definition of the class field)

1. Let $\mathfrak{p} \nmid f$, $N_{K/\mathbb{Q}}\mathfrak{p} = p$ and $\bar{\mathfrak{a}} \in CL(\mathfrak{O}_f)$. Then

$$J_p(j(\bar{\mathfrak{a}}), j(\bar{\mathfrak{a}}\bar{\mathfrak{p}}_f^{-1})) = 0$$

since $\mathfrak{a}\mathfrak{p}_f^{-1} = (\alpha_1, \alpha_2)$ implies that $\mathfrak{a}$ is generated by $P\begin{pmatrix}\alpha_1\\ \alpha_2\end{pmatrix}$ with $P \in A_p$. If $\mathfrak{p}_f$ is a principal module, then $j(\mathfrak{a}\mathfrak{p}_f^{-1}) = j(\mathfrak{a})$. Hence $j(\mathfrak{a})$ is a root of the polynomial $J_p(x, x)$. Proposition 2.28 implies that $j(\mathfrak{a})$ is an algebraic integer.

2. Proposition 2.29 implies

$$(j(\mathfrak{a}) - j(\mathfrak{a}\mathfrak{p}_f^{-1})^p)(j(\mathfrak{a})^p - j(\mathfrak{a}\mathfrak{p}_f^{-1})) \equiv 0 \pmod{\mathfrak{p}}.$$

Hence for any prime divisor $\mathfrak{P}$ of $\mathfrak{p}$ in $K(j(\mathfrak{a}), j(\mathfrak{a}\mathfrak{p}_f^{-1}))$

$$j(\mathfrak{a}) \equiv j(\mathfrak{a}\mathfrak{p}_f^{-1})^p \pmod{\mathfrak{P}} \quad \text{or} \quad j(\mathfrak{a})^p \equiv j(\mathfrak{a}\mathfrak{p}_f^{-1}) \pmod{\mathfrak{P}}. \tag{2.29}$$

If $\mathfrak{p}_f$ is a principal module, then $j(\mathfrak{a})^p \equiv j(\mathfrak{a}) \pmod{\mathfrak{P}}$ and if $\mathfrak{p}$ is prime to the discriminant of $j(\mathfrak{a})$ it follows that $\mathfrak{p}$ splits completely in $K(j(\mathfrak{a}))/K$.

If $\mathfrak{p}$ splits completely in $K(j(\mathfrak{a}))/K$, then

$$j(\mathfrak{a})^p \equiv j(\mathfrak{a}) \pmod{\mathfrak{P}}$$

and (2.13) implies

$$j(\mathfrak{a}) \equiv j(\mathfrak{a}\mathfrak{p}_f^{-1}) \pmod{\mathfrak{P}}.$$

Now we assume $\mathfrak{P} \nmid j(\bar{\mathfrak{a}}) - j(\bar{\mathfrak{b}})$ for arbitrary $\bar{\mathfrak{b}} \in CL(\mathfrak{O}_f)$. This excludes only finitely many $\mathfrak{p}$. Then $j(\mathfrak{a}) = j(\mathfrak{a}\mathfrak{p}_f^{-1})$. Hence $\mathfrak{p}_f$ is a principal module. □

2.3. The Reciprocity Law

Theorem 2.31. *The Artin symbol for a class $\bar{\mathfrak{a}} \in CL(\mathfrak{O}_f)$ is given by*

$$\left(\frac{\bar{\mathfrak{a}}}{K_f/K}\right) j(\bar{\mathfrak{b}}) = j(\bar{\mathfrak{b}}\bar{\mathfrak{a}}^{-1}) \qquad \text{for } \bar{\mathfrak{b}} \in CL(\mathfrak{O}_f). \ \boxtimes$$

The proof is based on the congruence

$$j(\mathfrak{a})^{N(\mathfrak{p})} \equiv j(\mathfrak{a}\mathfrak{p}_f^{-1}) \pmod{\mathfrak{p}} \tag{2.30}$$

for prime ideals $\mathfrak{p}$ with $\mathfrak{p} \nmid f$.

For the proof of (2.30) one uses the function $\varphi_S(z)$ for $S \in A_s$, which is defined as follows:

$$\varphi_S(z) := s^{12} \frac{\Delta(Sz)}{\Delta(z)}$$

where $\Delta(z)$ is the discriminant, i.e. a non trivial parabolic modular form of weight 12.

One of the main advantages of the function $\varphi_S(z)$ consists in our knowledge of its ideal theoretical structure at singular arguments: $\sqrt[24]{\Delta(z)} = \sqrt{2\pi}\eta(z)$ is a regular function in $\operatorname{Im} z > 0$. $\eta(z)$ is called the Dedekind function. Hence $\varphi_S(z)$ is regular in $\operatorname{Im} z > 0$, too.

Theorem 2.32. *Let $\mathfrak{b}$ be an ideal of K which is prime to $6f$. Furthermore let α_1, α_2, $\operatorname{Im} \alpha_1/\alpha_2 > 0$, be a basis of a module $\mathfrak{a}_f$ with order $\mathfrak{O}_f$ and let B be a rational matrix which transforms α_1, α_2 in a basis of $\mathfrak{a}_f g \mathfrak{b}_f^{-2}$ where g denotes complex conjugation. Then $\beta := \varphi_B(\alpha_1/\alpha_2)$ is a number in K_f and $(\beta) = \mathfrak{b}$.* ⊠

If $f = 1$, this is the principal ideal theorem (§ 1.11).

2.4. The Construction of the Ray Class Field. Kronecker conjectured that every abelian extensions of an imaginary-quadratic number field K is a subfield of a field of the form $K(j(\alpha), e^{2\pi i\beta})$ for some $\alpha \in K$, $\operatorname{Im} \alpha > 0$ and $\beta \in \mathbb{Q}$. This is the special form of his conjecture which he called "seinen liebsten Jugendtraum". It is wrong since the ray class field $K(f)$ mod f is in general a proper extension of $K_0(f) := K_f(e^{2\pi i/f})$. But $G(K(f)/K_0(f))$ is an elementary 2-group (Hasse (1970), I. § 10). Therefore the conjecture is true for abelian extensions of odd degree.

For the generation of the full ray class field mod f^∞ one uses the values of elliptic functions at division points.

Let $\mathfrak{a} = (\omega_1, \omega_2)$, $\operatorname{Im} \omega_1/\omega_2 > 0$, be an ideal of K and $\wp(z, \mathfrak{a})$ the corresponding Weierstrass $\wp$-function. For the purposes of number theory one has to multiply $\wp(z, \mathfrak{a})$ by a factor $g^{(e)}$ where e is the number of roots of unity in $\mathfrak{O}_K$, i.e. $e = 6$ if $\mathfrak{O}_K = \mathbb{Z}[(1 + \sqrt{-3})/2]$, $e = 4$ if $\mathfrak{O}_K = \mathbb{Z}[\sqrt{-1}]$ and $e = 2$ in all other cases.

$$g^{(2)} := -2^7 3^5 g_2(\omega) g_3(\omega) \omega_2^2 / \Delta(\omega),$$
$$g^{(4)} := 2^8 3^4 g_2^2(\omega) \omega_2^4 / \Delta(\omega),$$
$$g^{(6)} := -2^9 3^6 g_3(\omega) \omega_2^6 / \Delta(\omega)$$

where as usual $g_2(\omega)$, $g_3(\omega)$ denote the Eisenstein series such that

$$\wp'(z, \mathfrak{a})^2 = 4\wp(z, \mathfrak{a})^3 - g_2(\omega)\wp(z, \mathfrak{a}) - g_3(\omega), \qquad \omega = \omega_1/\omega_2,$$
$$\Delta(\omega) = g_2(\omega)^3 - 27 g_3(\omega)^2.$$

The function

$$\tau(z, \mathfrak{a}) := g^{(e)} \wp(z, \mathfrak{a})^{e/2}$$

is called *Weber τ-function* of the ideal $\mathfrak{a}$. The factor $g^{(e)}$ is chosen such that $\tau(z, \mathfrak{a})$ has a Fourier expansion

$$(q^{e/2} + \cdots + t_\nu q^\nu + \cdots)\left(1 + 12u(1-u)^{-2} + 12 \sum_{n,m=1}^{\infty} n q^{nm}(u^n + u^{-n} - 2)\right)^{e/2}$$

in $q = e^{2\pi i}$, $u = e^{2\pi i z/\omega_1}$ with integral coefficients t_ν.

Let N be a natural number. Then

$$\tau\left(\frac{x_1\omega_1 + x_2\omega_2}{N}, \mathfrak{a}\right) \qquad \text{with } x_1, x_2 \in \mathbb{Z}, (x_1, x_2) \not\equiv (0, 0) \pmod N,$$

is called an N-th division value of $\tau(z, \mathfrak{a})$. By evaluating $\tau\left(\frac{x_1\omega_1 + x_2\omega_2}{N}, \mathfrak{a}\right)$ in a Laurent series of $q^{1/N}$ one verifies that

$$T_N(x, j(\omega)) := \prod_{\substack{x_1, x_2 \bmod N \\ (x_1, x_2, N) = 1}} \left(x - \tau\left(\frac{x_1\omega_1 + x_2\omega_2}{N}, \mathfrak{a}\right)\right)$$

has rational coefficients which are integers if N is not a prime power and which become integers after multiplication by l^e if N is a power of the prime l. It follows that

$$\tau(\gamma, \mathfrak{a}) = \tau(1, \mathfrak{a}\gamma^{-1}) \qquad \text{for } \gamma \in K, \gamma \notin \mathfrak{a},$$

is an algebraic number.

If $\mathfrak{m}$ and $\mathfrak{r}$ are integral ideals of K and $\mathfrak{r}$ is prime to $\mathfrak{m}$, then $\tau(1, \mathfrak{m}\mathfrak{r}^{-1})$ depends only on the ray class of $\mathfrak{r}^{-1}$ mod $\mathfrak{m}$. In fact let $\mathfrak{r}' = \lambda\mathfrak{r}$ with $\lambda \equiv 1 \pmod{\mathfrak{m}}$, then

$$\tau(1, \mathfrak{m}\mathfrak{r}'^{-1}) = \tau(\lambda, \mathfrak{m}\mathfrak{r}^{-1}) = \tau(1, \mathfrak{m}\mathfrak{r}^{-1}).$$

If $\mathfrak{r}^{-1}$ lies in the ray class $\mathfrak{A}$, we write $\tau(\mathfrak{A}) := \tau(1, \mathfrak{m}\mathfrak{r}^{-1})$.

Proposition 2.33. *Let $\mathfrak{A}$ and $\mathfrak{A}'$ be ray classes* mod $\mathfrak{m}$ *such that $\mathfrak{A}'\mathfrak{A}^{-1}$ consists of principal ideals. Then $\mathfrak{A} = \mathfrak{A}'$ if and only if $\tau(\mathfrak{A}) = \tau(\mathfrak{A}')$.* ⊠

The following theorem is called the second main theorem.

Theorem 2.34. *Let $\mathfrak{m}$ be an ideal of $\mathfrak{O}_K$ and $\mathfrak{A}$ a ray class* mod $\mathfrak{m}$. *Then $K_1(\tau(\mathfrak{A}))$ is the ray class field* mod $\mathfrak{m}$. ⊠

The proof is analogous to the proof of Theorem 2.30. It is based on the following congruence:

Theorem 2.35. *Let $\mathfrak{p}$ be a prime ideal of K of degree one which is prime to $6N\mathfrak{m}d_{K/\mathbb{Q}}$. Then*

$$\tau(\mathfrak{A}\mathfrak{p}^{-1}) \equiv \tau(\mathfrak{A})^{N(\mathfrak{p})} \pmod{\mathfrak{P}}$$

for any prime divisor $\mathfrak{P}$ of $\mathfrak{p}$ in $K_1(\tau(\mathfrak{A}), \tau(\mathfrak{A}\mathfrak{p}^{-1}))$. ⊠

From this congruence one also derives the reciprocity law:

Theorem 2.36. Let $\mathfrak{A}$ and $\mathfrak{B}$ be ray classes mod $\mathfrak{m}$ and $\Omega_{\mathfrak{m}}$ the ray class field mod $\mathfrak{m}$. Then

$$\left(\frac{\mathfrak{A}}{\Omega_{\mathfrak{m}}/K}\right) j(\mathfrak{B}) = j(\mathfrak{B}\mathfrak{A}^{-1}),$$

$$\left(\frac{\mathfrak{A}}{\Omega_{\mathfrak{m}}/K}\right) \tau(\mathfrak{B}) = \tau(\mathfrak{B}\mathfrak{A}^{-1}). \text{ ⊠}$$

2.5. Algebraic Theory of Complex Multiplication. (Main reference: Shimura (1971), Chap. 4). The theory of complex multiplication can also be developed algebraically as part of the theory of elliptic curves. If E is an elliptic curve defined over $\mathbb{C}$, then $\mathrm{End}(E)$ is isomorphic to $\mathbb{Z}$ or to an order in an imaginary-quadratic number field. In the latter case one says that E has complex multiplication.

The connection with the analytic theory is given by the following theorem.

Theorem 2.37. *Let E be an elliptic curve defined over $\mathbb{C}$ such that* $\mathrm{End}_{\mathbb{Q}}(E) = \mathrm{End}(E) \otimes_{\mathbb{Z}} \mathbb{Q}$ *is isomorphic to an imaginary-quadratic number field K and let $\mathfrak{O}$ be the order in K corresponding to* $\mathrm{End}(E)$. *Then E is isomorphic (as analytical group) to the torus $\mathbb{C}/\mathfrak{a}$ for some module $\mathfrak{a}$ with order $\mathfrak{O}$. On the other hand* $\mathrm{End}(\mathbb{C}/\mathfrak{a})$ *is isomorphic to $\mathfrak{O}$ for any module $\mathfrak{a}$ with order $\mathfrak{O}$. If $\mathfrak{a}$ and $\mathfrak{b}$ are modules with order $\mathfrak{O}$, then the curves $\mathbb{C}/\mathfrak{a}$ and $\mathbb{C}/\mathfrak{b}$ are isomorphic if and only if $\mathfrak{b} = \mu\mathfrak{a}$ for some $\mu \in K^{\times}$.* ⊠

The invariant of E is equal to $j(\mathfrak{a})$. Hence E is defined over the algebraic number field $\mathbb{Q}(j(\mathfrak{a}))$. By means of the reduction theory of E one obtains a beautiful proof of Theorem 2.31.

For more details see Shimura (1971), Chap. 4, or Lang (1973), Part 2.

2.6. Generalization. The generation of class fields by the values of transcendental functions has already been emphasized by Hilbert in his twelfth problem (Hilbert (1976)). He writes: *The extension of Kronecker's theorem to the case that in place of the realm of rational numbers or of the imaginary quadratic field any algebraic field whatever is laid down as realm of rationality, seems to me of the greatest importance. I regard this problem as one of the most profound and farreaching in the theory of numbers and of functions.*

From the point of view of complex multiplication the most natural generalization is to replace elliptic curves by abelian varieties. Such a generalization was established by Shimura and Taniyama (1961) for CM-fields, i.e. totally complex quadratic extensions of totally real number fields, but only with partial success since one obtains not all abelian extensions. The further development of this investigations led to the notion of Shimura variety (Deligne (1971), Borel Casselman (1975), Part 4) which proposes to be of the greatest importance in arithmetical algebraic geometry and in the theory of automorphic L-functions.

An approach to Hilbert's twelfth problem of quite another nature was proposed by Stark (1976)–(1982), see also Tate (1984), who connects units of abelian extensions with the value of Artin L-functions (Chap. 5.1.12) at $s = 0$.

§ 3. Cohomology of Groups

(Main references: Serre (1962), Part 3; Koch (1970), § 3)

The cohomology of groups is a convenient mathematical language for the proofs of the theorems of class field theory. It leads to a further development of this theory called Galois cohomology, which will be considered in the next chapter. We present the cohomology of groups here in its easiest form as a theory of factor systems. For the more general point of view of derived functors the reader is referred to the classical book of Cartan-Eilenberg (1956).

3.1. Definition of Cohomology Groups. Let G be a group, $\Lambda = \mathbb{Z}[G]$ its group ring over $\mathbb{Z}$, a (left) G-module is a unitary (left) Λ-module. For $n \geqslant 1$ and the G-module A we denote by $K^n(G, A)$ the set of mappings of the n-fold product of G into A. We put $K^0(G, A) := A$ and we transfer the addition in A to $K^n(G, A)$. An element f of $K^n(G, A)$ is called an *n-dimensional cochain.*

$$\begin{aligned}(df)(x_1, \ldots, x_{n+1}) &= x_1 f(x_2, \ldots, x_{n+1}) \\ &\quad + \sum_{i=1}^{n} (-1)^i f(x_1, \ldots, x_{i-1}, x_i x_{i+1}, x_{i+2}, \ldots, x_{n+1}) \\ &\quad + (-1)^{n+1} f(x_1, \ldots, x_n)\end{aligned}$$

defines a homomorphism $d = d_n$ from $K^n(G, A)$ into $K^{n+1}(G, A)$ with $d_n d_{n-1} = 0$ for $n \geqslant 1$. Hence $K(G, A) := \sum_{n=0}^{\infty} K^n(G, A)$ with the endomorphism $\sum_{n=0}^{\infty} d_n$ is a complex (Appendix 1.4).

A *cocycle* or *factor system* is a cochain f with $df = 0$, a *coboundary* or *splitting factor system* is a cochain f with $f \in \operatorname{Im} d$. Of course every coboundary is a cocycle.

We define the *cohomology groups* $H^n(G, A)$ of G and A as the cohomology groups of the complex $K(G, A)$, i.e.

$$H^n(G, A) := H^n(K(G, A)) = \begin{cases} \operatorname{Ker} d_n / \operatorname{Im} d_{n-1} & \text{if } n \geqslant 1 \\ \operatorname{Ker} d_n & \text{if } n = 0. \end{cases}$$

Example 10. $H^0(G, A) = A^G$ is the group of elements in A which are fixed by G. □

Example 11. Let L/K be a finite separable normal extension of fields. Then L is a $G(L/K)$-module. E. Noether s generalization of Hilbert's Satz 90 (Proposition 2.6) means $H^1(G(L/K), L^\times) = \{0\}$. □

Example 12. If G acts trivial on A, then $H^1(G, A) = \operatorname{Hom}(G, A)$ is the group of homomorphisms of G into A. □

Example 13. A two dimensional factor system defines a group extension of A with G. The equivalence classes of such extensions are in one-to-one correspondence with the elements of $H^2(G, A)$ (Hall (1959), Chap. 15). □

Let X be an abelian group and $M_G(X)$ the group of mappings from G into X. We define a G-module structure for $M_G(X)$ as follows:

$$gf(x) = f(xg) \qquad \text{for } g \in G, f \in M_G(X).$$

$M_G(X)$ is called induced module.

Proposition 2.38. $H^n(G, M_G(X)) = \{0\}$ *for* $n \geqslant 1$. □

3.2. Functoriality and the Long Exact Sequence. Let G and H be groups, A a G-module and B an H-module. A morphism of A into B is a pair $[\varphi, \psi]$ of group homomorphisms $\varphi\colon H \to G$, $\psi\colon A \to B$ such that

$$\psi(\varphi(h)a) = h\psi(a) \qquad \text{for } h \in H, a \in A.$$

If $H = G$ and $\varphi = id$, then $[\varphi, \psi]$ is a homomorphism of G-modules.

A morphism $[\varphi, \psi]$ of A into B induces a morphism of $K(G, A)$ into $K(H, B)$ and therefore a homomorphism $[\varphi, \psi]_*$ of $H^n(G; A)$ into $H^n(H, B)$.

Example 14. Let H be a subgroup of G, $B = A$, $\psi = id$ and φ the embedding of H in G. Then $\operatorname{Res} := [\varphi, \psi]_* : H^n(G, A) \to H^n(H, A)$ is called *restriction* from G to H. □

Example 15. Let H be a normal subgroup of G. Then A^H is a G/H-module and we have the natural morphism $[\varphi, \psi]$ with $\psi\colon A^H \to A$, $\varphi\colon G \to G/H$. The corresponding homomorphism $\operatorname{Inf}\colon H^n(G/H, A^H) \to H^n(G, A)$ is called *inflation* from G/H to G. □

Let G and H be groups. A commutative diagram

$$\begin{array}{ccccccccc} \{0\} & \to & A & \to & B & \to & C & \to & \{0\} \\ & & \downarrow & & \downarrow & & \downarrow & & \\ \{0\} & \to & A' & \to & B' & \to & C' & \to & \{0\}, \end{array}$$

where the first row is an exact sequence of G-modules and the second row is an exact sequence of H-modules, induces an exact and commutative diagram

$$\begin{array}{ccccccccc} \{0\} & \to & K(G, A) & \to & K(G, B) & \to & K(G, C) & \to & \{0\} \\ & & \downarrow & & \downarrow & & \downarrow & & \\ \{0\} & \to & K(H, A') & \to & K(H, B') & \to & K(H, C') & \to & \{0\} \end{array}$$

of complexes. Hence one has an exact and commutative diagram

$$\begin{array}{ccccccccccc} \{0\} \to H^0(G, A) & \to \cdots \to & H^n(G, C) & \xrightarrow[\Delta_n]{} & H^{n+1}(G, A) & \to & H^{n+1}(G, B) & \to & H^{n+1}(G, C) & \to \cdots \\ \downarrow & & \downarrow & & \downarrow & & \downarrow & & \downarrow & \\ \{0\} \to H^0(H, A') & \to \cdots \to & H^n(H, C') & \xrightarrow[\Delta_n]{} & H^{n+1}(H, A') & \to & H^{n+1}(H, B') & \to & H^{n+1}(H, C') & \to \cdots \end{array} \tag{2.31}$$

where Δ_n is the n-th connecting homomorphism for the cohomology groups of the complexes. The rows of (2.31) are called long exact sequences.

3.3. Dimension Shifting. A G-module A can be embedded in $M_G(A)$ by means of the mapping which associates to $a \in A$ the function $f_a(x) = xa$ for $x \in G$. Let

$$\{0\} \to A \to M_G(A) \to C \to \{0\}$$

be the corresponding exact sequence. By Proposition 2.38 the associated connecting homomorphism

$$H^n(G, C) \to H^{n+1}(G, A)$$

is an isomorphism for $n \geqslant 1$ and a surjection for $n = 0$.

By means of this *dimension shifting* it is possible to proof theorems about cohomology groups by reducing them to small dimensions. This method is based on the following principle:

Proposition 2.39. *Let G and H be groups and F an exact covariant functor from the category of H-modules into the category of G-modules which transforms induced modules in induced modules. Furthermore let $m \geqslant 0$ be an integer and λ_m a functorial morphism defined for every H-module A such that $\lambda_m(A)$ is a homomorphism of $H^m(G, FA)$ into $H^m(H, A)$*

Then there is one and only one family $\{\lambda_n | n = m, m + 1, \ldots\}$ of functorial morphisms such that for all exact sequences of H-modules

$$\{0\} \to A \to B \to C \to \{0\}$$

and all $n \geqslant m$ the diagram

$$\begin{array}{ccc} H^n(G, FC) & \xrightarrow{\lambda_n(C)} & H^n(H, C) \\ \downarrow \Delta_n & & \downarrow \Delta_n \\ H^{n+1}(G, FA) & \xrightarrow{\lambda_{n+1}(A)} & H^{n+1}(H, A) \end{array}$$

is commutative. If λ_m is an isomorphism, then also λ_n is an isomorphism for $n \geqslant m$. □

In the following sections we give some applications of Proposition 2.39.

3.4. Shapiro's Lemma. Let H be a subgroup of G and A an H-module. The G-module $M_G^H(A)$ is the submodule of $M_G(A)$ consisting of the $f \in M_G(A)$ such that

$$hf(x) = f(hx) \qquad \text{for } h \in H, x \in G.$$

Take $F := M_G^H$ and let $\lambda_n(A) := \psi_{n*}(A)$ be the homomorphism which is induced from the morphism $[\varphi, \psi]: [G, M_G^H(A)] \to [H, A]$ where φ is the injection $H \to G$ and $\psi(f) = f(1)$ for $f \in M_G^H(A)$.

It is easy to see that the assumptions of Proposition 2.39 are satisfied. Since $\psi_{0*}(A): (M_G^H(A))^G = A^H \to A^H$ is the identity, we have proved

Proposition 2.40 (Shapiro's lemma). *$\psi_{n*}(A)$ is an isomorphism.* □

3.5. Corestriction. Let H be a subgroup of G of finite index. In Example 14 we constructed a homomorphism, called restriction, from $H^n(G, A)$ into $H^n(H, A)$. Now we want to construct a homomorphism from $H^n(H, A)$ into $H^n(G, A)$, called *corestriction.*

Take $F(A) = A$ and let $\lambda_0(A) = \mathrm{Cor}$ be given by

$$\mathrm{Cor}(a) = \sum_{r \in G/H} ra \qquad \text{for } a \in A^H.$$

λ_0 is a functorial morphism and can be extended according to Proposition 2.39 to higher dimensions (observe that the roles of G and H are exchanged). The resulting morphism is the corestriction Cor.

Proposition 2.41. $\mathrm{Cor} \cdot \mathrm{Res} = [G:H]$.

Proof. This is trivial in dimension 0 and follows in higher dimension from Proposition 2.39. □

Proposition 2.42. *Let G be a finite group and let A be a G-module. Then $H^n(G, A)$ is annihilated by $|G|$ for $n \geqslant 1$.*

Proof. Take $H = \{1\}$ in Proposition 2.41. □

Let G be a finite group and H a subgroup of G. We consider $\mathbb{Q}/\mathbb{Z}$ as G-module with trivial action of G. Then

$$\mathrm{Cor}: H^1(H, \mathbb{Q}/\mathbb{Z}) \to H^1(G, \mathbb{Q}/\mathbb{Z})$$

induces a homomorphism $G/[G, G] \to H/[H, H]$, which is the transfer map defined in § 1.6.

3.6. The Transgression and the Hochschild-Serre-Sequence. Let H be a normal subgroup of G and A a G-module. For every $g \in G$ we define an automorphism $\tilde{g}$ of $H^n(H, A)$: $\tilde{g}$ is induced by the morphism $[\varphi, \psi] : \psi(a) = ga$, $\varphi(h) = g^{-1}hg$ for $a \in A$, $h \in H$.

By means of this action of $\tilde{g}$ the group $H^n(H, A)$ becomes a G-module. Dimension shifting shows that $\tilde{g}$ is the identity for $g \in H$. Therefore $H^n(H, A)$ can be considered as a G/H-module. The image of the restriction map $H^n(G, A) \to H^n(H, A)$ lies in $H^n(H, A)^{G/H}$. The *transgression* is a functorial homomorphism from $H^n(H, A)^{G/H}$ into $H^{n+1}(G/H, A^H)$, which is defined only if $H^i(H, A) = \{0\}$ for $i = 1, \ldots, n-1$.

For $n = 1$ the transgression is defined as follows: let $\bar{a} \in H^1(H, A)^{G/H}$, $da = 0$. We extend a to an element $b \in K^1(G, A)$ with the properties $gb(g^{-1}hg) - b(h) = hb(g) - b(g)$, $b(hg) = b(h) - hb(g)$ for $g \in G$, $h \in H$. This is possible since a is invariant by G. Now we put

$$f(g_1, g_2) := db(g_1, g_2) \qquad \text{for } g_1, g_2 \in G.$$

Then $f(g_1, g_2)$ depends only on the classes of g_1, g_2 mod H and has values in A^H. Hence it defines a class $\operatorname{Tra} \bar{a}$ in $H^2(G/H, A^H)$, the transgression.

One checks directly that the following sequence is exact:

$$\{0\} \to H^1(G/H, A^H) \xrightarrow[\text{Inf}]{} H^1(G, A) \xrightarrow[\text{Res}]{} H^1(H, A)^{G/H} \xrightarrow[\text{Tra}]{} H^2(G/H, A^H) \xrightarrow[\text{Inf}]{} H^2(G, A).$$

In the general case one defines the transgression and proves the following proposition by dimension shifting.

Proposition 2.43. *Let H be a normal subgroup of G and suppose that $H^i(H, A) = \{0\}$ for $i = 1, \ldots, n-1$. Then the following Hochschild-Serre-sequence is exact:*

$$\{0\} \to H^n(G/H, A^H) \xrightarrow[\text{Inf}]{} H^n(G, A) \xrightarrow[\text{Res}]{} H^n(H, A)^{G/H} \xrightarrow[\text{Tra}]{} H^{n+1}(G/H, A^H) \xrightarrow[\text{Inf}]{} H^{n+1}(G, A). \boxtimes$$

One can establish the transgression and Proposition 2.43 also by means of the Hochschild-Serre spectral sequence $H^p(G/H, H^q(H, A)) \Rightarrow H^n(G, A)$ (Cartan-Eilenberg (1956), Chap. 16).

As application we prove a proposition which will be useful in the following:

Proposition 2.44. *Let G be a finite group, A a G-module and $n, m \in \mathbb{Z}$ with $n > 0$, $m \geqslant 0$. We assume that the following two conditions are satisfied. 1) $H^i(H, A) = \{0\}$ for all $0 < i < n$ and all subgroups H of G. 2) If $H \subseteq H' \subseteq G$ where H is a normal subgroup of H' and H'/H is cyclic of prime order, then the order of $H^n(H, A)$ divides $[H' : H]^m$.*

Then $H^n(G, A)$ is of order dividing $|G|^m$.

Proof. Let p be a prime and G_p a p-Sylow group of G. Proposition 2.41 shows that the restriction $H^n(G, A) \to H^n(G_p, A)$ is injective on the p-component of $H^n(G, A)$. Hence it is sufficient to show the assertion for p-groups G. Let $|G| = p^s$. Then induction over s and application of Proposition 2.43 proves that $H^n(G, A)$ divides $|G|^m$. □

3.7. Cup Product. Let G be a group and let A, B, C be G-modules. Furthermore let $\circ$ be a bilinear map from $A \times B$ into C with

$$g(a \circ b) = ga \circ gb \qquad \text{for } g \in G,\ a \in A,\ b \in B.$$

$\circ$ is called a *pairing* from $A \times B$ into C.

Example 16. Let A be a ring with trivial action of G. Then the multiplication in A is a pairing from $A \times A$ into A. □

Example 17. Let A and B be arbitrary G-modules. The tensor product $A \otimes_{\mathbb{Z}} B$ is defined as G-module by

$$g(a \otimes b) = ga \otimes gb \qquad \text{for } g \in G,\ a \in A,\ b \in B. \tag{2.32}$$

Then $a, b \to a \otimes b$ defines a pairing from $A \times B$ into $A \otimes_{\mathbb{Z}} B$. This is the universal pairing by definition of the tensor product. □

A pairing from $A \times B$ into C induces a bilinear map

$$H^m(G, A) \times H^n(G, B) \circlearrowright H^{n+m}(G, C)$$

for every $n, m \geqslant 0$, called cup product, which is defined as follows:

For $f \in K^m(G, A)$, $f' \in K^n(G, B)$ we put

$$(f \circ f')(x_1, \ldots, x_{m+n}) = f(x_1, \ldots, x_n) \circ x_1 \ldots x_m f'(x_{m+1}, \ldots, x_{m+n}).$$

Then

$$d(f \circ f') = df \circ f' + (-1)^m f \circ df'.$$

It follows that

$$\bar{f} \cup \bar{f}' := \overline{f \circ f'}$$

defines a bilinear map of $H^m(G, A) \times H^n(G, B)$ into $H^{m+n}(G, C)$, the *cup product*.

Since every pairing from $A \times B$ into C induces a homomorphism of $A \otimes_{\mathbb{Z}} B$ into C, the cup product of $H^m(G, A) \times H^n(G, B)$ into $H^{m+n}(G, C)$ is the composition of the cup product of $H^m(G, A) \times H^n(G, B)$ into $H^{m+n}(G, A \otimes_{\mathbb{Z}} B)$ and the homomorphism $H^{m+n}(G, A \otimes_{\mathbb{Z}} B) \to H^{m+n}(G, C)$. Therefore it is sufficient to consider the pairing from $A \times B$ into $A \otimes_{\mathbb{Z}} B$.

The following properties of the cup product follow immediately from the definition.

Proposition 2.45. *Let A and B be G-modules. Then $\alpha \cup \beta = \alpha \otimes \beta$ for $\alpha \in H^0(G, A)$, $\beta \in H^0(G, B)$.* □

Proposition 2.46. *Let*

$$\{0\} \to A \to A' \to A'' \to \{0\}$$

be an exact sequence of G-modules and let B be a G-module such that the sequence

$$\{0\} \to A \otimes_{\mathbb{Z}} B \to A' \otimes_{\mathbb{Z}} B \to A'' \otimes_{\mathbb{Z}} B \to \{0\}$$

is exact. Then

$$(\Delta_m \alpha) \cup \beta = \Delta_{m+n}(\alpha \cup \beta) \qquad \text{for } \alpha \in H^m(G, A''),\ \beta \in H^n(G, B). \ \square$$

Proposition 2.47. *Let*

$$\{0\} \to B \to B' \to B'' \to \{0\}$$

be an exact sequence of G-modules and let A be a G-module such that the sequence

$$\{0\} \to A \otimes_{\mathbb{Z}} B \to A \otimes_{\mathbb{Z}} B' \to A \otimes_{\mathbb{Z}} B'' \to \{0\}$$

is exact. Then

$$\alpha \cup (\Delta_n \beta) = (-1)^m \Delta_{m+n}(\alpha \cup \beta) \qquad \text{for } \alpha \in H^m(G, A),\ \beta \in H^n(G, B'').$$

Propositions 2.46–2.47 show that the method of dimension shifting is applicable to the cup product: With

$$\{0\} \to A \to M_G(A) \to C \to \{0\}$$

also

$$\{0\} \to A \otimes_{\mathbb{Z}} B \to M_G(A) \otimes_{\mathbb{Z}} B \to C \otimes_{\mathbb{Z}} B \to \{0\}$$

is exact. This in particular can be applied to the proof of the following propositions.

Proposition 2.48. *Let G and H be groups and let* $[\varphi, \psi_A]$, $[\varphi, \psi_B]$ *be morphisms of the G-modules A, B into the H-modules A′, B′. Then*

$$(\psi_A \otimes \psi_B)^*(\alpha \cup \beta) = \psi_A^* \alpha \cup \psi_B^* \beta \qquad \text{for } \alpha \in H^m(G, A),\ \beta \in H^n(G, B). \ \square \tag{2.33}$$

(2.33) together with Propositions 2.45–2.47 determine the cup product uniquely up to isomorphy.

Proposition 2.49. *Let A, B, C be G-modules with pairings* $(A \circ B) \circ C$ *and* $A \circ (B \circ C)$ *such that*

$$(a \circ b) \circ c = a \circ (b \circ c) \qquad \text{for } a \in A,\ b \in B,\ c \in C.$$

Then

$$(\alpha \cup \beta) \cup \gamma = \alpha \cup (\beta \cup \gamma) \qquad \text{for } \alpha \in H^l(G, A),\ \beta \in H^m(G, B),\ \alpha \in H^n(G, C). \ \square$$

Proposition 2.50. *Let A, B be G-modules and let* $A \circ B$ *and* $B \circ A$ *be pairings such that*

$$a \circ b = b \circ a \qquad \text{for } a \in A,\ b \in B.$$

Then

$$\alpha \cup \beta = (-1)^{mn}(\beta \cup \alpha) \qquad \textit{for } \alpha \in H^m(G, A),\, \beta \in H^n(G, B). \ \square$$

Proposition 2.51. *Let A, B be G-modules, let H be a subgroup of finite index in G and $A \circ B$ a pairing.* Res *denotes the restriction from G to H and* Cor *denotes the corestriction from H to G. Then*

$$\operatorname{Res}(\alpha \cup \beta) = \operatorname{Res} \alpha \cup \operatorname{Res} \beta \qquad \textit{for } \alpha \in H^m(G, A),\, \beta \in H^n(G, B), \tag{2.34}$$

$$\operatorname{Cor}(\operatorname{Res} \alpha \cup \beta) = \alpha \cup \operatorname{Cor} \beta \qquad \textit{for } \alpha \in H^m(G, A),\, \beta \in H^n(H, B). \ \square \tag{2.35}$$

3.8. Modified Cohomology for Finite Groups. Now let G be a finite group and A a G-module. Then $M_G(A)$ is isomorphic to $\mathbb{Z}[G] \otimes_{\mathbb{Z}} A$ (see (2.32) for the action of G). The isomorphism is given by $f \to \sum_{g \in G} g \otimes gf(g^{-1})$. There is an injection of the G-module $\mathbb{Z}$ with trivial action of G in $\mathbb{Z}[G]$ defined by $h \to \sum_{g \in G} hg$ for $h \in \mathbb{Z}$. Let $J_G := \mathbb{Z}[G]/\mathbb{Z}$. Then the exact sequence

$$\{0\} \to A = \mathbb{Z} \otimes_{\mathbb{Z}} A \to \mathbb{Z}[G] \otimes_{\mathbb{Z}} A \to J_G \otimes_{\mathbb{Z}} A \to \{0\}$$

implies the isomorphism

$$H^n(G, J_G \otimes A) \cong H^{n+1}(G, A) \qquad \text{for } n \geqslant 1.$$

On the other hand we have a surjection $\mathbb{Z}[G] \to \mathbb{Z}$ given by $\sum_{g \in G} n_g g \to \sum_{g \in G} n_g$. Let I_G be its kernel. Then we have the isomorphism

$$H^n(G, A) = H^{n+1}(G, I_G \otimes A) \qquad \text{for } n \geqslant 1.$$

For $n = 0$ we get a surjection

$$H^0(G, A) \to H^1(G, I_G \otimes A)$$

with kernel

$$tr_G A := \left\{ \sum_{g \in G} ga \,\middle|\, a \in A \right\}.$$

The group $\hat{H}^0(G, A) := H^1(G, I_G \otimes A) \cong A^G / tr_G A$ is called the *modified zero dimensional cohomology group.*

We write J_G^n, I_G^n for the n-fold tensor product of J_G, I_G respectively. Then we have functorial isomorphisms

$$H^m(G, J_G^n \otimes I_G^1 \otimes A) \cong H^{m-n+1}(G, A) \qquad \text{for } m - n + 1 \geqslant 1.$$

We define the *modified cohomology groups* for higher dimensions by

$$\hat{H}^m(G, A) := \hat{H}^0(G, I_G^{-m} \otimes A) \qquad \text{for } m < 0,$$

$$\hat{H}^m(G, A) := \hat{H}^0(G, J_G^{m} \otimes A) \qquad \text{for } m > 0.$$

With this definition it is clear from the principle of dimension shifting that we can formulate the definitions and propositions of the sections § 3.2–§ 3.7 almost without changes for the groups $\hat{H}^m(G, A)$, $m \in \mathbb{Z}$. We leave this to the reader.

Example 18. The elements of $(I_G \otimes A)^G$ have the form $\sum_{g \in G} g \otimes ga$ for $a \in A$ with $\sum_{g \in G} ga = 0$. $a \to \sum_{g \in G} g \otimes ga$ induces an isomorphism

$$\left\{\alpha \in A \,\middle|\, \sum_{g \in G} ga = 0\right\} \Big/ I_G A \cong \hat{H}^{-1}(G, A)$$

where $I_G A := (ua | u \in I_G, a \in A) = (ga - a | g \in G, a \in A)$. □

Example 19. $\hat{H}^{-2}(G, \mathbb{Z}) \cong \hat{H}^{-1}(G, I_G) \cong I_G/I_G I_G \cong G/[G, G]$. The isomorphism $G/[G, G] \to I_G/I_G I_G$ is given by $g[G, G] \to g - 1 + I_G I_G$. □

Proposition 2.52. *The cup product*

$$\hat{H}^n(G, \mathbb{Z}) \times \hat{H}^{-n}(G, \mathbb{Z}) \to \hat{H}^0(G, \mathbb{Z}) \cong \mathbb{Z}/|G|\mathbb{Z}$$

defines a perfect pairing, i.e. the induced homomorphism

$$\hat{H}^{-n}(G, \mathbb{Z}) \cong \operatorname{Hom}(\hat{H}^n(G, \mathbb{Z}), \mathbb{Z}/|G|\mathbb{Z}) \tag{2.36}$$

is an isomorphism (*compare* Proposition 2.42). □

Let $\mathbb{Q}$ be the G-module with trivial action of G. Then $\hat{H}^n(G, \mathbb{Q}) = \{0\}$ (Proposition 2.42). Therefore the exact sequence

$$\{0\} \to \mathbb{Z} \to \mathbb{Q} \to \mathbb{Q}/\mathbb{Z} \to \{0\}$$

induces isomorphisms $\hat{H}^n(G, \mathbb{Q}/\mathbb{Z}) \cong \hat{H}^{n+1}(G, \mathbb{Z})$, $n \in \mathbb{Z}$. In particular $\hat{H}^2(G, \mathbb{Z}) \cong \hat{H}^1(G, \mathbb{Q}/\mathbb{Z}) \cong \operatorname{Hom}(G, \mathbb{Q}/\mathbb{Z})$. Hence Proposition 2.52 implies the isomorphism $\hat{H}^{-2}(G, \mathbb{Z}) \cong G/[G, G]$ of Example 19. (2.35) shows that the transfer map (§ 1.6, § 3.5) corresponds to the restriction $\hat{H}^{-2}(G, \mathbb{Z}) \to \hat{H}^{-2}(H, \mathbb{Z})$ (see Serre (1962), Chap. 7.8, for a direct proof).

3.9. Cohomology for Cyclic Groups. Let G be a finite cyclic group of order m with generator s. The isomorphism of I_G onto J_G given by

$$s^{i+1} - s^i \to s^i + \mathbb{Z} \qquad \text{for } i = 1, \ldots, m-1$$

induces isomorphisms

$$\delta_n\colon \hat{H}^n(G, A) \to \hat{H}^{n+1}(G, I_G \otimes_{\mathbb{Z}} A) \to \hat{H}^{n+1}(G, J_G \otimes_{\mathbb{Z}} A) \to \hat{H}^{n+2}(G, A) \quad \text{for } n \in \mathbb{Z}.$$

δ_n can also be described as follows: Let $\chi \in H^1(G, \mathbb{Q}/\mathbb{Z})$ be the character with $\chi(s) = \dfrac{1}{m} + \mathbb{Z}$ and let $\Delta_1\colon \hat{H}^1(G, \mathbb{Q}/\mathbb{Z}) \to \hat{H}^2(G, \mathbb{Z})$. Then

$$\delta_n(\alpha) = \alpha \cup \Delta_1(\chi) \qquad \text{for } \alpha \in \hat{H}^n(G, A). \tag{2.37}$$

A G-module A such that $\hat{H}^0(G, A)$ and $\hat{H}^{-1}(G, A)$ (and therefore $\hat{H}^n(G, A)$ for $n \in \mathbb{Z}$) are finite is called *Herbrand module*. The quotient

$$h(A) := |\hat{H}^0(G, A)|/|\hat{H}^{-1}(G, A)|$$

is called *Herbrand quotient*.

Let $\{0\} \to A \to B \to C \to \{0\}$ be an exact sequence of G-modules, two of which are Herbrand modules. The long exact sequence (2.31) implies that also the third module is an Herbrand module and that

$$h(B) = h(A)h(C). \tag{2.38}$$

Proposition 2.53. *Let A be a finite G-module. Then $h(A) = 1$.* □

For a G-module A we denote by $h_0(A)$ the Herbrand quotient of the abelian group A with trivial action of G if it exists. If A is finitely generated, then $h_0(A) = |G|^{rk A}$.

Proposition 2.54. *Let p be a prime, G a group of order p, and A a G-module such that $h_0(A)$ exists. Then also $h_0(A^G)$ and $h(A)$ exist and $h(A)$ is given by*

$$h(A)^{p-1} = h_0(A^G)^p / h_0(A). \boxtimes \tag{2.39}$$

3.10. The Theorem of Tate

Theorem 2.55 (Theorem of Tate). *Let G be a finite group and let A be a G-module such that $H^1(H, A) = \{0\}$ and $H^2(H, A)$ is cyclic of order $|H|$ for every subgroup H of G.*

Let ζ be a generator of $H^2(G, A)$. Then the homomorphism $\varphi_n\colon \hat{H}^n(G, \mathbb{Z}) \to \hat{H}^{n+2}(G, A)$ given by $\varphi_n(\alpha) = \alpha \cup \zeta$ for $\alpha \in \hat{H}^n(G, \mathbb{Z})$ is an isomorphism for all $n \in \mathbb{Z}$. ⊠

§4. Proof of the Main Theorems of Class Field Theory

(Main reference: Cassels, Fröhlich (1967), Chap. 6, 7; Chevalley (1954))

In this paragraph we show how one can prove the main theorems of class field theory using cohomology of groups.

4.1. Application of the Theorem of Tate to Class Field Theory. In § 1.6 we have defined the notion of a class module A_K of a local or global field K. From the point of view of cohomology the following theorem contains the fundamental facts about class modules.

Theorem 2.56. *Let L/K be a finite normal extension of local or global fields with Galois group G. Then $H^1(G, A_L) = \{0\}$ and $H^2(G, A_L)$ is a cyclic group of order $|G|$ with a canonical generator $\zeta_{L/K}$ called the canonical class.*

For local fields, $H^1(G, A_L) = \{0\}$ is E. Noether's generalization of Hilbert's Satz 90 (Proposition 2.6). All other statements of Theorem 2.56 are deep results, being an essential part of class field theory. The proof of Theorem 2.56 will occupy us in the following sections.

Combining Theorem 2.55 with Theorem 2.56 we get

Theorem 2.57. *Let L/K be a finite normal extension of local or global fields with Galois group G. Then there is a canonical isomorphism between the groups $\hat{H}^n(G, \mathbb{Z})$ and $\hat{H}^{n+2}(G, A_L)$ for any $n \in \mathbb{Z}$.*

In particular for $n = -2$ we have

$$G/[G,G] \cong \hat{H}^{-2}(G,\mathbb{Z}) \cong \hat{H}^{0}(G,A_L) \cong A_K/N_{L/K}A_L. \square \tag{2.40}$$

This proves a part of the main theorem of class field theory in the local and global case. It turns out, that the homomorphism from A_K onto $G/[G,G]$ induced by (2.40) is the norm symbol (§ 1.3, § 1.5).

The canonical class induces a monomorphism inv: $H^2(G, A_L) \to \mathbb{Q}/\mathbb{Z}$, called the invariant, which is defined by $\operatorname{inv}\zeta_{L/K} = \dfrac{1}{[L:K]} + \mathbb{Z}$.

One has the following characterization of the norm symbol: If $\chi \in H^1(G, \mathbb{Q}/\mathbb{Z})$ and Δ_1 is the connecting homomorphism $H^1(G, \mathbb{Q}/\mathbb{Z}) \to H^2(G, \mathbb{Z})$, then

$$\chi(a, L/K) = \operatorname{inv}(a \cup \Delta_1 \chi) \qquad \text{for } a \in A_K \tag{2.41}$$

(Serre (1962), Chap. 11.3).

The behavior of the cohomological operations deliver us the functorial properties of the norm symbol (§ 1.6).

Since we consider in this paragraph arbitrary normal extensions L/K we get something more than the results stated in § 1. For example one has the following theorem about the norm:

Theorem 2.58. *Let L/K be a finite extension of local or global fields and let L'/K be the maximal abelian subextension of L/K. Then $N_{L/K}A_L = N_{L'/K}A_{L'}$.*

Proof. Let E be a normal extension of K containing L. We put $G := G(E/K)$, $H := G(E/L)$, then $G(E/L') = [G,G]H$. Let $a \in N_{L'/K}(A_{L'})$ and therefore $(a, L'/K) = 1$. Hence $(a, E/K)$ is in the image of $H^{ab} \to G^{ab}$. Since $A_L \to H^{ab}$ is surjective, (2.4) shows that there exists an $\alpha \in A_L$ such that $(N_{L/K}\alpha, F/K) = (a, F/K)$. It follows now from (2.40) that there exists an $\alpha' \in A_E$ such that $N_{E/K}\alpha' = N_{L/K}\alpha/a$, hence $a = N_{L/K}(\alpha/N_{E/L}\alpha')$. $\square$

4.2. Class Formations. We will collect a part of the needed information about class modules in the form of axioms and deduce from them the main theorems of class field theory. It is customary to do this in a rather abstract manner. But for our purposes we restrict ourselves to field formations:

Let F be an arbitrary field and $\bar{F}$ a separable algebraic closure of F. A *field formation* over F is a covariant functor A from the category $\mathfrak{K}_F$ of finite extensions K/F with $K \subseteq \bar{F}$ into the category of abelian topological groups with the following property:

Let $K, L \in \mathfrak{K}_F$ with $K \subseteq L$. Then the homomorphism $\iota\colon A_K \to A_L$ corresponding to the injection $K \to L$ is an immersion, i.e. ι is an injection and $\operatorname{Im}\iota$ is closed in A_L. In the following we consider A_K as a subgroup of A_L. If L/K is a normal extension with Galois group G, then $A_L^G = A_K$.

It is clear that the class modules for local (resp. global) fields form a field formation over $\mathbb{Q}_p$ (resp. $\mathbb{Q}$) and these are the only field formations which we need in this article.

Let L/K be a normal extension with $L, K \in \mathfrak{K}_F$ and let K' be an intermediate field of L/K. We introduce the notation $H^n(L/K) := H^n(G(L/K), A_L)$, $\mathrm{Res}_{K \to K'}$ for the restriction from $G(L/K)$ to $G(L/K')$, and similar notation for the inflation and corestriction map.

A field formation A over F is called a *class formation* if the following axioms are fulfilled:

1. *Let L/K be a normal extension of prime degree of fields in $\mathfrak{K}_F$. Then $H^1(L/K) = \{0\}$ and the order of $H^2(L/K)$ divides $[L:K]$.*
2. *For every natural number m and any field $K \in \mathfrak{K}_F$ there exists a finite abelian extension $E \in \mathfrak{K}_F$ with $m \mid [E:K]$ and $H^2(E/K)$ is a cyclic group of order $[E:K]$ with a canonical generator $\zeta_{E/K}$. We call E/K a special extension.*
3. *If E/K is a special extension and $K' \in \mathfrak{K}_F$ is an arbitrary extension of K, then EK'/K' is a special extension and the homomorphism $H^2(E/K) \to H^2(EK'/K')$ induced from the projection $G(EK'/K') \to G(E/K)$ maps $\zeta_{E/K}$ onto $[EK':E]\zeta_{EK'/K'}$.*
4. *If E/K and E'/K are special extensions and v and v' are natural numbers with $[E:K]/[E':K] = v/v'$, then*

$$v \,\mathrm{Inf}_{E \to EE'} \zeta_{E/K} = v' \,\mathrm{Inf}_{E' \to EE'} \zeta_{E'/K}.$$

We shall see in the next sections that unramified extensions in the local case and cyclotomic extensions in the global case serve as special extensions.

In the rest of this section we show that our axioms allow us to give a full picture of the structure of $H^1(L/K)$ and $H^2(L/K)$ for arbitrary normal extensions L/K in $\mathfrak{K}_F$.

Proposition 2.59. *Let L/K be a normal extension in $\mathfrak{K}_F$. Then $H^1(L/K) = \{0\}$ and the order of $H^2(L/K)$ divides $[L:K]$.*

Proof. This follows from axiom 1 and Proposition 2.44. □

Proposition 2.60. *Let L/K and L'/K be normal extensions in $\mathfrak{K}_F$ with $L' \subseteq L$. Then $\mathrm{Inf}_{L' \to L}\colon H^2(L'/K) \to H^2(L/K)$ is injective.*

Proof. This follows from Proposition 2.59 and Proposition 2.43. □

We are going now to construct a canonical generator of $H^2(L/K)$ for arbitrary normal extensions L/K in $\mathfrak{K}_F$:

Let E/K be a special extension with $[L:K] \mid [E:K]$. We put $v = [E:K]/[L:K]$. Then $[EL:E]\zeta_{EL/L} = 0$ since the order of $\zeta_{EL/L}$ is $[EL:L]$ and $v[EL:E] = [EL:L]$. Therefore axiom 3 implies

$$\mathrm{Res}_{K \to L}(v \,\mathrm{Inf}_{E \to EL} \zeta_{E/K}) = v[EL:E]\zeta_{EL/L} = 0.$$

Hence the exact sequence

$$\{0\} \longrightarrow H^2(L/K) \xrightarrow[\mathrm{Inf}]{} H^2(EL/K) \xrightarrow[\mathrm{Res}]{} H^2(EL/L) \qquad \text{(Proposition 2.43)}$$

shows that there is a unique element $\zeta_{L/K}$ in $H^2(L/K)$ such that

$$\mathrm{Inf}_{L \to EL} \zeta_{L/K} = \mathrm{Inf}_{E \to EL} \zeta_{E/K}.$$

Since inflation is injective, $\zeta_{L/K}$ has order $[L:K]$. It follows easily from axiom 4 that $\zeta_{L/K}$ does not depend on the choice of the special extension E.

Proposition 2.61. *Let L/K be a normal extension in $\mathfrak{K}_F$ and let K' be an intermediate field of L/K. Then $H^2(L/K)$ is cyclic of order $[L:K]$ with the canonical generator $\zeta_{L/K}$. Moreover*

$$\operatorname{Res}_{K\to K'} \zeta_{L/K} = \zeta_{L/K'}, \tag{2.42}$$

$$\operatorname{Cor}_{K'\to K} \zeta_{L/K'} = [K':K]\zeta_{L/K} \tag{2.43}$$

and

$$\operatorname{Inf}_{K'\to L} \zeta_{K'/K} = [L:K']\zeta_{L/K} \qquad \text{if } K'/K \text{ is normal.} \tag{2.44}$$

Proof. The first assertion follows from the construction of $\zeta_{L/K}$ and Proposition 2.59, (2.42) is proved by means of axiom 3, (2.43) follows from (2.42) and Proposition 2.41, and (2.44) follows from the construction of $\zeta_{L/K}$. □

The use of special extensions in the construction of the canonical generator of $H^2(L/K)$ is called in German *Durchkreuzung*. This method was discovered by Chebotarev, who applied it for the proof of his density theorem (Chap. 1.6.7). Later on it was applied by Artin in the proof of his reciprocity law (§ 1.6). The method can be used to prove that the norm symbol is characterized by the properties (i)–(iii) (§ 1.2). This is the starting point of Neukirch (1986) in his approach to class field theory.

Let L/K be a normal extension in $\mathfrak{K}_F$ and g an isomorphism of L onto gL. Then g induces the morphism $[\varphi, \psi]: [G(L/K), A_L] \to [G(gL/gK), A_{gL}]$ given by $\psi(a) = ga$, $\varphi(h) = g^{-1}hg$ for $a \in A_L$, $h \in G(gL/gK)$. The resulting homomorphism from $H^2(L/K)$ to $H^2(gL/gK)$ will be denoted by g^*.

Proposition 2.62. $g^*\zeta_{L/K} = \zeta_{gL/gK}$. □

If we knew that the class modules for local and global fields form a class formation, for the proof of the main theorems of class field theory (Theorem 2.3, 2.8) it remained to prove that ϕ is surjective. In the scope of class formation this can be formulated as follows:

Axiom 5. *For every field $K \in \mathfrak{K}_F$ and every closed subgroup U of finite index in A_K there exists an abelian extension $L \in \mathfrak{K}_F$ of K such that $N_{L/K}A_L = U$.*

We have also to show that the norm symbol constructed in § 1 by means of Lubin-Tate extensions and a local-global principle coincides with the map given by (2.40). The functorial properties of the norm symbol (§ 1.7) then follow from (2.42)–(2.44) and Proposition 2.62.

4.3. Cohomology of Local Fields. Let K be a finite extension of the field $\mathbb{Q}_p$ of rational p-adic numbers. We want to show that $A_K = K^\times$ defines a class formation over $\mathbb{Q}_p$.

Proposition 2.63. *Let L/K be a cyclic extension of local fields of degree n. Then the order of $H^2(L/K)$ is equal to n.*

Proof. By means of the exponential function (Chap. 1.4.9) one constructs a submodule V of finite index in the group of units U_L with $H^m(G(L/K), V) = \{0\}$ for $m \in \mathbb{Z}$. It follows that the Herbrand quotient $h(U_L)$ is equal to 1(§3.9) and $h(L^\times) = n$, hence $|H^2(L/K)| = n$ since $|H^1(L/K)| = 1$ (Proposition 2.6). □

As special extensions of K we take the unramified extensions. We know already that unramified extensions are cyclic and for every $n \in N$ there exists one and only one unramified extension of K in $\overline{\mathbb{Q}}_p$ of degree n (Chap. 1.4.6).

Proposition 2.64. *Let E/K be a unramified extension of degree n. Then $\hat{H}^m(G(E/K), U_E) = \{0\}$ for $m \in \mathbb{Z}$.*

Proof. Since $n = |H^2(E/K)| = |\hat{H}^0(E/K)| = [K^\times : N_{E/K}E^\times]$ and $N_{E/K}E^\times = N_{E/K}U_{E^\times}(\pi^n)$ for a prime element π of K, we have $N_{E/K}U_E = U_K$ hence $H^0(G(E/K), U_E) = \{0\}$. Now $h(U_E) = 1$ implies $H^1(G(E/K), U_E) = \{0\}$. □

The exact sequence

$$\{1\} \to U_E \to E^\times \underset{v_E}{\to} \mathbb{Z} \to \{0\}$$

shows together with Proposition 2.64 that $v_E^*\colon H^2(E/K) \to H^2(G(E/K), \mathbb{Z})$ is an isomorphism. Furthermore $\Delta_1\colon H^1(G(E/K), \mathbb{Q}/\mathbb{Z}) \to H^2(G(E/K), \mathbb{Z})$ is an isomorphism. Let χ be the character of $G(E/K)$ such that $\chi(F) = \frac{1}{n} + \mathbb{Z}$ where F denotes the Frobenius automorphism of E/K.

Now we define the canonical class $\zeta_{E/K}$ by means of the equation

$$v_E^*\zeta_{E/K} = \Delta_1\chi.$$

It is not difficult to verify the axioms 2–4 and to show that the homomorphism of $K^\times$ onto $G(E/K)$ given by (2.40) maps π onto F, i.e. coincides with the norm symbol.

One can use Lubin-Tate theory to show that axiom 5 is fulfilled. (Serre (1967), 3.7–3.8).

4.4. Cohomology of Ideles and Idele Classes. The theorems of this section represent the hard kernel of class field theory. The proof of Theorem 2.72 is rather long and technical. It shows the importance of Kummer theory (Chap. 1.4.7) in class field theory. The reader who is not interested in details and believes that Axiom 1 is true for global formation modules may skip §4.4 after Proposition 2.67.

Let K be an algebraic number field and L a finite normal extension of K (L/K is a normal extension of $\mathfrak{K}_{\mathbb{Q}}$) with Galois group G. The places of K will be denoted by v and the places of L by w. G_w denotes the decomposition group of w with respect to L/K, i.e. G_w is the image of $G(L_w/K_v)$ in G if $w|v$. For every place v of K we choose one place w_v of L above v. The set of all places of K will be denoted by P_K.

G acts on $\prod_{w|v} L_w$ (§1.4). We define an isomorphism φ from $\prod_{w|v} L_w$ onto $M_G^{G_{w_v}}(L_{w_v})$ by

$$\varphi\left(\prod_{w|v}\alpha_w\right)(x) = x\alpha_{x^{-1}w_v} \qquad \text{for } x \in G \ (\S\,3.4).$$

Shapiro's lemma implies

Proposition 2.65. *$\hat{H}^m(G, \prod_{w|v} L_w^\times) \cong \hat{H}^m(G_{w_v}, L_{w_v}^\times)$, $\hat{H}^m(G, \prod_{w|v} U_w) \cong \hat{H}^m(G_{w_v}, U_{w_v})$ for all $m \in \mathbb{Z}$ where U_w is the group of units of L_w.* □

For the proof of the following proposition we need a principle which will be considered in much more general form in Chap. 3.1.9:

Let $A_1 \subseteq A_2 \subseteq \cdots$ be submodules of a G-module A such that $\bigcup_{i=1}^\infty A_i = A$ and suppose that $A_i \subseteq A_{i+1}$ induces injections $\hat{H}^m(G, A_i) \to \hat{H}^m(G, A_{i+1})$ so that we can consider $\hat{H}^m(G, A_i)$ as a subgroup of $\hat{H}^m(G, A_{i+1})$, $i = 1, 2, \ldots$. Then $\hat{H}^m(G, A) = \bigcup_{i=1}^\infty \hat{H}^m(G, A_i)$.

Proposition 2.66. *$\hat{H}^m(G, J(L)) \cong \sum_{v \in P_K} \hat{H}^m(G_{w_v}, L_{w_v}^\times)$ for all $m \in \mathbb{Z}$.*

Proof. Let S be a finite set of places of K and

$$J_S(L) = J_S = \prod_{v \in S}\prod_{w|v} L_w^\times \cdot \prod_{v \notin S}\prod_{w|v} U_w$$

Then

$$J(L) = \bigcup_S J_S$$

and Proposition 2.64 together with Proposition 2.65 imply

$$\hat{H}^m(G, J_S) \cong \sum_{v \in S} \hat{H}^m(G_{w_v}, L_{w_v}^\times) \tag{2.45}$$

if S contains all ramified places.

Now we consider (2.45) for a sequence S_i, $i = 1, 2, \ldots$, with $\bigcup_{i=1}^\infty S_i = P_K$. The above principle proves the assertion. □

According to § 4.3, § 4.4 for every place v of K we have a canonical isomorphism $H^2(G_w, L_w^\times) \to \frac{1}{n_v}\mathbb{Z}/\mathbb{Z}$ defined by $\zeta_{L_w/K_v} \to \frac{1}{n_v} + \mathbb{Z}$ where $n_v := [L_w : K_v]$. Since $H^1(G_w, L_w^\times) = \{0\}$, we have

Proposition 2.67. $H^1(G, J_L) = \{0\}$, $H^2(G, J_L) = \sum_v \frac{1}{n_v}\mathbb{Z}/\mathbb{Z}$. □

We come now to the main problem of this section, the proof of axiom 1 for the formation module $\mathfrak{C}(K) = J(K)/K^\times$. We need some preparations.

Theorem 2.68. *Let L/K be a cyclic extension of prime degree p. Then $h(\mathfrak{C}(L)) = p$.*

Proof. Let S be a finite set of places of K including the infinite places, the places ramified in L and a set of places such that the places of L above them generate the ideal class group of L. We denote by T the set of places of L lying above places in S. Then

$$J(L) = L^\times J_S, \qquad \mathfrak{C}(L) = L^\times J_S/L^\times \cong J_S/(J_S \cap L^\times).$$

Since $h(J_S) = \prod_{v \in S} n_v$ ((2.45), Proposition 2.63), it remains to show

$$ph(L_T) = \prod_{v \in S} n_v$$

where $L_T = J_S \cap L^\times$ is the group of T-units in L (Chap. 1.3.8).

L_T is a finitely generated group of rank $|T| - 1$ and $L_T^G = K_S$ is the group of S-units in K which has rank $|S| - 1$ (Theorem 1.58). Hence Proposition 2.53 implies

$$h(L_T)^{p-1} = p^{(|S|-1)p}/p^{|T|-1} = \prod_{v \in S} n_v^{p-1}/p^{p-1}. \ \square$$

As a corollary of Theorem 2.68 we get immediately

Theorem 2.69 (First inequality). *Let L/K be a cyclic extension of prime degree p. Then*

$$|\hat{H}^0(G, \mathfrak{C}(L))| = [J(K) : K^\times N_{L/K} J(L)] \geqslant p. \ \square$$

The next theorem is a consequence of Theorem 1.110. We mention it here because we wish to give a purely algebraic proof of the main theorem.

Theorem 2.70. *Let L/K be a cyclic extension of prime degree. There are infinitely many places of K which do not split in L.*

Proof. Let S be the set of all places of K which do not split in L. We assume that S is finite. Let πa_v be an arbitrary idele in $J(K)$. There exists an $x \in K$ such that $x^{-1} a_v \in K_v^p$ for $v \in S$ (Proposition 1.68). Then $x^{-1}\pi a_v \in N_{L/K} J(L)$ hence $K^\times N_{L/K} J(L) = J(K)$ in contradiction to Theorem 2.69. $\square$

Theorem 2.71. *Let p be a prime with $\mu_p \subset K$ and S a set of places of K containing all infinite places, all prime divisors of p, and a set of places which generate the ideal class group of K. Furthermore let S' be a set of places of K with $S' \cap S \neq \varnothing$ such that the natural map*

$$K_S \to \prod_{v \in S'} U_v/U_v^p$$

is surjective.

Then any $x \in K$ with $x \in J(K)^p \prod_{v \notin S} U_v \prod_{v \in S'} K_v^\times$ is a p-th power in $K^\times$.

Proof. Let $K' = K(\sqrt[p]{x})$. We have to prove $K' = K$.

$$D := \prod_{v \in S} K_v^\times \prod_{v \in S'} U_v^p \prod_{v \notin S \cup S'} U_v \subseteq N_{K'/K}(J(K'))$$

by local class field theory (§ 4.3). Since

$$K_S \to \prod_{v \in S'} U_v/U_v^p \cong J_S(K)/D$$

is surjective, $J(K) = K^\times J_S(K) = K^\times D$ hence $J(K) = K^\times N_{K'/K}(J(K'))$. This implies $K' = K$ by Theorem 2.69. $\square$

We look now at the special case of Theorem 2.71 that S' is empty. We put

$$H := J(K)^p \prod_{v \notin S} U_v, \qquad N := |S|.$$

Then $K_S \cap H = K_S^p$ hence

$$[HK_S : H] = [K_S : K_S \cap H] = p^N.$$

Furthermore

$$[J_S(K) : H] = \prod_{v \in S} p^2/p^{v(p)} = p^{2N}$$

by Proposition 1.73, (1.26) (v is here to be understood as the normalized valuation). Therefore

$$[J(K) : HK^\times] = [J_S(K)K^\times : HK^\times] = [J_S(K) : HK_S] = p^N. \tag{2.46}$$

After this preparations we are able to prove axiom 1 for the field formation $\mathfrak{K}_{\mathbb{Q}}$.

Theorem 2.72. *Let L/K be a cyclic extension of prime degree p. Then $H^1(L/K) = \{0\}$ and $|H^2(L/K)| = p$.*

Proof. By means of Theorem 2.68 it is sufficient to prove

$$H^0(L/K) = [J(K) : K^\times N_{L/K}(J(L))] \leqslant p. \tag{2.47}$$

(2.47) is called the second inequality.

With some cohomological machinery one reduces (2.47) to the case that K contains the p-th roots of unity. Then let $L = K(\sqrt[p]{a})$ and let S be a finite set of places of K satisfying the conditions of Theorem 2.71 and the extra condition that S contains all places v for which a is not a unit in K_v. The extension $E_S := K(\sqrt[p]{x} \mid x \in K_S)$ of K has degree p^N (Chap. 1.4.7). Let $g_1, \ldots, g_N$ be a generator system of $G(E/K)$ such that L is the fixed field of g_N. We denote the fixed field of g_i by E_i. By Theorem 2.70 there exists a place w_i of L_i which does not split in E; let v_i be the place of K with $w_i | v_i$, $i = 1, \ldots, N$. Then

$$\varphi : K_S \to \prod_{i=1}^{N} U_{v_i}/U_{v_i}^p$$

is surjective. In fact, E_i is the decomposition field of v_i in E_S/K (Chap. 1.3.7). If $x \in \operatorname{Ker} \varphi$ then v_i splits in $K(\sqrt[p]{x})/K$ hence $K(\sqrt[p]{x}) \subseteq E_i$, $i = 1, \ldots, N$, which implies $K(\sqrt[p]{x}) = K$. Therefore

$$p^N = [K_S : K_S^p] \,\Big|\, \prod_{i=1}^{N} [U_{v_i} : U_{v_i}^p] = p^N.$$

This proves the assertion.

We apply Theorem 2.71 to the set $S' = \{v_1, \ldots, v_N\}$ and get

$$K^\times \cap H \prod_{v \in S'} K_v^\times = K^\times \cap J(K)^p \prod_{v \in S} U_v \prod_{v \in S'} K_v^\times = K^{\times p}.$$

This implies

$$\left[K^\times H \prod_{i=1}^{n} K_{v_i} : K^\times H\right] = p^n \qquad \text{for } n = 1, \ldots, N. \tag{2.48}$$

Now we are looking for ideles which are in $N_{L/K}(J(L))$. We find

$$J(K)^p \prod_{v \notin S} U_v \prod_{i=1}^{N-1} K_{v_i}^\times = H \prod_{i=1}^{N-1} K_{v_i}^\times \subseteq N_{L/K}(J(L))$$

and therefore by means of (2.48)

$$[J(K) : K^\times N_{L/K}(J(L))] \leqslant \left[J(K) : K^\times H \prod_{i=1}^{N-1} K_{v_i}^\times\right] = p. \ \square$$

4.5. Analytical Proof of the Second Inequality (Main references: Goldstein (1971), Chap. 11; Hasse (1970), Part I). In this section we show how to prove (2.47) rather quickly with the methods of Chap. 1.6.6. We will consider functions of the real variable $s > 1$. By $g_1(s), g_2(s), \ldots$, we will denote such functions which are bounded for $s \to 1$.

Let L/K be a normal extension of degree n. We want to show

$$[J(K) : K^\times N_{L/K}(J(L))] \leqslant n. \tag{2.49}$$

By local consideration it is easy to see that there is a defining module $\mathfrak{m}$ such that $K^\times N_{L/K}(J(L)) \subseteq K^\times U_{\mathfrak{m}}$ (Chap. 1.5.5). Hence there is a subgroup $\mathfrak{B}_{\mathfrak{m}}$ of $\mathfrak{A}_{\mathfrak{m}}$ containing the ray group $R_{\mathfrak{m}}$ such that

$$J(K)/K^\times N_{L/K}(J(L)) \cong \mathfrak{A}_{\mathfrak{m}}/\mathfrak{B}_{\mathfrak{m}}.$$

$\mathfrak{B}_{\mathfrak{m}}$ contains all prime ideals in $\mathfrak{A}_{\mathfrak{m}}$ which split completely in L/K. Let C be the set of this prime ideals. Then we have

$$\sum_{\chi} \log L(s, \chi) = [\mathfrak{A}_{\mathfrak{m}} : \mathfrak{B}_{\mathfrak{m}}] \sum_{\mathfrak{p} \in \mathfrak{B}_m} N(\mathfrak{p})^{-s} + g_1(s) \geqslant [\mathfrak{A}_{\mathfrak{m}} : \mathfrak{B}_{\mathfrak{m}}] \sum_{\mathfrak{p} \in C} N(\mathfrak{p})^{-s} + g_1(s)$$

where χ runs over all characters of $\mathfrak{A}_{\mathfrak{m}}/\mathfrak{B}_{\mathfrak{m}}$.

Since $\log L(s, \chi)$ is bounded for $s \to 1$ if $\chi \neq \chi_0$ and

$$\log L(s, \chi_0) = -\log(s-1) + g_2(s)$$

we have

$$-[\mathfrak{A}_{\mathfrak{m}} : \mathfrak{B}_{\mathfrak{m}}]^{-1} \log(s-1) \geqslant \sum_{\mathfrak{p} \in C} N(\mathfrak{p})^{-s} + g_3(s). \tag{2.50}$$

On the other hand we have

$$\log \zeta_L(s) = \sum_{\mathfrak{P}} N(\mathfrak{P})^{-s} + g_4(s) = -\log(s-1) + g_5(s) \tag{2.51}$$

where the sum runs over all prime ideals $\mathfrak{P}$ of L which split completely in L/K (compare Theorem 1.112).

(2.50) and (2.51) and imply

$$-[\mathfrak{A}_{\mathfrak{m}} : \mathfrak{B}_{\mathfrak{m}}]^{-1} \log(s-1) \geqslant -n^{-1} \log(s-1) + g_6(s).$$

We divide by $-\log(s-1)$. Then for $s \to 1$ one gets

$$[\mathfrak{A}_{\mathfrak{m}} : \mathfrak{B}_{\mathfrak{m}}]^{-1} \geqslant n^{-1}. \square$$

4.6. The Canonical Class for Global Extensions. Now we consider the axioms 2–4 for $\mathfrak{K}_{\mathbb{Q}}$. A cyclotomic extension of K is a subextension of some extension $K(\zeta)/K$ where ζ is a root of unity. As special extensions of K we take the cyclic cyclotomic extensions of K. We have to construct the canonical class for such extensions. We proceed as follows:

For an arbitrary normal extension L/K with Galois group G we define the symbol $\langle a, \chi\rangle \in H^2(L/K)$ for $a \in C(K) = H^0(L/K)$, $\chi \in H^1(G, \mathbb{Q}/\mathbb{Z})$ by

$$\langle a, \chi\rangle := a \cup \Delta_1 \chi$$

where Δ_1 is the connecting homomorphism $H^1(G, \mathbb{Q}/\mathbb{Z}) \to H^2(G, \mathbb{Z})$. It is easy to show that $\langle a, \chi\rangle$ has the following degeneracy properties:

$$\langle a, \chi\rangle = 0 \qquad \text{for all } \chi \in H^1(G, \mathbb{Q}/\mathbb{Z}) \text{ if and only if } \quad a \in \bigcap_M N_{M/K}\mathfrak{C}(M)$$

where M runs over all cyclic subextensions of L/K.

$$\langle a, \chi\rangle = 0 \qquad \text{for all } a \in \mathfrak{C}(K) \text{ if and only if } \chi = \chi_0.$$

Furthermore we use our good knowledge of cyclotomic fields (§ 1.1). Let ζ be a primitive m-th root of unity. We define $(\mathbb{Q}(\zeta)/\mathbb{Q}, a) \in G(\mathbb{Q}(\zeta)/\mathbb{Q})$ for $a \in J(\mathbb{Q})$ by

$$(\mathbb{Q}(\zeta)/\mathbb{Q}, a)\zeta = \zeta^u$$

where $\bar{u} \in (\mathbb{Z}/m\mathbb{Z})^\times$ is the class corresponding to a under the isomorphism

$$\varphi: J(\mathbb{Q})/\mathbb{Q}^\times U_{m\infty} \to (\mathbb{Z}/m\mathbb{Z})^\times \qquad \text{(Chap. 1.5.6)}$$

which is constructed as follows: $\bar{a} \in \mathfrak{C}(\mathbb{Q})$ has a unique representative $\prod u_p$ in $U_\infty = \mathbb{R}_+ \prod_{p\,\text{finite}} U_p$. Furthermore there is one and only one $\bar{u} \in (\mathbb{Z}/m\mathbb{Z})^\times$ such that

$$\prod_p u_p \equiv u \pmod{U_{m\infty}}.$$

We put $\varphi(\bar{a}) = \bar{u}$.

For an arbitrary algebraic number field K and $a \in J(K)$ one shows that $(\mathbb{Q}(\zeta)/\mathbb{Q}, N_{K/\mathbb{Q}}a)$ is in the image of the injection $G(K(\zeta)/K) \to G(\mathbb{Q}(\zeta)/\mathbb{Q})$. One defines $(K(\zeta)/K, a)$ as the preimage of $(\mathbb{Q}(\zeta)/\mathbb{Q}, N_{K/\mathbb{Q}}a)$. If E/K is an arbitrary cyclotomic extension, $(E/K, a)$ is defined as the restriction of $(K(\zeta)/K, a)$ for some ζ with $E \subseteq K(\zeta)$. Then $(E/K, a)$ does not depend on the choice of ζ.

Theorem 2.73. *Let E/K be a cyclic cyclotomic extension. Then $\bar{a} \to (E/K, a)$ is an epimorphism of $\mathfrak{C}(K)$ onto $G(E/K)$ with kernel $N_{E/K}\mathfrak{C}(E)$.* ⊠

Let E/K be a cyclic cyclotomic extension of degree n with Galois group G. We define the canonical class of E/K as follows. Let s be a generator of G and $\chi \in H^1(G, \mathbb{Q}/\mathbb{Z})$ with $\chi(s) = \dfrac{1}{n} + \mathbb{Z}$, and let $a \in J(K)$ such that $(E/K, a) = s$. Then we put

$$\zeta_{E/K} := \langle a, \chi \rangle.$$

It is easy to see that $\zeta_{E/K}$ does not depend on the choice of s and a.

It remains to show that $\zeta_{E/K}$ satisfies the axioms 2–4 (Chevalley (1954), § 15–16).

Now let L/K be an arbitrary normal extension in $K_{\mathbb{Q}}$. The canonical class $\zeta_{L/K}$ defined by means of § 4.2 leads to a symbol $(a, L/K) \in G(L/K)^{ab}$ for $a \in J(K)$ (2.40). One shows that for cyclotomic extensions E/K one has $(a, E/K) = (E/K, a)$.

Furthermore let v be a place of K, w a place of L above v, $G_v = G(L_w/K_v)$ the decomposition group of w, and K_w the decomposition field of w. We denote by ψ_w the map from $H^2(L_w/K_v)$ to $H^2(L/K_w)$ induced by the injection $L_w^\times \to C(L)$. Since the definition of the local and global canonical classes correspond to each other, we have

$$\psi_w \zeta_{L_w/K_v} = \zeta_{L/K_w}. \tag{2.52}$$

This shows that $(a, L/K)$ is the norm symbol defined in § 1.5.

To finish the proof of the main theorem of global class field theory one has to proof axiom 5. This is done by reduction to the case that the closed subgroup U of finite index in $\mathfrak{C}(K)$ has the property that $\mathfrak{C}(K)/U$ has prime exponent p and K contains the p-th roots of unity. Then one uses (2.46): On one hand every U contains a group $H = J(K)^p \prod_{v \notin S} U_v$ for some S. On the other hand the extension $E_S = K(\sqrt[p]{x} | x \in K_S)$ is the class field of the group $J(K)/HK^\times$:

$$K^\times N_{E_S/K}(J(E_S)) = K^\times H.$$

§ 5. Simple Algebras

The theory of simple algebras over local and global fields is intimately related to class field theory. On one hand, the cohomological method is an outgrowth of the earlier use of simple algebras in class field theory. On the other hand, the structure of simple algebras over local and global fields can be determined by means of results of class field theory.

5.1. Simple Algebras over Arbitrary Fields (Main reference: Weil (1967), Chap. 9). Let K be an arbitrary field. We consider only algebras A over K with unit element being finite dimensional and central over K, i.e. A is a finite dimensional vector space over K and a ring with unit element 1 such that K is the center of A.

An algebra A over K is *simple* if it has only the trivial twosided ideals $\{0\}$ and A. With A and B also $A \otimes_K B$ is a simple algebra over K. Let A be a simple algebra over K and L a field extension of K. Then $A \otimes_K L$ is a simple algebra over L.

Theorem 2.74. *Let D be a division algebra over K and n a natural number. Then the matrice algebra $M_n(D)$ is a simple algebra over K. An arbitrary simple algebra A over K is isomorphic to $M_n(D)$ for some division algebra D and some natural number n. For a given simple algebra A the number n is uniquely determined and D is uniquely determined up to isomorphism.* ☒

For an algebra A over K the *opposite algebra* $A^{\bullet}$ is the algebra which coincides with A as vector space over K and which has multiplication $(x, y) \to yx$ for $x, y \in A$.

Proposition 2.75. *The dimension of a simple algebra A over K is equal to n^2 for some natural n.*

Proof. Let L be the algebraic closure of K. Then $\dim_L A \otimes_K L = \dim_K A$. Since L is algebraically closed a division algebra over L is equal to L. Therefore $A \otimes_K L = M_n(L)$ for some number n by Theorem 2.74. □

Proposition 2.76. *Let A be a simple algebra over K and*

$$\varphi\colon A \otimes_K A^{\bullet} \to \mathrm{End}_K(A)$$

the linear map which sends $a \otimes b$ to the endomorphism $x \to axb$. Then φ is an isomorphism of algebras. ☒

Proposition 2.77. *Let A be a simple algebra over K and let α be an automorphism of A which fixes K. Then $\alpha(x) = a^{-1}xa$ for some $a \in A$.*

Let B, C be subalgebras of A with centers containing K and let β be an isomorphism of B onto C. Then β extends to an automorphism of A. ☒

5.2. The Reduced Trace and Norm. Let L be the algebraic closure of K and A a simple algebra over K. Furthermore let φ be an algebra isomorphism of A into $M_n(L)$. Such isomorphisms always exist (see the proof of Proposition 2.75).

Proposition 2.78. *The functions*

$$\tau(a) = \mathrm{tr}(\varphi(a)) \quad \textit{and} \quad \nu(a) = \det(\varphi(a)) \qquad \textit{for } a \in A$$

have values in K and they are independent of the choice of φ. ☒

$\tau(a)$ is called the *reduced trace* and $\nu(a)$ is called the *reduced norm* of a. Obviously

$$\tau(a + b) = \tau(a) + \tau(b), \qquad \tau(\alpha a) = \alpha\tau(a), \qquad \tau(ab) = \tau(ba),$$

$$\nu(ab) = \nu(a)\nu(b) \qquad \text{for } a, b \in A,\ \alpha \in K.$$

Let A be a simple algebra of degree n^2 over K and $L \subseteq A$ a field extension of K of degree m. Then m divides n and

$$\tau(a) = \frac{n}{m}\,\mathrm{tr}_{L/K}\, a, \qquad \nu(a) = (N_{L/K}a)^{n/m} \qquad \text{for } a \in L.$$

5.3. Splitting Fields. Let A be a simple algebra over K of dimension n. A field extension L of K is called *splitting field* of A if $A \otimes_K L \cong M_n(L)$. Thus the algebra of K is a splitting field of A.

Proposition 2.79. *For every simple algebra A over K there exists a separable normal splitting field of finite degree over K.* □

Proposition 2.80. *Every maximal subfield of a division algebra D over K is a splitting field of D. Every splitting field of D is a splitting field of $M_n(D)$.* ⊠

5.4. The Brauer Group. Two simple algebras A and B over K are called similar if $A \cong M_m(D)$, $B \cong M_n(D)$ with the same division algebra D. The class of algebras similar to A will be denoted by $[A]$. We define the product of two classes $[A_1]$ and $[A_2]$ by

$$[A_1]\cdot[A_2] = [A_1 \otimes_K A_2]. \tag{2.53}$$

This product is well defined, commutative and associative. Furthermore $[A]\cdot[K] = [A]$ and $[A]\cdot[A^\cdot] = [K]$. Therefore the set of classes of simple algebras over K form an abelian group $B(K)$ under the multiplication (2.53) with unit element $[K]$ called the *Brauer group* of K.

Let L be a field extension of K. Then the classes of simple algebras over K with splitting field L form a subgroup $B(L/K)$ of $B(K)$ called the Brauer group of L/K.

Now let L/K be a finite normal separable extension and D a division algebra over K with splitting field L. Then there is a natural number r such that L is isomorphic to a maximal subfield in $A = M_r(D)$. We identify L with its image in A. By Proposition 2.77 there exists $u_g \in A$ for $g \in G := G(L/K)$ with

$$g\alpha = u_g \alpha u_g^{-1} \qquad \text{for } \alpha \in L. \tag{2.54}$$

The elements u_g are uniquely determined by g up to a factor in $L^\times$ and $\{u_g | g \in G\}$ is a basis of the vector space A over L. Furthermore

$$f(g_1, g_2) = u_{g_1} u_{g_2} u_{g_1 g_2}^{-1}, \qquad g_1, g_2 \in G, \tag{2.55}$$

is a cocycle in the sense of § 3.1, which multiplies by a coboundary if we multiply the u_g by some elements $f_1(g)$ in $L^\times$. Therefore we have a map ϕ of $B(L/K)$ into $H^2(G, L^\times)$, defined by $\phi([D]) = \bar{f}$.

On the other hand any $\bar{f} \in H^2(G, L^\times)$ gives rise to a simple algebra

$$A = \left\{ \sum_{g \in G} \alpha_g u_g | \alpha_g \in L \right\}$$

of formal linear combinations with multiplication defined by (2.54) and (2.55).

Theorem 2.81. *ϕ is an isomorphism of $B(L/K)$ onto $H^2(G, L^\times)$.* ⊠

In view of Theorem 2.81 the group $H^2(G, L^\times)$ is called the Brauer group of L/K, too.

If G is cyclic, $\chi \in \operatorname{Hom}(G, \mathbb{Q}/\mathbb{Z}) = H^1(G, \mathbb{Q}/\mathbb{Z})$ and $\alpha \in K^\times$. Then $\alpha \cup \Delta_1 \chi \in H^2(G, L^\times)$ where $\Delta_1: H^1(G, \mathbb{Q}/\mathbb{Z}) \to H^2(G, \mathbb{Z})$ is the connecting homomorphism corresponding to the exact sequence

$$\{0\} \to \mathbb{Z} \to \mathbb{Q} \to \mathbb{Q}/\mathbb{Z} \to \{0\} \qquad (\S\, 3.8).$$

The corresponding simple algebra is called cyclic algebra. It will be denoted by $[L/K; \chi, \alpha]$. If s is a generator of G, then

$$[L/K; \chi, \alpha] = \left\{ \sum_{i=0}^{n-1} \alpha_i u_s^i \middle| \alpha_i \in L \right\}, \qquad u_s^n = \alpha, \qquad n := |G|. \tag{2.56}$$

5.5. Simple Algebras over Local Fields (Main reference: Weil (1967), Chap. 10). Let K be a finite extension of $\mathbb{Q}_p$ and let D be a division algebra over K of dimension n^2.

Theorem 2.82. *The integral elements of D form a ring $\mathfrak{O}_D$, which is the maximal compact subring of D. Any maximal compact subring of $M_m(D)$ is conjugated to $M_m(\mathfrak{O}_D)$ under an inner automorphism of $M_m(D)$.* ⊠

Theorem 2.83. *D contains an isomorphic image of any field extension L/K with $[L:K]|n$.* ⊠

In particular let L/K be an unramified extension of degree n in D. Let χ be the character of $G := G(L/K)$ which sends the Frobenius automorphism to $\frac{1}{n} + \mathbb{Z}$ and let π be a prime element of K. Then D is isomorphic to the cyclic algebra $[L/K; \chi, \pi^r]$ for some $r \in \mathbb{Z}$ with $1 \leqslant r \leqslant n$, $(r, n) = 1$ (§ 4.3). The class of r/n in $\mathbb{Q}/\mathbb{Z}$ is uniquely determined by D and is called the Hasse invariant of $[D]$. It will be denoted by $h(D)$. (2.41) implies

$$h(D) = \operatorname{inv} \phi([D]). \tag{2.57}$$

Theorem 2.84. *$h(D)$ determines D up to isomorphism over K* (§ 4.3). □

Proposition 2.85. *Let A be a simple algebra over K and let ν be the reduced norm in A. Then $\nu(A^\times) = K^\times$.* ⊠

Now let K be the field $\mathbb{R}$ of real numbers. The algebra $\mathbb{H}$ of quaternions is the unique (central) division algebra over $\mathbb{R}$ distinct from $\mathbb{R}$. As cyclic algebra $\mathbb{H}$ equals $[\mathbb{C}/\mathbb{R}; \chi, -1]$ where χ is the non trivial character in $\operatorname{Hom}(G(\mathbb{C}/\mathbb{R}), \mathbb{Q}/\mathbb{Z})$. The reduced norm maps $M_m(\mathbb{H})$ onto $\mathbb{R}_+$.

$\mathbb{H}/\mathbb{R}$ is generated by $1, i, j, k$ with $i^2 = j^2 = k^2 = -1$, $ij = -ji = k$, $jk = -kj = i$, $ki = -ik = j$. In the following we put

$$x' = \alpha_1 - \alpha_i i - \alpha_j j - \alpha_k k \qquad \text{for } x = \alpha_1 + \alpha_i i + \alpha_j j + \alpha_k k.$$

Then

$$\tau(x) = x + x', \qquad \nu(x) = x'x. \tag{2.58}$$

ι is called the main involution of $\mathbb{H}$.

5.6. The Structure of the Brauer Group of an Algebraic Number Field. Let L be a finite normal extension of an algebraic number field K. We keep the notation of §4.4. The exact sequence

$$\{1\} \to L^\times \to J(L) \to \mathfrak{C}(L) \to \{1\}$$

induces the exact sequence

$$H^1(G, \mathfrak{C}(L)) \to H^2(G, L^\times) \to H^2(G, J(L)) \to H^2(G, \mathfrak{C}(L)). \tag{2.59}$$

By means of Theorem 2.56 and Proposition 2.66 we derive from (2.59) the exact sequence

$$\{0\} \to H^2(G, L^\times) \to \sum_{v \in P_K} H^2(G_{w_v}, L_{w_v}^\times) \underset{\psi}{\to} H^2(G, \mathfrak{C}(L)).$$

The definition of ψ and (2.36) imply

$$\psi(\zeta_{L_{w_v}/K_v}) = \operatorname{Cor} \zeta_{L/K_{w_v}}$$

where K_{w_v} denotes the decomposition field of w_v. Hence

$$\operatorname{inv}(\psi(a)) = \operatorname{inv} a \qquad \text{for } a \in H^2(G_{w_v}, L_{w_v}^\times).$$

This shows that the map

$$\psi' \colon \sum_{v \in P_K} H^2(G_{w_v}, L_{w_v}) \underset{\psi}{\longrightarrow} H^2(G, \mathfrak{C}(L)) \underset{\text{inv}}{\longrightarrow} \mathbb{Q}/\mathbb{Z}$$

is given by

$$\psi'\left(\sum_v a_v\right) = \sum_v \operatorname{inv} a_v. \tag{2.60}$$

Therefore we have the following description of the Brauer group of L/K:

Theorem 2.86. *Let L/K be a finite normal extension with Galois group G. Then with the notation of* §4.4 *one has the exact sequence*

$$\{0\} \to H^2(G, L^\times) \to \sum_{v \in P_K} H^2(G_{w_v}, L_{w_v}^\times) \underset{\psi'}{\to} \mathbb{Q}/\mathbb{Z}. \boxtimes$$

If G is cyclic, then $H^2(G, L^\times) = \hat{H}^0(G, L^\times)$. This proves the following Hasse local global principle for the norm:

Theorem 2.87. *Let L/K be a finite cyclic extension and let v_0 be a fixed place of K. An element a of $K^\times$ is contained in $N_{L/K}L^\times$ if and only if a is contained in $N_{L_w/K_v}L_w^\times$ for all places $v \neq v_0$ of K.* □

Theorem 2.87 is not true for non cyclic normal extensions in general, e.g. let $K = \mathbb{Q}$, $L = \mathbb{Q}(\sqrt{13}, \sqrt{17})$. Then all decomposition groups are cyclic. Furthermore the theorem of Tate (§4.1) implies $\hat{H}^{-1}(G, \mathfrak{C}(L)) \cong \hat{H}^{-3}(G, \mathbb{Z})$. Since $\hat{H}^{-3}(G, \mathbb{Z}) \cong H^2(G, \mathbb{Q}/\mathbb{Z}) \cong \mathbb{Z}/2\mathbb{Z}$ (2.36) for the group $G = \mathbb{Z}/2\mathbb{Z} \times \mathbb{Z}/2\mathbb{Z}$ and the sequence

$$\{0\} \to \hat{H}^{-1}(G, \mathfrak{C}(L)) \to \hat{H}^0(G, L^\times) \to \sum_v \hat{H}^0(G_{w_v}, L_{w_v}^\times)$$

is exact, there exists an $a \in K^\times$ which is a local norm at all places of K but is not a global norm.

The following (strong) Hasse local global principle for quadratic forms can be rather easily deduced from Theorem 2.87. One considers first the case of two variables (comp. Chap. 1.1.6) and then one proves the theorem by induction (Borevich, Shafarevich (1985), Chap. 1.7; Serre (1970), Chap. 4).

Theorem 2.88. *Let $f(x_1, \ldots, x_n) = \sum_{i,j=1}^{n} a_{ij}x_ix_j$ be a non degenerate quadratic form in n variables with coefficients a_{ij} in K. Then a number a of K can be represented by f over K if and only if a can be represented by f over all completions K_v of K.* ☒

There is a corresponding theorem about the equivalence of forms called the weak Hasse local global principle.

The Hasse local global principle is in general not true for forms of higher degree. This follows from the example above. See Borevich, Shafarevich (1985), Chap. 1.7.6 for more information about this question.

Furthermore Theorem 2.77 together with Theorem 2.23 implies the following theorems.

Theorem 2.89. *Every simple algebra over K has a cyclic cyclotomic splitting field.* ☒

Theorem 2.90 (Brauer-Hasse-Noether theorem). *Every simple algebra over K is cyclic.* ☒

It arises the question whether any simple algebra over an arbitrary field is cyclic. Albert (1932) showed that this is not true in general and Amitsur, Rowen, Tignol (1979) gave an example of a simple algebra over the field $\mathbb{Q}(t)$ which is not isomorphic to the tensor product of simple algebras. But Merkuriev, Suslin (1982) obtained a positive result:

Theorem 2.91. *Let n be a natural number and let F be a field of characteristic 0 containing the n-th roots of unity. Then every simple algebra over F with exponent n in the Brauer group is similar to the product of cyclic algebras of dimension n^2.* ☒

5.7. Simple Algebras over Algebraic Number Fields (Main reference: Deuring (1968)). We keep the notation of the last section. Let A be a simple algebra over K with splitting field L and v a place of K. The localization of A at v is the simple algebra $A_v := A \otimes_K K_v$ over K_v. We say that A is unramified at v if A_v is similar to K_v, i.e. $A_v \in [K_v]$.

$\phi([A_v]) \in H^2(G_{w_v}, L_{w_v}^\times)$ is the localization of $\phi([A]) \in H^2(G, L^\times)$. Therefore Theorem 2.86 implies that A is similar to K if and only if A is unramified at all places of K.

We denote the ring of integers of K_v by $\mathfrak{O}_v$ for finite places v.

Theorem 2.92. *Let A be a simple algebra over K and let M be a finite subset of A containing a basis of A over K. For each finite place v of K, let M_v be the*

$\mathfrak{O}_v$-module generated by M in A_v. Then, for almost all v, A_v is trivial over K_v and M_v is a maximal compact subring of A_v. ⊠

It is possible to develop an ideal theory for simple algebras over algebraic number fields which generalizes the ideal theory for algebraic number fields (Deuring (1968)). We give here a few basic results of this theory.

An order of the simple algebra A over K is a subring of A containing the unit element of A which forms a lattice in A when A is viewed as a vector space over $\mathbb{Q}$.

Among the orders of an algebraic number field there is a unique maximal order (Chap. 1.1). The same is true for the orders of division algebras over local fields (§5.5). For simple algebras over algebraic number fields one has only the following result:

Proposition 2.93. *Let A be a simple algebra over K. Then there are maximal orders in A. These are $\mathfrak{O}_K$-modules in A, and a $\mathfrak{O}_K$-module M in A is a maximal order if and only if its closure M_v in A_v is a maximal compact subring in A_v for every finite place v of K. Every order in A is contained in a maximal order.* ⊠

Example 19. Let $K = \mathbb{Q}$ and $A = a_0 + a_1 i + a_2 j + a_3 k$, $a_0, a_1, a_2, a_3 \in \mathbb{Q}$ with i, j, k as in §5.5. $a_0 + a_1 i + a_2 j + a_3 k$ is integral if and only if $2a_0 \in \mathbb{Z}$ and $a_0^2 + a_1^2 + a_2^2 + a_3^2 \in \mathbb{Z}$. Hence i and $\frac{3}{5}i + \frac{4}{5}j$ are integral but their sum is not integral. □

Let $\mathfrak{m}$ be a finite generated $\mathfrak{O}_K$-module in A which generates A as K-module. Then the *left* (resp. *right*) *order* of $\mathfrak{m}$ is defined by $\mathfrak{O}_1(\mathfrak{m}) = \{\alpha \in A | \alpha\mathfrak{m} \subseteq \mathfrak{m}$ (resp. $\mathfrak{O}_r(\mathfrak{m}) = \{\alpha \in A | \mathfrak{m}\alpha \subseteq \mathfrak{m}\}$). $\mathfrak{m}$ is called a *normal ideal* if its left and right orders are maximal orders. $\mathfrak{m}$ is called integral if it is contained in its left order and hence in its right order. $\mathfrak{m}$ is called *equilateral* if its left order is equal to its right order.

Theorem 2.94 (theorem of Brandt). *The normal ideals of A form a gruppoid with respect to ideal multiplication with the maximal orders as units (the product of two normal ideals $\mathfrak{a}$ and $\mathfrak{b}$ is admissible only if the right order of $\mathfrak{a}$ is equal to the left order of* $\mathfrak{b}$). ⊠

Theorem 2.95. *The equilateral normal ideals of a maximal order $\mathfrak{O}$ form an abelian group $G(\mathfrak{O})$ and the integral ideals in $G(\mathfrak{O})$ decompose uniquely in a product of prime ideals.* ⊠

§6. Explicit Reciprocity Laws and Symbols

As mentioned above (§1.8) the solution of the ninth problem of Hilbert was given by Shafarevich and more explicitly by Brückner and Vostokov. In view of Theorem 2.15 one has to compute the symbols $\left(\frac{\alpha, \beta}{v}\right)$ for infinite valuations v

and for $v|m$. Since $\left(\frac{\alpha, \beta}{v}\right)$ is given by (2.17) if v is real and $\left(\frac{\beta, \alpha}{v}\right) = 1$ if v is complex, it remains to compute $\left(\frac{\beta, \alpha}{v}\right)$ for $v|m$. According to Theorem 2.14.7 it is sufficient to consider the case that m is a power of a prime p. Therefore in the following an explicit reciprocity law means an explicit formula for $(\beta, \alpha)_{p^\kappa}$ where $\alpha, \beta \neq 0$ are in a local field K of residue characteristic p and $\mu_{p^\kappa} \subset K$.

6.1. The Explicit Reciprocity Law of Shafarevich (Main reference: Kneser (1951)). We begin with the explanation of the explicit reciprocity law of Shafarevich (1950) and we use the following notations: $\mathfrak{p}$ is the prime ideal and π a fixed prime element of K. Ω is the group of *p^κ-primary numbers ω in K*, i.e.

$$\Omega := \{\omega \in K^\times | K(\sqrt[p^\kappa]{\omega})/K \text{ unramified}\}.$$

T is the inertia field of K, i.e. $T/\mathbb{Q}_p$ is the maximal unramified subextension of $K/\mathbb{Q}_p$.

$\tilde{T}$ is the completion of the maximal unramified subextension of $\bar{K}$. R (resp. $\tilde{R}$) is the system of representatives of $\mathfrak{O}_T/(p)$ (resp. $\mathfrak{O}_{\tilde{T}}/(p)$) in $\mathfrak{O}_T$ (resp. $\mathfrak{O}_{\tilde{T}}$) consisting of the roots of unity of order prime to p in $\mathfrak{O}_T$ (resp. $\mathfrak{O}_{\tilde{T}}$) and of 0. e is the ramification index of $K/\mathbb{Q}_p$. F is the Frobenius automorphism of $T/\mathbb{Q}_p$.

The map $\mathscr{P}: \mathfrak{O}_{\tilde{T}} \to \mathfrak{O}_{\tilde{T}}$ given by $\mathscr{P}(\alpha) = F(\alpha) - \alpha$, is an endomorphism of $\mathfrak{O}_{\tilde{T}}^+$ onto $\mathfrak{O}_{\tilde{T}}^+$. In particular for every $\alpha \in \mathfrak{O}_T$ there is an $\tilde{\alpha} \in \mathfrak{O}_{\tilde{T}}$ with $\mathscr{P}(\tilde{\alpha}) = \alpha$.

The reciprocity law of Shafarevich is based on the presentation of $\alpha \in K^\times$ by means of the Artin-Hasse function $E(\alpha)$ and the Shafarevich function $E(\alpha, x)$ with $\alpha \in \mathfrak{O}_T$, $x \in \mathfrak{p}$. We put

$$E(\alpha, x) := \prod_{\substack{m=1 \\ p \nmid m}}^{\infty} (1 - \alpha^m x^m)^{\mu(m)/m} \qquad \text{for } \alpha \in \tilde{R},\ x \in \mathfrak{p}$$

where μ denotes the Moebius function. $E(\alpha, x)$ is well defined since $\mu(m)/m \in \mathbb{Z}_p$. By the Moebius inversion formula,

$$1 - \alpha x = \prod_{\substack{m=1 \\ p \nmid m}}^{\infty} E(\alpha^m, x^m)^{1/m}.$$

For arbitrary $\alpha = \sum_{i=0}^{\infty} \alpha_i p^i \in \mathfrak{O}_{\tilde{T}}$, $\alpha_i \in \tilde{R}$, we put

$$E(\alpha, x) = \prod_{i=0}^{\infty} E(\alpha_i, x)^{p^i}.$$

Proposition 2.95. $E(\alpha + \beta, x) = E(\alpha, x)E(\beta, x)$ *for* $\alpha, \beta \in \mathfrak{O}_{\tilde{T}}$. □

Let ζ be a primitive p^κ-th root of unity. There exists a unique $\tau \in \mathfrak{p}$ with $\zeta = E(1, \tau)$. Furthermore if $\alpha \in \mathfrak{O}_T$, let $\tilde{\alpha} \in \mathfrak{O}_{\tilde{T}}$ with $\mathscr{P}(\tilde{\alpha}) = \alpha$. We put

$$E(\alpha) := E(p^\kappa \tilde{\alpha}, \tau) = E(\tilde{\alpha}, \tau)^{p^\kappa}.$$

Theorem 2.96. 1. *$E(\alpha)$ is independent of the choice of $\tilde{\alpha}$ and it is a principal unit of K.*

2. $E(\alpha+\beta) = E(\alpha)E(\beta)$ *for* $\alpha, \beta \in \mathfrak{O}_T$.
3. $\{E(\alpha) | \alpha \in \mathfrak{O}_T\} K^{\times p^\kappa} = \Omega$. ⊠

In the following we are interested in the group $K^\times / K^{\times p^\kappa}$. For $\xi, \eta \in K^\times$ we write $\xi \sim \eta$ if $\xi\eta^{-1} \in K^{\times p^\kappa}$.

Theorem 2.97. 1. *Every* $\xi \in K^\times$ *has a presentation*

$$\xi \sim \pi^a \prod_i E(\alpha_i, \pi^i) E(\alpha) \qquad \text{with } a \in \mathbb{Z}, \alpha_i, \alpha \in \mathfrak{O}_T$$

where the product runs over all i with $1 \leqslant i < pe/(p-1)$, $p \nmid i$ (compare Chap. 1.4.4).

2. *If*

$$\pi^a \prod_i E(\alpha_i, \pi^i) E(\alpha) \sim \pi^b \prod_i E(\beta_i, \pi^i) E(\beta),$$

then

$$a \equiv b \ (\mathrm{mod}\ p^\kappa),\ E(\alpha_i, \pi^i) \sim E(\beta_i, \pi^i),\ E(\alpha) \sim E(\beta). \ ⊠$$

Now we are able to formulate the explicit reciprocity law of Shafarevich:

Theorem 2.98. 1. $(\pi, E(\alpha, \pi^i)) = 1$ *for* $p \nmid i$, $\alpha \in \mathfrak{O}_T$.
2. $(\pi, E(\alpha)) = \zeta^{\mathrm{Tr}\,\alpha}$ *if* $\alpha \in \mathfrak{O}_T$, $\mathrm{Tr}\,\alpha := \mathrm{Tr}_{T/\mathbb{Q}_p} \alpha$.
3. $(\pi, \pi) = (\pi, -1)$.
4. $(E(\alpha), \varepsilon) = 1$ *for* $\alpha \in \mathfrak{O}_T$, $\varepsilon \in \mathfrak{O}_K^\times$.
5. *If* $p \neq 2$, *then* $(E(\alpha, \pi^i), E(\beta, \pi^j)) = (\pi^j, E(\alpha\beta, \pi^{i+j}))$,
If $p = 2$, *then* $(E(\alpha, \pi^i), E(\beta, \pi^j)) =$
$(-\pi^j, E(\alpha\beta, \pi^{i+j})) \prod_{s=1}^\infty (-1, E(\alpha F^s(\beta), \pi^{i+jp^s})) \prod_{r=1}^\infty (-1, E(F^r(\alpha)\beta, \pi^{ip^r+j}))$
for $\alpha, \beta \in \mathfrak{O}_T$, $p \nmid i$, $p \nmid j$.
6. *If* $p = 2$, *then* $(-1, E(\alpha, \pi^i)) = \prod_{s=0}^\infty (\pi, E(i2^s F^{s+1}(\alpha), \pi^{i2^{s+1}}))$ *for* $\alpha \in \mathfrak{O}_T$, $p \nmid i$. ⊠

The functions $E(\alpha, x)$ and $E(\alpha)$ have simple transformation properties with respect to the Galois group of a tamely ramified normal extension K/K_0. This will be used together with Theorem 2.98 in Chap. 3.2.4.

6.2. The Explicit Reciprocity Law of Brückner and Vostokov (Main reference: Brückner (1979)). The reciprocity law of Shafarevich is not as explicit as one would like since if one wants to compute (ξ, η), one has to write ξ, η in the form

$$\xi \sim \pi^a \prod_i E(\alpha_i, \pi^i) E(\alpha), \qquad \eta \sim \pi^b \prod_j E(\beta_j, \pi^j) E(\beta)$$

and then one has to write $E(\alpha_i\beta_j, \pi^{i+j})$ in such a form if $p | (i+j)$.

A more explicit general reciprocity law was found by Brückner (1967) and Vostokov (1978).

We keep the notations of § 6.1. For simplicity we assume $p \neq 2$. For the case $p = 2$ see Brückner (1979a). Let $\mathfrak{O}_T((x))$ be the ring of formal power series $\sum_{i \geqslant i_0} \alpha_i x^i$, $i_0 \in \mathbb{Z}$, with coefficients in $\mathfrak{O}_T$. We consider $\mathfrak{O}_T((x))$ as topological ring with a system of neighbourhoods of 0 given by the subgroups $x^i \mathfrak{O}_T((x)) +$

$p^j\mathfrak{O}_T((x))$ for $i \in \mathbb{Z}, j \in \mathbb{N}$. The Frobenius automorphism F extends by

$$F\left(\sum_i \alpha_i x^i\right) := \sum_i F(\alpha_i) x^{pi}$$

to a continuous endomorphism of $\mathfrak{O}_T((x))$. For $f \in \mathfrak{O}_T((x))$, the quotient f^p/Ff lies in $1 + p\mathfrak{O}_T[[x]]$. We define $\log g$ for $g \in 1 + p\mathfrak{O}_T[[x]]$ as usually by

$$\log g := \sum_{i=1}^{\infty} (-1)^{i+1}(g-1)^i/i.$$

Then $\log\colon 1 + p\mathfrak{O}_T[[x]] \to p\mathfrak{O}_T[[x]]$ is a continuous group homomorphism. Furthermore we put for arbitrary $f \in \mathfrak{O}_T((x))^\times$

$$d\log f := \frac{f'}{f}\,dx$$

where f' is the formal derivative of f. The residue of $h\,dx = \sum_i \alpha_i x^i\,dx$ is as usually defined by $\operatorname{res} h\,dx := \alpha_{-1}$.

For the fixed prime element π of K and the fixed primitive p^κ-th root of unity ζ we define $t \in \mathfrak{O}_T((x))$ as follows: let $1 - \zeta = \pi^{e_1}\varepsilon$ with $\varepsilon \in \mathfrak{O}_K^\times$ and $u \in \mathfrak{O}_T[[x]]$ such that $u(\pi) = \varepsilon$. Then

$$t := 1 - (1 - x^{e_1}u)^{p^\kappa} = (x^{e_1}u)^{p^\kappa}(1 - v) \qquad \text{with } v \in p\mathfrak{O}_T((x)).$$

$1/t$ is a Laurent series $\sum_{i=-\infty}^{\infty} \alpha_i x^i$ with $\lim_{i\to-\infty} \alpha_i = 0$. It lies in the completion $\widehat{\mathfrak{O}_T((x))}$ of $\mathfrak{O}_T((x))$ with respect to the topology defined above.

Let s_π be the map from $\mathfrak{O}_T[[x]]d\log x$ to $\mathbb{Z}_p$ defined by

$$s_\pi\omega = \operatorname{tr}_{K/\mathbb{Q}_p} \operatorname{res}\left(\frac{1}{t} - \frac{1}{2}\right)\omega \qquad \text{for } \omega \in \mathfrak{O}_T[[x]]d\log x.$$

Furthermore for $f, g \in \mathfrak{O}_T((x))^\times$ we put

$$(f, g) := \left(\frac{1}{p}\log\frac{f^p}{Ff}\right)(d\log g) - \left(\frac{1}{p}\log\frac{g^p}{Fg}\right)\left(\frac{1}{p}d\log Ff\right) \in \mathfrak{O}_T[[x]]d\log x.$$

Now we can formulate the reciprocity law of Brückner and Vostokov in the form of Brückner (1979).

Theorem 2.99. *Let $p \neq 2$, $\xi, \eta \in K$ and $f, g \in \mathfrak{O}_T((x))^\times$ with $f(\pi) = \xi$, $g(\pi) = \eta$. Then*

$$(\xi, \eta) = \zeta^{w(f,g)}$$

with

$$w(f, g) = s_\pi(f, g). \boxtimes$$

Example 20. (Artin, Hasse (1928), Artin, Tate (1968), Chap. 12). In special cases the explicit reciprocity law looks considerably simpler, than in the form of Brückner and Vostokov. Let $K := \mathbb{Q}_p(\zeta)$ with the fixed primitive p-th root of unity ζ. $(p) = (1 - \zeta)^{p-1}$ is the prime ideal decomposition of p in K. We put

$\lambda := 1 - \zeta$. Then

$$(\alpha, \beta) = \zeta^{\operatorname{tr}(\zeta \log \alpha D \log \beta)/p} \qquad \text{if } \alpha \equiv 1 \pmod{\lambda^2},\ \beta \equiv 1 \pmod{\lambda},$$

$$(\zeta, \alpha) = \zeta^{\operatorname{tr}(\log \alpha)/p} \qquad \text{if } \alpha \equiv 1 \pmod{\lambda},$$

$$(\alpha, \lambda) = \zeta^{\operatorname{tr}(\zeta \log \alpha)/p} \qquad \text{if } \alpha \equiv 1 \pmod{\lambda},$$

where $D \log \beta$ means the logarithmic derivative of β considered as power series in λ and tr is the trace from K to $\mathbb{Q}_p$. Since the Hilbert symbol is trivial for $p-1$-th roots of unity and since $\zeta \not\equiv 1 \pmod{\lambda^2}$ the symbol (α, β) for arbitrary $\alpha, \beta \in K^\times$ is easily reduced to the three cases above. ⊠

The Hilbert symbol can be considered as the norm symbol for Kummer extensions. More general one can define a "Hilbert symbol" for extensions generated by division points of formal groups of dimension one. Kolyvagin (1979) establishes explicit reciprocity laws for such extensions. See also Wiles (1978) and a series of papers of Vostokov partially with other authors (Vostokov, Fesenko (1983)) for the case of Lubin-Tate extensions (Chap. 1.4.10). Vostokov also got an explicit reciprocity law for higher dimensional local fields.

6.3. Application to Fermat's Last Theorem II (Main reference: Hasse (1970), II, §22). Furtwängler noticed that the reciprocity law for the p-th power residue symbol can be used to get interesting sufficient conditions for the first case of Fermat's last theorem (Chap. 1.3.10). In the following we keep the notations of Chap. 1.3.10. Since $x^p + y^p = z^p$, g.c.d. $(x, y, z) = 1$, $p \nmid xyz$ implies $(x + \zeta^i y) = \mathfrak{a}_i^p$ for $i = 0, \ldots, p-1$ and ideals $\mathfrak{a}_i$ of $\mathbb{Z}[\zeta]$, the quotient $\alpha = \dfrac{x + \zeta y}{x + y}$ is the p-th power of a fractional ideal. Moreoever $\alpha = 1 - \dfrac{y\lambda}{x+y} \equiv 1 \pmod{\lambda}$, $\lambda := 1 - \zeta$. Hence $\left(\dfrac{\beta}{\alpha}\right) = 1$ for all $\beta \in \mathbb{Z}[\zeta]$ which are prime to α, i.e. g.c.d. $(\beta, \operatorname{supp} \alpha) = 1$.

We give as an example of the application of condition $\left(\dfrac{\beta}{\alpha}\right) = 1$ a sketch of the proof of the following theorem of Furtwängler, for more information see Hasse (1970), II, §22.

Theorem 2.100. *Let x, y, z be natural numbers with $x^p + y^p = z^p$,* g.c.d. *$(x, y, z) = 1$ and $p \nmid xyz$. Moreover let q be a prime with $q | xyz$. Then $q^{p-1} \equiv 1$* $\pmod{p^2}$.

Proof. One uses the following formula of Hasse (1970), II, §16.1.

$$\left(\frac{\beta}{\alpha}\right)\left(\frac{\alpha}{\beta}\right)^{-1} = \zeta^{\operatorname{tr} \eta} \qquad \text{with } \eta = \frac{\beta - 1}{p} \cdot \frac{\alpha - 1}{\lambda} \tag{2.61}$$

for all $\alpha, \beta \in \mathbb{Q}(\zeta)$ with g.c.d. $(\alpha, \beta) = 1$, $\alpha \equiv 1 \pmod{\lambda}$, $\beta \equiv 1 \pmod{p}$, where tr denotes the trace from $\mathbb{Q}(\zeta)$ to $\mathbb{Q}$. (2.61) follows from the explicit reciprocity law of Artin, Hasse (Example 20).

We apply (2.61) to $\beta := q^{p-1} \equiv 1 \pmod{p}$. Without loss of generality let $q|y$ hence $\alpha \equiv 1 \pmod{q}$ and $\left(\frac{\alpha}{q}\right) = 1$. Thus

$$\frac{q^{p-1}-1}{p}\operatorname{tr}\left(\frac{\alpha-1}{\lambda}\right) \equiv 0 \pmod{p},$$

but

$$\operatorname{tr}\left(\frac{\alpha-1}{\lambda}\right) = -(p-1)\frac{y}{x+y} \not\equiv 0 \pmod{p}. \square$$

Since always $2|xyz$, Theorem 2.100 implies the Wieferich criterion $2^{p-1} \equiv 1 \pmod{p^2}$.

The smallest primes p with $2^{p-1} \equiv 1 \pmod{p^2}$ are 1093 and 3511 and there are no other such primes p with $p < 3 \cdot 10^9$. Mirimanoff proved that the first case holds if $3^{p-1} \not\equiv 1 \pmod{p^2}$. Since this is fulfilled for 1093 and 3511 the first case holds for all $p < 3 \cdot 10^9$. Adleman, Heath-Brown (1985) and Fouvry (1985) showed that the first case holds for infinitely many primes.

6.4. Symbols (Main reference: Milnor (1971)). The proof of Theorem 2.99 in Brückner (1979) is based on the theory of symbols. In this section we explain the connection of Theorem 2.99 with symbols and consider symbols on local fields which serve as a background for all considerations about explicit formulas for the Hilbert symbol.

Let R be a commutative ring with unit element and A an abelian group. A *symbol* of R with values in A is a bimultiplicative map $c: R^\times \times R^\times \to A$ with

1. $c(-a, a) = 0$ if $a \in R^\times$,
2. $c(1-a, a) = 0$ if $a \in R^\times$, $1-a \in R^\times$.

Proposition 2.101. $c(b, a) = -c(a, b)$ *if* $a, b \in R^\times$.

Proof. $c(b, a) = c(b, -ab) = c(b(ab)^{-1}, -ab) = c(a^{-1}, -ab) = -c(a, b)$. $\square$

If R is a field, then 1. follows from 2.: Let $a \neq 1$, then

$$0 = c(1-a, a) = c(-a(1-a^{-1}), a) = c(-a, a) + c(1-a^{-1}, a) = c(-a, a).$$

Proposition 2.102. *Every symbol of a finite field* $\mathbb{F}_q$ *is trivial.*

Proof. If u is a generator of $F_q^\times$, then $c(u, u) = c(-1, u)$ therefore $2(a, b) = 0$ for all $a, b \in \mathbb{F}_q^\times$. Furthermore there exists $a, b \in \mathbb{F}_q^\times$ with $1 = (a^2 + b^2)u$ hence

$$0 = c(ua^2, 1-ua^2) = c(ua^2, ub^2) = c(u, u). \square$$

Proposition 2.103. *Let R be a field. Then for $a, b, c \in R^\times$ with $a + b = c$,*

$$c(a, b) = c(a, c) + c(c, b) + c(-1, c).$$

Proof. $0 = c(bc^{-1}, ac^{-1}) = c(b, a) - c(b, c) - c(c, a) + c(-1, c)$. $\square$

A symbol $c\colon \mathfrak{O}_T((x)) \times \mathfrak{O}_T((x)) \to A$ is called *admissible* if A is a complete Hausdorff topological group with subgroup topology, if c is continuous, $c(\mathfrak{O}_T, \mathfrak{O}_T) = \{0\}$, and $c(\mu_{q-1}, \mathfrak{O}_T((x))) = \{0\}$.

The Hilbert norm symbol $(\ ,\ )_{p^\kappa}$ induces on $\mathfrak{O}_T((x)) \times \mathfrak{O}_T((x))$ a symbol $\hat{c}$:

$$\hat{c}(f, g) = (f(\pi), g(\pi))$$

which is admissible.

Theorem 2.104. *$f, g \to (f, g) + d\mathfrak{O}_T[[x]]$ is a universal admissible symbol of $\mathfrak{O}_T((x))$ with values in $\mathfrak{O}_T[[x]]\, d\log x/d\mathfrak{O}_T[[x]]$.* ⊠

6.5. Symbols of $\mathfrak{p}$-adic Number Fields. Now let K be a $\mathfrak{p}$-adic number field and $U_1 := 1 + \mathfrak{p}$. Then $K^\times = (\pi) \times \mu_{q-1} \times U_1$ where q is the number of elements in the residue class field of K. The Hilbert norm symbol $(\ ,\ )_m$ is a symbol of K with values in μ_m. Let p be the residue characteristic of K.

Proposition 2.105. *Let A be a Hausdorff complete topological abelian group and $c\colon K^\times \times K^\times \to A$ a continuous symbol. Then*

1. $c(\mu_{q-1}, U_1) = \{0\}$,
2. $c(\mu_{q-1}, \mu_{q-1}) = \{0\}$,
3. $c(U_1, U_1) \subseteq c(U_1, \pi)$.

Proof. 1. follows from the fact that U_1 is a $\mathbb{Z}_p$-module (Chap. 1.4.4) hence $1/(q-1) \in \mathbb{Z}_p$. 1. implies that $c\colon \mu_{q-1} \times \mu_{q-1} \to A$ factors through $\mu_{q-1}U_1/U_1$. Therefore 2. follows from Proposition 2.102. For the proof of 3. one uses that for $i, j \in N$, $\sigma, \rho \in \mu_{q-1}$ one has by Proposition 2.103

$$\begin{aligned} c(1-\rho\pi^i, 1-\sigma\pi^j) &= c(1-\rho\pi^i, \rho\pi^i - \sigma\rho\pi^{i+j}) = c(1-\rho\pi^i, 1-\sigma\rho\pi^{i+j}) \\ &\quad + c(1-\sigma\rho\pi^{i+j}, 1-\sigma\pi^j) + ic(1-\sigma\rho\pi^{i+j}, \pi) \\ &\quad + c(-1, 1-\sigma\rho\pi^{i+j}). \end{aligned}$$

By iteration one finds that $c(U_1, U_1) \subseteq c(U_1, \pi) \subseteq (U_1, 1+\mathfrak{p}^h)$ for arbitrary large h. Since c is continuous, this implies $c(U_1, U_1) \subseteq c(U_1, \pi)$. □

It remains to know $c(K^\times, \pi)$. Let $i = i_0 p^\nu$ with $p \nmid i_0$, then

$$p^\nu c(1-\rho\pi^i, \pi) = (1/i_0)c(1-\rho\pi^i, \rho\pi^i) = 0. \tag{2.62}$$

If the p-th roots of unity do not belong to K, then as $\mathbb{Z}_p$-module, U_1 is generated by elements of the form $1-\rho\pi^i$ with $\rho \in \mu_{q-1}$ (Chap. 1.4.4) and therefore $c(U_1, \pi) = 0$.

If $\mu_p \subset K^\times$ and n is maximal with $\mu_{p^n} \subset K$, then U_1 is generated by $1-\rho\pi^i$ for $\rho \in \mu_{q-1}$, $1 \leqslant i < e_0 p^n$, $p \nmid i$ and one element of the form

$$\delta := 1 - \rho\pi^{e_0 p^n} \qquad \text{with } \rho \in \mu_{q-1},\ \operatorname{tr}_{T/\mathbb{Q}_p} \rho \equiv 0 \pmod{p}$$

where e_0 is the ramification index of $K/\mathbb{Q}_p(\mu_{p^n})$ (Chap. 1.4.4). In the following we fix such an element δ.

Proposition 2.106. *Let ζ be a primitive root of unity of order p^n in K, $n \geqslant 1$. Then $c(U_1, \pi)$ is generated by $c(\alpha, \zeta)$ for some $\alpha \in K^\times$.*

Proof. We know already that $c(U_1, \pi)$ is a cyclic $\mathbb{Z}_p$-module generated by (δ, π). Therefore it is sufficient to find $\alpha \in K^\times$ such that $c(\alpha, \zeta)$ generates $c(U_1, \pi)/pc(U_1, \pi)$. Let $m \in N$ be maximal with $\zeta \equiv u \pmod{U_1^p}$ for some $u \in U_m$. Then $m \leqslant e_0 p^n$. If $m = e_0 p^n$, then we can take $\alpha := \pi$. If $m < e_0 p^n$, then m is prime to p. We take $\alpha := 1 - \dfrac{u}{\delta} \equiv \sigma\pi^m \pmod{U_1}$ for some $\sigma \in \mu_{q-1}$. Then $c\left(\alpha, \dfrac{u}{\delta}\right) = 0$ and

$$c(\alpha, \zeta) \equiv c(\alpha, u) \equiv c(c, \delta) \equiv mc(\pi, \delta) \pmod{c(U_1, \pi)}. \ \square$$

6.6. Tame and Wild Symbols. A symbol c of the $\mathfrak{p}$-adic number field K is called *tame* (resp. *wild*) if $c(U_1, K^\times) = \{0\}$ (resp. $c(\mu_{q-1}, K^\times) = \{0\}$). The Hilbert norm symbol $(\ ,\ )_m$ is tame for $p \nmid m$ and wild for $m = p^s$.

Proposition 2.107. *If K does not contain the p-th roots of unity, then every continuous wild symbol of K is trivial.*

Proof. (2.62). $\square$

Proposition 2.108. *Every continuous symbol of K can be uniquely represented as sum of a tame and a wild symbol.*

Proof. Let $c = t + w$ where t is a tame and w a wild symbol. Put $\alpha = \pi^a \rho u$, $\beta = \pi^b \sigma v \in K^\times$ with $a, b \in \mathbb{Z}$, $\rho, \sigma \in \mu_{q-1}$, $u, v \in U_1$. Then

$$t(\alpha, \beta) = \begin{cases} abc(-1, \pi) - ac(\sigma, \pi) + bc(\rho, \pi) & \text{if } p \neq 2, \\ -ac(\sigma, \pi) + bc(\rho, \pi) & \text{if } p = 2. \end{cases} \tag{2.63}$$

This proves the uniqueness. On the other hand (2.63) defines for arbitrary c a tame symbol and $c - t$ is a wild symbol. $\square$

A continuous symbol c of K is called *universal tame* (resp. *wild*) *symbol* if for every tame (resp. wild) symbol $c'\colon K^\times \times K^\times \to A'$ there exists a homomorphism $f\colon A \to A'$ such that $c' = fc$.

Theorem 2.109 (Theorem of Moore). *Let K be a $\mathfrak{p}$-adic number field with $\mu_{p^n} \subset K$, $\mu_{p^{n+1}} \not\subset K$. Then the Hilbert norm symbol $(\ ,\ )_{p^n}$ is a universal wild symbol of K.*

Proof. This follows from § 6.5. $\square$

According to (2.63) a universal tame symbol $t\colon K^\times \times K^\times \to \mu_{q-1}$ is given by

$$t(\alpha, \beta) = (-1)^{v(\alpha)v(\beta)} \alpha^{v(\beta)} \beta^{-v(\alpha)} \pmod{\mathfrak{p}} \text{ for } \alpha, \beta \in K^\times.$$

6.7. Remarks about Milnor's *K*-Theory (Main reference: Milnor (1970)). The notion of symbol leads to *Milnor's K-theory*: For any field F one defines the graded ring $K_*(F) = \bigoplus_{i=0} K_i(F)$ as the factor ring of the tensor algebra

$$\mathbb{Z} \dotplus F^\times \dotplus F^\times \otimes_{\mathbb{Z}} F^\times \dotplus F^\times \otimes_{\mathbb{Z}} F^\times \otimes_{\mathbb{Z}} F^\times \dotplus \cdots \tag{2.64}$$

with respect to the ideal generated by the elements $a \otimes (1 - a)$ for all $a \in F^\times$, $a \neq 1$. If F is a topological field, then (2.64) has a natural topology and one defines $K_*^{\mathrm{top}}(F)$ as the factor ring of (2.64) with respect to the closed ideal generated by the elements $a \otimes (1 - a)$ for all $a \in F^\times$, $a \neq 1$. The results about symbols on F can be reformulated as results about $K_2(F)$ or $K_2^{\mathrm{top}}(F)$.

A one dimensional analogue to symbols is the exponential valuation $v: F^\times \to \mathbb{Z}$. Parshin had the fruitful idea that, since $K_1(F) = F^\times$ is responsible for the class field theory of local fields, $K_n(F)$ should be in the same way responsible for the theory of abelian extensions of "local fields of dimension n". This is in fact true for local fields of dimension 0, i.e. finite fields, and it turns out to be true in general if one defines a local field of dimension n by induction as follows: A local field of dimension $n + 1$ is a discrete valuated field such that the residue class field is a local field of dimension n.

§ 7. Further Results of Class Field Theory

7.1. The Theorem of Shafarevich-Weil (Main reference: Artin-Tate (1968), Chap. 13, 15). Let F be a local or global field. We consider a finite normal extension K/F and a finite abelian extension L/K which is given by the corresponding subgroup H of the class module A_K (§ 1.6). According to Theorem 2.11, H is invariant under $G(K/F)$ if and only if L/K is normal. If we assume this, we have a group extension

$$\{1\} \to A_K/H \to G(L/F) \to G(K/F) \to \{1\} \tag{2.65}$$

where $G(K/F)$ acts on A_K/H in the natural way.

The question arises which class in $H^2(G(K/F), A_K/H)$ belongs to the group extension (2.65) (§ 3.1 example 9). If there is a prestabilized harmony in mathematics one may hope that the class of (2.65) is connected with the canonical class in $H^2(G(K/F), A_K)$ (Theorem 2.56). This is indeed true:

Theorem 2.110 (theorem of Shafarevich-Weil). *The class of the group extension* (2.65) *is the image of the canonical class under the map* $H^2(G(K/F), A_K) \to H^2(G(K/F), A_K/H)$. ⊠

7.2. Universal Norms (Main reference: Artin-Tate (1968), Chap. 6.5, Chap. 9). Let L/K be a finite normal extension of local or global fields. The kernel of the norm symbol $\alpha \to (\alpha, L/K)$ is the group $N_{L/K}A_L$ (§ 1). A universal norm is by definition an element of A_K which is a norm for all finite normal extensions L/K. Hence the group of universal norms is the group

$$\mathfrak{U}_K := \bigcap_L N_{L/K}A_L \subseteq A_K$$

where the intersection runs over all normal extensions L of K.

In the local case it is easy to see that $\mathfrak{U}_K = \{1\}$. In the global case one has the following result:

Theorem 2.111. *Let K be an algebraic number field. Then $\mathfrak{U}_K$ is the connected component of the idele class group $\mathfrak{C}(K)$. The group $\mathfrak{U}_K$ is also characterized as the group of elements in $\mathfrak{C}(K)$ which are infinitely divisible.*

Let V be the group of integral adeles of $\mathbb{Q}$, i.e. the components are integers at all finite places (Chap. 1.5.4). *Let r_1 (resp. r_2) be the number of real* (*resp. complex*) *places of K. Then $\mathfrak{U}_K$ is topological isomorphic to the group*

$$\mathbb{R}^+ \times (V/\mathbb{Z})^{r_1+r_2-1} \times (\mathbb{R}/\mathbb{Z})^{r_2}. \boxtimes$$

7.3. On the Structure of the Ideal Class Group. In this section we give some applications of the isomorphism between the ideal class group $CL(K)$ and the Galois group of the Hilbert class field $H(K)$ of an algebraic number field K (§ 1.2).

Theorem 2.112. *Let L/K be a finite extension of number fields which contains no subextension F/K with $F \neq K$ which is unramified at all places. Then $h(K)|h(L)$. In fact,*

$$N_{L/K}CL(L) = CL(K).$$

Proof. According to (2.4) we have the commutative diagram

$$\begin{array}{ccc} CL(L) & \xrightarrow[\cong]{} & G(H(L)/L) \\ \Big\downarrow{\scriptstyle N} & & \Big\downarrow{\scriptstyle \kappa} \\ CL(K) & \xrightarrow[\cong]{} & G(H(K)/K). \end{array}$$

By our assumption, $H(K) \cap L = K$. Therefore κ is surjective, hence the norm map is surjective, too. □

Theorem 2.113. *Let p be a prime and let L/K be a finite p-extension, i.e. L/K is normal of p-power degree. We assume that there is at most one place which ramifies in L/K. Then $p|h(L)$ implies $p|h(K)$.*

If K has no abelian extension which is unramified at finite places (*e.g.* $K = \mathbb{Q}$), *then it is sufficient to assume that K has at most one finite place which ramifies in L/K.*

Proof. We assume $p|h(L)$. Let H be the maximal unramified abelian p-extension of L. Then H/K is normal. We put $G := G(H/K)$. Let v be the place of K (if it exists) which ramifies in L/K, let w be a place of H above v, and let V_0 be the inertia group of w with respect to H/K. By assumption, $V_0 \neq G$. Since G is a p-group, if follows that V_0 fixes a cyclic subextension E of degree p of H/K. Then E/K is unramified at all places. □

Let $\mathfrak{m}$ be the product of all real places of K and $CL_0(K)$ the ray class group mod $\mathfrak{m}$ (Chap. 1.5.6). $CL_0(K)$ is called the ideal class group in the narrow sense. If K is totally complex, i.e. K has no real places, then $CL_0(K) = CL(K)$. The class field of $CL_0(K)$ is the maximal abelian extension of K which is unramified at finite places.

Example 21. $K = \mathbb{Q}(\sqrt{6})$ has class number one, but $\mathbb{Q}(\sqrt{-2}, \sqrt{-3})/\mathbb{Q}(\sqrt{6})$ is unramified at finite places, $|CL_0(K)| = 2$. □

The following theorem goes back to Gauss, who formulated it in the language of quadratic forms (Chap. 1.1.6).

Theorem 2.114. *Let K be a quadratic extension of $\mathbb{Q}$ with t ramified finite places. Then $[CL_0(K) : CL_0(K)^2] = 2^{t-1}$.*

Proof. Let d be the discriminant of $K/\mathbb{Q}$. Then d has a unique representation

$$d = p_1^* \dots p_t^*$$

where p_i^* is a prime discriminant, i.e. $\mathbb{Q}(\sqrt{p_i^*})$ has one ramified prime p_i, $i = 1, \dots, t$. A prime discriminant p^* has the form

$$p^* = (-1)^{(p-1)/2} p \quad \text{if} \quad p \neq 2 \quad \text{and} \quad p^* = -4, 8, -8 \quad \text{if} \quad p = 2.$$

The class field $H_0(K)$ of $CL_0(K)$ contains the field $E = \mathbb{Q}(\sqrt{p_1^*}, \dots, \sqrt{p_t^*})$. On the other hand let E'/K be the maximal 2-elementary subextension of $H_0(K)/K$. Then $E'/\mathbb{Q}$ is normal. The ramification group V_0 of a prime divisor of p_1 in E' has order 2 and is not contained in $G(E'/K)$. Therefore $G(E'/\mathbb{Q})$ contains no element of order 4. This implies that $G(E/\mathbb{Q})$ is an abelian 2-elementary group, hence $E = E'$. □

The classes of $CL_0(K)/CL_0(K)^2$ are called the genera of K and $CL_0(K)^2$ is called the principal genus. If $\mathfrak{a}$ is an ideal of K which is prime to d, then by (2.4)

$$\left(\frac{\mathfrak{a}}{E/K}\right) = \left(\frac{N(\mathfrak{a})}{E/\mathbb{Q}}\right).$$

This proves the following theorem:

Theorem 2.115. *Let K be a quadratic extension of $\mathbb{Q}$ with discriminant d. Then the norm from K to $\mathbb{Q}$ induces an injection*

$$\iota\colon CL_0(K)/CL_0(K)^2 \to (\mathbb{Z}/d\mathbb{Z})^\times/(\mathbb{Z}/d\mathbb{Z})^{\times 2}.$$

The image of ι equals $U(K)/(\mathbb{Z}/d\mathbb{Z})^{\times 2}$ where $U(K)$ is the subgroup of $(\mathbb{Z}/d\mathbb{Z})^\times$ corresponding to K in the sense of § 1.1. □

For a generalization of Theorem 2.114 and Theorem 2.115 see Leopoldt (1953). For more information about the 2-component of $CL_0(K)$ see Chap. 3.

7.4. Leopoldt's Spiegelungssatz (Main reference: Washington (1982), Chap. 10.2). In this section we use some results of the representation theory of semi-simple algebras over fields (Curtis-Reiner (1962)).

Let p be a fixed prime number and L a finite normal extension of an algebraic number field K such that $\mu_p \subset L$ and $p \nmid [L : K]$. We put $G := G(L/K)$. The p-Sylow group A of the ideal class group of L is a $\mathbb{Z}_p[G]$-module.

The group ring $\mathbb{Q}_p[G]$ is the direct sum of simple algebras $\mathfrak{A}_\phi$ which correspond to the irreducible characters ϕ of $\mathbb{Q}_p[G]$. The algebra $\mathfrak{A}_\phi$ is the full matrix algebra over its centre Z_ϕ. The rank of $\mathfrak{A}_\phi$ over Z_ϕ is the degree $\chi(1)$ of any absolute irreducible component χ of ϕ. The extension $Z_\phi/\mathbb{Q}_p$ is unramified of degree $\phi(1)/\chi(1)$. To ϕ corresponds the primitive idempotent

$$1_\phi := \frac{\chi(1)}{|G|} \sum_{g \in G} \phi(g^{-1})g \in \mathbb{Z}_p[G]$$

with $\mathfrak{A}_\phi = \mathbb{Q}_p[G]1_\phi$ and

$$1_\phi 1_{\phi'} = 0 \qquad \text{if } \phi \neq \phi', \quad 1 = \sum_\phi 1_\phi. \tag{2.66}$$

(2.66) implies the direct product decomposition

$$A = \prod_\phi A_\phi \qquad \text{with } A_\phi := A^{1_\phi}.$$

The index $[A_\phi : A_\phi^p]$ is a power $p^{\phi(1)e_\phi}$ of $p^{\phi(1)}$. The number e_ϕ is called the G-rank of A_ϕ.

Let $\omega\colon G \to \mu_{p-1} \subset \mathbb{Z}_p$ be the character with

$$\zeta^g = \zeta^{\omega(g)} \qquad \text{for } g \in G, \zeta \in \mu_p.$$

The *reflection* (*Spiegelung*) $\bar{\phi}$ of ϕ is the character defined by

$$\bar{\phi}(g) = \omega(g)\phi(g^{-1}) \qquad \text{for } g \in G.$$

one has $\bar{\bar{\phi}} = \phi$.

Let E be the unit group of K and $\mathfrak{E} = E/E^p$. A class εE^p of $\mathfrak{E}$ is called primary if $K(\sqrt[p]{\varepsilon})/K$ is unramified for all places of K above p. Let $\mathfrak{E}_0$ be the group of primary classes. Since $\mathfrak{E}_0$ is a $\mathbb{Z}_p[G]$-module one has the direct product decomposition

$$\mathfrak{E}_0 = \prod_\phi \mathfrak{E}_{0\phi} \qquad \text{with } \mathfrak{E}_{0\phi} = E_0^{1_\phi}.$$

Theorem 2.116 (Spiegelungssatz of Leopoldt). *Let e_ϕ (resp. δ_ϕ) be the G-rank of A_ϕ (resp. $\mathfrak{E}_{0\phi}$). Then*

$$\delta_\phi \geqslant e_{\bar{\phi}} - e_\phi \geqslant -\delta_{\bar{\phi}}. \boxtimes \tag{2.67}$$

(Leopoldt (1958)).

The proof combines our knowledge from class field theory about the maximal unramified p-elementary extension of K with the Kummer theory of this extensions.

In the special case that L is a CM-field, i.e. a totally complex quadratic extension of a totally real extension of K and ι denotes the complex conjugation one has $\phi(\iota) = \pm\phi(1)$. The characters ϕ with $\phi(\iota) = \phi(1)$ (resp. $\phi(\iota) = -\phi(1)$) are called even (resp. odd). Since $\delta_\phi = 0$ for odd characters, (2.67) has in this case the form

$$\delta_\phi \geqslant e_{\bar{\phi}} - e_\phi \geqslant 0 \qquad \text{if } \phi \text{ is even.} \tag{2.68}$$

This applies in particular to the case $K = \mathbb{Q}$, $L = \mathbb{Q}(\mu_p)$. In this case Theorem 2.116 was proved by Pollaczek and strengthens a result of Kummer.

Another special case is the following *theorem of Scholz.*

Theorem 2.117. *Let $d > 1$ be square-free, let r be the 3-rank of the ideal class group of $\mathbb{Q}(\sqrt{d})$ and s the 3-rank of the ideal class group of $\mathbb{Q}(\sqrt{-3d})$. Then*

$$r \leqslant s \leqslant r + 1.$$

Proof. We apply Theorem 2.116 in the case $p = 3$, $K = \mathbb{Q}$, $L = \mathbb{Q}(\sqrt{-3}, \sqrt{d})$. We have $G = (\iota, \sigma)$ with $\sigma(\sqrt{d}) = -\sqrt{d}$, $\sigma(\sqrt{-3}) = \sqrt{-3}$. Let χ be the non trivial even character of G. Then $r = e_\chi$, $s = e_{\bar{\chi}}$, $\delta_\chi \leqslant 1$. Hence the claim follows from (2.67). □

7.5. The Cohomology of the Multiplicative Group (Main reference: Tate (1967), 11.4). We keep the notations of §5.6 where we determined the structure of $H^2(G, L^\times)$.

In general the exact sequence

$$\{1\} \to L^\times \to J_L \to \mathfrak{C}_L \to \{1\}$$

implies the exact sequence

$$\to \hat{H}^{n-1}(G, J_L) \underset{\psi}{\to} \hat{H}^{n-1}(G, \mathfrak{C}_L) \to \hat{H}^n(G, L^\times) \underset{\varphi}{\to} \hat{H}^n(G, J_L) \to .$$

We have

$$\hat{H}^{n-1}(G, J_L) \cong \sum_{v \in P_K} \hat{H}^{n-1}(G_{w_v}, L_{w_v}^\times) \cong \sum_{v \in P_K} \hat{H}^{n-3}(G_{w_v}, \mathbb{Z})$$

and

$$\hat{H}^{n-1}(G, \mathfrak{C}_L) \cong \hat{H}^{n-3}(G, \mathbb{Z})$$

((2.45), Theorem 2.57). Therefore Ker φ is isomorphic to the cokernel of the map

$$\psi_1\colon \sum_{v \in P_K} \hat{H}^{n-3}(G_{w_v}, \mathbb{Z}) \to \hat{H}^{n-3}(G, \mathbb{Z})$$

given by

$$\psi_1\left(\sum_v \beta_v\right) = \sum_v \operatorname{Cor}_{G_{w_v}, G} \beta_v.$$

By Proposition 2.52 Coker ψ_1 is the dual of the kernel of the map

$$\psi_2\colon \hat{H}^{3-n}(G, \mathbb{Z}) \to \prod_{v \in P_K} \hat{H}^{3-n}(G_{w_v}, \mathbb{Z})$$

given by

$$\psi_2(\beta) = \prod_v \operatorname{Res}_{G, G_{w_v}} \beta.$$

We emphasize the two cases $n = 0$ and $n = 3$.

Theorem 2.118. *Let*

$$N_1 := \{\alpha \in K^\times \mid \alpha \in N_{L_{w_v}/K_v}(L_{w_v}^\times) \qquad \text{for } v \in P_K\}.$$

Then $N_1/N_{L/K}(L^\times)$ *is in duality to the kernel of the restriction map*

$$H^3(G, \mathbb{Z}) \to \prod_{v \in P_K} H^3(G_{w_v}, \mathbb{Z}). \boxtimes$$

Example 22. If there is a place v of K such that $G = G_{w_v}$, then every $\alpha \in K^\times$ which is a local norm for all $v \in P_K$ is a global norm. □

Theorem 2.119. $H^3(G, L^\times)$ *is cyclic of order* $[L:K]/\text{l.c.m.}\,\{[L_{w_v}:K_v] | v \in P_K\}$.

Proof. Theorem 2.57 implies $H^3(G_{w_v}, L_{w_v}^\times) \cong H^1(G_{w_v}, \mathbb{Z}) = \{0\}$. □

For more information see Tate (1952).

Chapter 3
Galois Groups

So far we have studied class field theory mainly as a theory of finite abelian extensions of local and global fields. In this chapter the main interest consists in the study of the Galois group of (mostly) infinite normal extensions on the basis of the cohomological treatment of class field theory.

The Galois group of a normal infinite field extension was introduced in Krull (1928). The importance of Galois groups of infinite extensions with certain maximality conditions became clear by the work of Shafarevich (1947), Iwasawa (1953), and Kawada (1954), about local fields and the work of Tate (1962) and Serre (1962) about global fields. The culminating point of this development was the solution of the class field tower problem by Golod, Shafarevich (1964).

The theory of field embeddings initiated by Faddejev and Hasse in a series of papers in the forties was applied in Shafarevich (1954) to the solution of the inverse problem of Galois theory over algebraic number fields for solvable groups. The theory of field embeddings was reformulated in terms of Galois cohomology by Bashmakov (1968) and Hoechsmann (1968).

A new direction in Galois theory has been created by Matzat (1977) and by Belyj (1979) who showed on the basis of Hilbert's irreducibility theorem that certain simple non commutative groups are Galois groups of normal extensions over abelian fields.

Chapter 3 is organized as follows: In § 1 we introduce the language of profinite groups and the cohomology of such groups. In the sections 1.12–1.16 we develop the theory of pro-p-groups including the classification of Demushkin groups and the theorem of Golod, Shafarevich. § 2 explains the most important results about Galois cohomology of local and global fields with application to the structure of the Galois group of the maximal p-extension of local fields (2.3), the algebraic closure of local fields (2.4), and the maximal p-extension with given ramification of global fields (2.6). § 3 contains the theory of field embeddings and results about extensions of global fields with given Galois group.

§ 1. Cohomology of Profinite Groups

(Main references: Serre (1964), Chap. 1, Koch (1970), § 1–7)

We begin with the explanation of the language of Galois cohomology, which is a topological version of the cohomology of groups (Chap. 2.3).

1.1. Inverse Limits of Groups and Rings. Let K be a field and L/K an infinite (separable algebraic) normal extension, i.e. L is the union of finite (separable) extensions L_i of K, where i runs through an index system I. The Galois group of L/K can be understood as the limit of the finite Galois groups $G(L_i/K)$. This limit is called the inverse or projective limit of the system $\{G(L_i/K) | i \in I\}$.

Inverse limits play a growing role not only in algebraic number theory but also in other domains of mathematics as algebraic geometry and topology (see e.g. Hartshorne (1977), Chap. 2.9, Sullivan (1970)).

We begin with the explanation of the notion of inverse limit in the scope of the theory of topological groups.

Let I be a *directed set* with respect to an order relation $\leqslant$, i.e. to every $i, j \in I$, there exists a $k \in I$ such that $i \leqslant k$, $j \leqslant k$. We consider I as a category with I as set of objects and with morphism sets $\mathrm{Hom}(i, j)$, $i, j \in I$, containing one element if $i \leqslant j$ and being empty otherwise.

An *inverse system* $P = \{I, G_i, \varphi_i^j\}$ of compact groups (rings) is a contravariant functor P from I in the category $\mathfrak{K}$ of compact groups (rings) such that $G_i = P(i)$ for $i \in I$ and $\varphi_i^j = P(i \leqslant j)$: $G_j \to G_i$. We write also $P = \{G_i | i \in I\}$ for short. $P(i \leqslant i)$ is the identity map.

Example 1. Let $I = N$ be the ordered set of natural numbers, $\{G_i | i \in N\}$ a family of compact groups, and φ_i^{i+1} arbitrary morphisms of G_{i+1} into G_i. For $i \leqslant j$ we define

$$\varphi_i^j := \varphi_i^{i+1} \varphi_{i+1}^{i+2} \cdots \varphi_{j-1}^j$$

Then $\{N, G_i, \varphi_i^j\}$ is an inverse system. □

For any directed set I and $G \in \mathfrak{K}$, the trivial inverse system P_G associates to $i \in I$ the group (ring) G and to $i \leqslant j$ the identity map. To any morphism φ of $G' \in \mathfrak{K}$ into G there corresponds a functor morphism of $P_{G'}$ into P_G again denoted by φ.

Let $P = \{I, G_i, \varphi_i^j\}$ be an inverse system. A group (ring) $G \in \mathfrak{K}$ together with a functor morphism ϕ of P_G to P is called *inverse limit* of P if for every $G' \in \mathfrak{K}$ and every functor morphism Φ' of $P_{G'}$ in P there exists one and only one morphism φ of G' in G such that the diagram

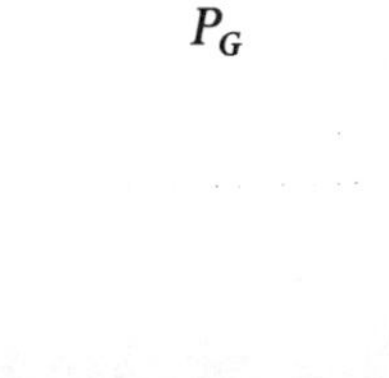

commutes. A functor morphism of P_G in P is given by a family $\{\varphi_i | i \in I\}$ of morphisms of G into G_i such that for $i \leqslant j$ the diagram

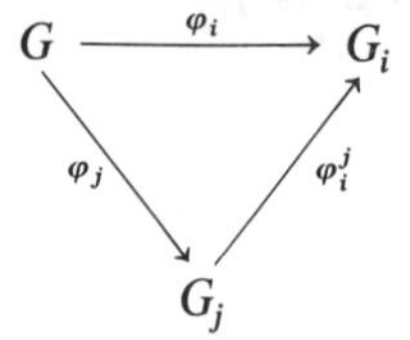

commutes.

It follows from the definition that the inverse limit is uniquely determined up to isomorphism.

Proposition 3.1. *For every inverse system $\{I, G_i, \varphi_i^j\}$ the inverse limit exists.*

Proof. We define G as the group of all elements $\prod_{i \in I} g_i$ of the direct product $\prod_{i \in I} G_i$ such that $g_i = \varphi_i^j g_j$ for all $i \leqslant j$. The morphism $\varphi_i \colon G \to G_i$ is the restriction of $\prod_{i \in I} g_i$ to g_i. □

The inverse limit of $\{I, G_i, \varphi_i^j\}$ will be denoted by $\varprojlim_{i \in I} G_i$ or $\varprojlim_i G_i$.

Example 2. Let G be an arbitrary topological group (ring) and I a set of closed normal subgroups (ideals) of finite index in G which is closed with respect to finite intersections. Let $i \leqslant j$ if and only if $i \supseteq j$ for $i, j \in I$. For $i \leqslant j$ let φ_i^j be the projection $G/j \to G/i$. Then $\{I, G/i, \varphi_i^j\}$ is an inverse system. The corresponding limit will be denoted by G^I. If I is the set of all closed normal subgroups (ideals) of finite index in G, then G^I is called *total completion* of G and is denoted by $\hat{G}$. Let p be a prime. If I is the set of all closed normal subgroups (ideals) of p-power index, then G^I is called *p-completion* of G. □

Example 3. The p-completion of the discrete ring $\mathbb{Z}$ of integers is isomorphic to the ring $\mathbb{Z}_p$ of p-adic integers, the total completion of $\mathbb{Z}$ is isomorphic to the direct product $\prod \mathbb{Z}_p$ taken over all primes p.

More general, let $\mathfrak{O}$ be a Dedekind ring (Chap. 1.3) and $\mathfrak{p}$ a prime ideal of finite index in $\mathfrak{O}$. Then the completion of $\mathfrak{O}$ with respect to the set I of the powers of $\mathfrak{p}$ is isomorphic to the ring $\mathfrak{O}_\mathfrak{p}$ of integers in the completion of the quotient field K of $\mathfrak{O}$ with respect to the valuation of K corresponding to $\mathfrak{p}$.

Let $\mathfrak{O}$ be a Dedekind ring such that $\mathfrak{O}/\mathfrak{p}$ is finite for all prime ideals $\mathfrak{p}(\neq \{0\})$. Then the total completion of $\mathfrak{O}$ is isomorphic to the direct product $\mathfrak{O}_\mathfrak{p}$ taken over all prime ideals $\mathfrak{p}$ of $\mathfrak{O}$. □

Let $P = \{I, G_i, \varphi_i^j\}$ and $Q = \{J, H_i, \psi_i^j\}$ be inverse systems. A morphism ψ of P into Q consists in a morphism ϕ of J into I and a family of morphisms ψ_j, $j \in J$, of $G_{\phi(j)}$ into H_j such that for $i \leqslant j$ the diagram

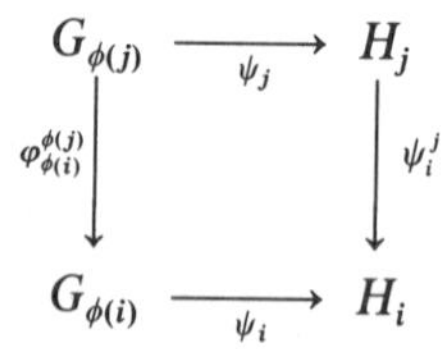

commutes.

We associate to ψ a morphism φ of $\varprojlim_i G_i$ into $\varprojlim_i H_i$ by means of

$$\varphi\left(\prod_{i\in I} g_i\right) = \prod_{j\in J} \psi_j(g_{\phi(j)}).$$

It is clear by the definitions above what one has to understand by the category $\mathfrak{K}_I$ of inverse systems over the directed set I. The functor $\varprojlim_{i\in I}$ of $\mathfrak{K}_I$ into $\mathfrak{K}$ is exact. With other words:

Proposition 3.2. *Let $P = \{I, F_i, \theta_i^j\}$, $Q = \{I, G_i, \varphi_i^j\}$, $R = \{I, H_i, \psi_i^j\}$ be inverse systems and let $\theta = \{\theta_i\}$ and $\varphi = \{\varphi_i\}$ be morphisms of P into Q and Q into R, respectively, such that for every $i \in I$ the sequence*

$$\{1\} \to F_i \to G_i \to H_i \to \{1\}$$

is exact. Then the sequence

$$\{1\} \to \varprojlim_i F_i \to \varprojlim_i G_i \to \varprojlim_i H_i \to \{1\}$$

is exact. □

1.2. Profinite Groups. A *profinite group* is a topological group which can be represented as inverse limit of finite groups.

Example 4. The groups G^I in example 2 are profinite groups, in particular $\mathbb{Z}_p^+$ and $\mathbb{Z}_p^\times$ are profinite groups. □

Example 5. An analytical group over $\mathbb{Q}_p$ is defined similar as an analytical group over $\mathbb{R}$. A compact analytical group over $\mathbb{Q}_p$ is a profinite group. In particular $GL_n(\mathbb{Z}_p)$, $SL_n(\mathbb{Z}_p)$ are profinite groups. □

Let G, $\{\varphi_i | i \in I\}$ be the inverse limit of the inverse system $\{I, G_i, \varphi_i^j\}$. Then $\{\operatorname{Ker} \varphi_i | i \in I\}$ is a full system of neighborhoods of the unit in the topology of G.

A subgroup of a profinite group is open if and only if it is closed and of finite index.

Now we give an inner characterization of profinite groups:

Proposition 3.3. *The following properties of compact groups G are equivalent: 1. G is profinite. 2. G is totally disconnected. 3. There exists a set $\mathfrak{U}$ of open normal subgroups of G which is a full system of neighborhoods of the unit of G.* □

The category of profinite groups (morphisms are continuous homomorphisms) is closed with respect to closed subgroups, factor groups, infinite direct products and inverse limits. The following proposition guarantees the existence of a continuous system of representatives for the cosets of subgroups:

Proposition 3.4. *Let H be a closed subgroup of the profinite group G. Then there is a continuous section σ of G/H into G with $\sigma(H) = 1$, i.e. a continuous map σ of G/H into G such that the composition of σ with the projection $G \to G/H$ is the identity.* □

Let G be an abelian profinite group and χ a continuous homomorphism of G into $\mathbb{R}/\mathbb{Z}$. With G also $\operatorname{Im}\chi$ is compact and totally disconnected. Therefore $\operatorname{Im}\chi$ is finite, hence cyclic and contained in $\mathbb{Q}/\mathbb{Z}$. This means that the Pontrjagin dual $X(G) := \operatorname{Hom}(G, \mathbb{R}/\mathbb{Z})$ of G is a discrete torsion group. $G \to X(G)$ is an exact functor from the category of abelian profinite groups onto the category of abelian discrete torsion groups.

Example 6. The dual group of the total completion of $\mathbb{Z}^+$ is $\mathbb{Q}/\mathbb{Z}$. □

The dual notion of the inverse limit of compact abelian groups is the notion of direct limit of discrete abelian groups (see § 1.4). Let $\{G_i | i \in I\}$ be an inverse system of compact abelian groups. Then

$$X\left(\varprojlim_i G_i\right) = \varinjlim_i X(G_i) \tag{3.1}$$

1.3. Supernatural Numbers. A *supernatural number* is a formal product $\prod_p p^{n_p}$ where p runs over all primes and n_p is a non negative integer or ∞. The product, g.c.d. and l.c.m. of any set of supernatural numbers is defined in obvious manner.

Let G be a profinite group and H a closed subgroup of G. The index $[G:H]$ of H in G is defined as l.c.m. of the finite indices $[G:HU]$ where U runs over all open normal subgroups of G.

Proposition 3.5.1. *Let $K \subseteq H \subseteq G$ be profinite groups, then*

$$[G:K] = [G:H][H:K].$$

2. *Let $\{H_i | i \in I\}$ be a descending filtering family of closed subgroups of G such that $H = \bigcap_{i \in I} H_i$. Then*

$$[G:H] = \text{l.c.m.}\{[G:H_i] | i \in I\}.$$

3. *A closed subgroup H is open in G if and only if $[G:H]$ is a natural number.* □

1.4. Pro-p-Groups and p-Sylow Groups. Let p be a prime. A profinite group H is a *pro-p-group* if it is the inverse limit of p-groups. This is the case if and only if the order of H is a finite or infinite p-power. A closed subgroup H of a profinite group G is called *p-Sylow group* if H is a pro-p-group and if the index $[G:H]$ is prime to p.

Proposition 3.6. *Let G be a profinite group and p a prime.*

1. *There exists a p-Sylow group of G and any two p-Sylow groups of G are conjugated.*
2. *Every pro-p-group contained in G is a subgroup of a p-Sylow group.*
3. *Let G' be the image of a morphism φ of G and H a p-Sylow group of G. Then $\varphi(H)$ is a p-Sylow group of G'.* ⊠

1.5. Free Profinite, Free Prosolvable, and Free Pro-p-Groups. Let I be a set and $F(I)$ the free discrete group with generator system $\{s_i | i \in I\}$. Furthermore let

$\mathfrak{N}$ be the system of open normal subgroups containing almost all s_i. Then $\varprojlim_{N \in \mathfrak{N}} F(I)/N$ is called the *free profinite group* $F = F_I$ with free generator system $\{s_i | i \in I\}$. If $\mathfrak{N}$ is the system of open subgroups with solvable factor group (resp. p-power index) containing almost all s_i, then $\varprojlim_{N \in \mathfrak{N}} F(I)/N$ is called the *free prosolvable* (resp. *free pro-p-group*) $F = F_I$ with free generator system $\{s_i | i \in I\}$.

These definitions are justified by the following facts which are valid in all three cases:

Proposition 3.7. 1. *The natural map* $F(I) \to F$ *is an injection.*

2. *Let G be a profinite (prosolvable or pro-p-) group and* $\{g_i | i \in I\}$ *a family of elements of G such that any neighborhood of* 1 *contains almost all* g_i. *Then* $s_i \to g_i$, $i \in I$, *defines a morphism of F into G.* ⊠

One can always find such a family $\{g_i | i \in I\}$ which generates G, i.e. the smallest closed subgroup of G containing $\{g_i | i \in I\}$ is equal to G. Then $\{g_i | i \in I\}$ is called a *generator system* of G. A generator system is called *minimal* if it has no proper subset generating G.

1.6. Discrete Modules. Let G be a profinite group and A a unitary G-module. We consider A as a topological space with the discrete topology. A is called *discrete G-module* if the action of G on A is continuous.

Proposition 3.8. *Let G be a profinite group and A a unitary G-module. The following conditions are equivalent:*

1. *A is a discrete G-module.*
2. *For every* $a \in A$, *the stabilizer* $G_a = \{g \in G | ga = a\}$ *is an open subgroup of G.*
3. *Let* $\mathfrak{U}$ *be the set of all open normal subgroups of G. Then*

$$A = \bigcup_{U \in \mathfrak{U}} A^U. \square$$

Example 7. Let A be an abelian group and G a profinite group acting trivially on A. Then A is a discrete G-module. □

Example 8. Multiplication by p-adic integers defines $\mathbb{Q}_p/\mathbb{Z}_p$ as discrete $\mathbb{Z}_p$-module. □

Let G be a profinite group, H a closed subgroup of G and A a discrete H-module. $M_G^H(A)$ denotes the set of all continuous maps f of G into A with

$$hf(x) = f(hx) \qquad \text{for } h \in H, x \in G \text{ (compare Chap. 2.3.4).}$$

The map f is continuous if and only if there exists an open normal subgroup U in G such that f depends only on the classes of G/U.

We define $M_G^H(A)$ as G-module by

$$(gf)(x) = f(xg) \qquad \text{for } g, x \in G.$$

It is easy to see that $M_G^H(A)$ is a discrete G-module. $M_G(A) := M_G^{(1)}(A)$ is called *induced module.*

In the following a G-module means always a discrete unitary G-module. Beside the category C_G of G-modules the following category C is important: The objects of C are the pairs $[G, A]$ where G is a profinite group and A a G-module. A morphism of $[G, A]$ into $[H, B]$ is a pair $[\varphi, \psi]$ such that φ is a morphism of H into G and ψ is a homomorphism of abelian groups of A into B with

$$\psi(\varphi(h)a) = h\psi(a) \qquad \text{for } h \in H,\ a \in A.$$

For finite groups G the category C coincides with the category considered in Chap. 2.3.2.

1.7. Inductive Limits in C. Let I be a directed set. A *direct system* (also called *inductive system*) $\{I, [G_i, A_i], [\varphi_i^j, \psi_i^j]\}$ in C is a covariant functor ϕ of the category I (§ 1.1) in C with $\phi(i) = [G_i, A_i]$ for $i \in I$ and $\phi(i \leqslant j) = [\varphi_i^j, \psi_i^j]$. Then $\{I, G_i, \varphi_i^j\}$ is an inverse system in the category of profinite groups and $\{I, A_i, \psi_i^j\}$ is a direct system of abelian groups. To every G-module A we associate the trivial direct system $D_{[G,A]} = \{I, [G, A], [\varphi_i^j, \varphi_i^j]\}$ where φ_i^j and ψ_i^j are the identical maps. To a morphism $[\varphi, \psi]$ of $[G, A]$ into $[G', A']$ we associate in an obvious way a morphism of $D_{[G,A]}$ into $D_{[G',A']}$ denoted by $[\varphi, \psi]$, too.

Let $D = \{I, [G_i, A_i], [\varphi_i^j, \psi_i^j]\}$ be a direct system. A G-module A together with a functor morphism ϕ of D into $D_{[G,A]}$ is called *direct limit* of D if for every G'-module A' and for every functor morphism ϕ' of D into $D_{[G',A']}$ there exists one and only one morphism $[\varphi, \psi]$ of $[G, A]$ into $[G', A']$ such that the diagram

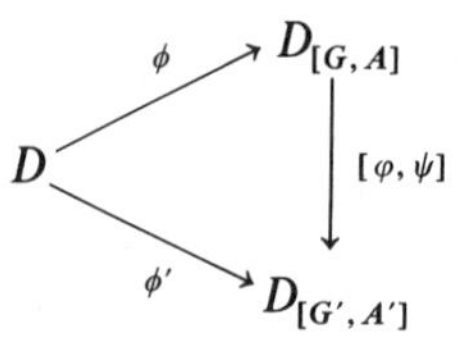

commutes.

A functor morphism of D into $D_{[G,A]}$ is given by a family $\{[\varphi_i, \psi_i] \mid i \in I\}$ of morphisms of $[G_i, A_i]$ into $[G, A]$ such that for $i \leqslant j$ the diagrams

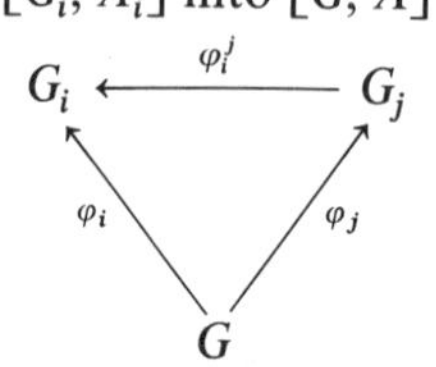

and

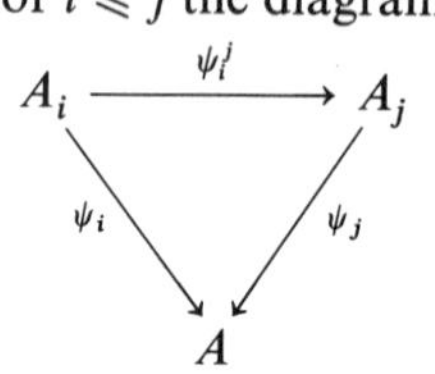

commute.

It follows from the definition that direct limits are uniquely determined up to isomorphism.

Proposition 3.9. *In the category C direct limits exist.*

Proof. Two elements $a_i \in A_i$ and $a_j \in A_j$ are called equivalent if there is a $k \in I$ such that $i \leqslant k$, $j \leqslant k$, and $\psi_i^k(a_i) = \psi_j^k(a_j)$.

One gets an equivalence relation in the disjunct union of the abelian groups A_i, $i \in I$. The classes with respect to this equivalence relation form an abelian group A in an obvious way, $A = \varinjlim_i A_i$ is the direct limit of the direct system $\{I, A_i, \psi_i^j\}$. A is a $G = \varprojlim_i G_i$-module and there is a functor morphism of D into $D_{[G,A]}$ defined in a natural way such that $[G, A]$, ϕ is the direct limit of D. The details are left to the reader. □

Let G be a profinite group and $\mathfrak{U}$ the set of open normal subgroups of G. Every pair $[G, A]$ is limit of the direct system $\{\mathfrak{U}, G/U, A^U\}$. Therefore every (discrete) G-module is the direct limit of G_i-modules with finite groups G_i.

1.8. Galois Theory of Infinite Algebraic Extensions. Let K be an arbitrary field and L an extension of K. We call L/K a normal extension if L is the union of finite normal separable algebraic extensions of K.

The Galois group $G(L/K)$ of the normal extension L/K is the topological group of all automorphisms of L which fix the elements of K. The topology of $G(L/K)$ is defined by means of the system

$$\mathfrak{U}(L/K) := \{G(L/N) | N \in \mathfrak{N}\}$$

of neighborhoods of the unit where $\mathfrak{N}$ is the set of all finite normal subextensions of L/K.

It is easy to see that $G(L/K)$ is the inverse limit of the inverse system $\{\mathfrak{N}, G(N/K), \varphi_N^M\}$ where for $N \subseteq M$, φ_N^M is the projection $G(M/K) \to G(N/K)$. Hence $G(L/K)$ is a profinite group.

Proposition 3.10 (Main theorem of Galois theory). *Let L/K be a normal extension. The map*

$$\phi : M \to G(L/M)$$

defines a one to one correspondence between all subextensions M/K of L/K and all closed subgroups of $G(L/K)$. The inverse of ϕ is the map which associates to a closed subgroup U of $G(L/K)$ the fixed field $K(U) := \{\alpha \in L | g\alpha = \alpha$ for $g \in U\}$. ⊠

Now let R be a Dedekind ring with quotient field K, v a valuation of K corresponding to a prime ideal of R, L a normal extension of K, and w an extension of v to L. A part of the considerations in Chap. 1.3.7 can be transferred to infinite extensions:

If $g \in G(L/K)$, we define gw by

$$gw(\alpha) = w(g^{-1}\alpha) \qquad \text{for } \alpha \in L.$$

Furthermore the decomposition group G_w and the inertia group T_w of w are defined by

$$G_w := \{g \in G | gw = w\},$$

$$T_w := \{g \in G_w | w(g\alpha - \alpha) < 1 \qquad \text{for } \alpha \in L, w(\alpha) \leqslant 1\}.$$

If the residue class field extension $\mathfrak{L}_w/\mathfrak{K}_v$ is separable, then there is a natural isomorphism of G_w/T_w onto $G(\mathfrak{L}_w/\mathfrak{K}_v)$. If $\mathfrak{K}_v$ is finite, then the Frobenius automorphism $F_w \in G_w/T_w$ is defined as in Proposition 1.52.5 and F_w generates G_w/T_w.

The higher ramification groups can be defined for the upper numeration by means of Theorem 1.55: If $y \in \mathbb{R}$, $y \geqslant 0$, we define

$$V^y(w, L/K) = \varprojlim_{N \in \mathfrak{N}} V^y(w|_N, N/K). \tag{3.2}$$

y is a jump if $V^y \neq V^{y+\varepsilon}$ for arbitrary $\varepsilon > 0$. Any non negative real number can be a jump. See Maus (1967)–(1973), Sen (1972), Gordeev (1981) for a detailed discussion of the filtration $\{V^y | y \in \mathbb{R}_+\}$ of $G(L/K)$ in the case of local fields.

Example 9. Let K be a finite field and L the algebraic closure of K. Then $G(L/K)$ is the total completion $\hat{\mathbb{Z}}^+$ of $\mathbb{Z}^+$. □

Example 10. Let K be a finite extension of $\mathbb{Q}_p$ and $L = K_{nr}$ the maximal unramified extension of K in an algebraic closure of K. Then $G(L/K) = \hat{\mathbb{Z}}^+$. □

Example 11. Let $K = \mathbb{Q}$ and let L be the union of all cyclotomic extensions of $\mathbb{Q}$ (Chap. 1.3.9). Then $G(L/K)$ is isomorphic to the multiplicative group of the ring $\hat{\mathbb{Z}}$. □

Example 12. Let K be a local or global field with class module A_K (Chap. 2.1.6), let $\bar{K}$ be a separable algebraic closure, and let K^{ab} be the maximal abelian extension of K in $\bar{K}$. Then by class field theory (Chap. 2.1) $G(K^{ab}/K) = G(\bar{K}/K)^{ab}$ is isomorphic to the total completion of A_K. This isomorphism is induced by the norm symbol $(\alpha, K) \in G(K^{ab}/K)$, which is defined by

$$(\alpha, K) := \varprojlim_{L} (\alpha, L/K) \qquad \text{for } \alpha \in A_K,$$

where L runs through the finite extensions of K in K^{ab} (compare Theorem 2.9). The functorial properties of this symbol are immediate consequences of the Theorems 2.10 and 2.11:

Let K'/K be a finite extension in $\bar{K}$. If H is an open subgroup of the profinite group G, the transfer from G into H is defined as in the case of finite groups. Let Ver be the transfer from $G(\bar{K}/K)$ into $G(\bar{K}/K')$. Then

$$\mathrm{Ver}(\alpha, K) = (\alpha, K') \qquad \text{for } \alpha \in A_K$$

and

$$\kappa(\alpha, K') = (N_{K'/K}\alpha, K) \qquad \text{for } \alpha \in A_{K'},$$

where κ denotes the homomorphism from $G(\bar{K}/K')^{ab}$ into $G(\bar{K}/K)^{ab}$ induced by the inclusion $G(\bar{K}/K') \subseteq G(\bar{K}/K)$.

Let g be an automorphism of $\bar{K}$, then

$$(g\alpha, gK) = g(\alpha, K)g^{-1} \qquad \text{for } \alpha \in A_K. \ \boxtimes$$

Example 13. Let F be the Riemann surface of the finite extension K of the function field $\mathbb{C}(z)$ and let S be a finite set of points of F. Let g be the genus of

F. Then the Galois group G of the maximal extension of K which is unramified outside S is isomorphic to the total completion of the fundamental group of $F - S$ (Springer (1957), Chap. 10.9). Corresponding to the well known structure of this group (see e.g. Seifert, Threlfall (1934), Chap. 7), G is isomorphic to the free profinite group with $2g + |S| - 1$ generators if $S \neq \emptyset$ and G is isomorphic to the profinite group with generators $s_1, t_1, \ldots, s_g, t_g$ and one relation

$$s_1^{-1} t_1^{-1} s_1 t_1 \ldots s_g^{-1} t_g^{-1} s_g t_g$$

if $S = \emptyset$. In particular if $g = 0$ and $S = \emptyset$, i.e. if K is of the form $\mathbb{C}(z)$, then $G = \{1\}$. This corresponds to Minkowski's discriminant theorem (Theorem 1.4.7). ⊠

Example 14. Let K be a $\mathfrak{p}$-adic number field with prime ideal $\mathfrak{p}|p$. The compositum of finite tamely ramified extensions of K is again tamely ramified (Proposition 1.78). Let L be the maximal tamely ramified extension of K in an algebraic closure of K. Then $G(L/K_{nr})$ is isomorphic to $\prod_{q \neq p} \mathbb{Z}_q$ where the product runs over all primes $q \neq p$. If s is an extension of the Frobenius automorphism of K_{nr}/K to L, then

$$sts^{-1} = t^{N(\mathfrak{p})} \qquad \text{for } t \in G(L/K_{nr}). \ \square \tag{3.3}$$

1.9. Cohomology of Profinite Groups. Let G be a profinite group and A a (discrete) G-module. The definition of the cohomology groups $H^n(G, A)$, $n = 0, 1, \ldots$, is similar to the definition of cohomology groups for abstract groups (Chap. 2.3). An n-dimensional cochain f of G with values in A is a continuous map of the n-fold product of G into A. Continuous means that there is an open subgroup U of G such that the values of f depend only on the cosets of G with respect to U. With this modification all what was said in Chap. 2.3.1–7 carries over to the case of profinite groups and we don't repeat it. We use the same notations as in Chap. 2.3.

Proposition 3.11. *Let $\{I, [G_i, A_i], [\varphi_i^j, \psi_i^j]\}$ be a direct system in C with limit $[G, A]$. Then*

$$H^n(G, A) = \varinjlim_i H^n(G_i, A_i) \qquad \text{for } n = 0, 1, \ldots. \ \square$$

Proposition 3.12. *Let G be a profinite group and A a G-module. Then $H^n(G, A)$ is a torsion group for $n > 0$.*

Proof. This follows from Proposition 2.42, Proposition 3.11, and § 1.7. □

1.10. Cohomological Dimension. Let G be a profinite group and p a prime. The *cohomological p-dimension* $cd_p(G)$ of G is the smallest integer n such that for all $m > n$ and for every G-module A being a torsion group, the p-primary component of $H^m(G, A)$ vanishes. If such an integer does not exists, we put $cd_p(G) = \infty$.

The *cohomological dimension* $cd(G)$ of G is the supremum of the p-dimensions.

Proposition 3.13. *Let $n \geqslant 0$ be an integer such that $H^{n+1}(G, A) = \{0\}$ for all simple G-modules A which are annihilated by p. Then $cd_p(G) \leqslant n$.* ⊠

An abelian group A is called p-divisible if $pA = A$.

Proposition 3.14. *Let $cd_p(G) \leqslant n$ and A a p-divisible G-module. Then the p-primary component of $H^m(G, A)$ vanishes for $m > n$.* □

Example 15. *$cd_p(G) \leqslant 1$ if G is a free pro-p-group (comp. § 1.12), a free profinite group, or a free prosolvable group. $cd_p(G) = 1$ if $G = \hat{\mathbb{Z}}$.* □

The *strict cohomological p-dimension* $scd_p(G)$ of G is the smallest integer n such that for all $m > n$ the p-primary component of $H^m(G, A)$ vanishes for every G-module A. The *strict cohomological dimension* $scd(G)$ is the supremum of the strict cohomological p-dimensions of G.

Proposition 3.15. *$scd_p(G)$ is equal to $cd_p(G)$ or $cd_p(G) + 1$.* ⊠

Proposition 3.16. *Let H be a closed subgroup of G. Then $cd_p(H) \leqslant cd_p(G)$, $scd_p(H) \leqslant scd_p(G)$. If H is a p-Sylow group of G, then $cd_p(H) = cd_p(G)$, $scd_p(H) = scd_p(G)$.* ⊠

Proposition 3.16 shows that for the question of cohomological p-dimension it is sufficient to consider torsion free pro-p-groups. (A pro-p-group with torsion has cohomological dimension ∞ since it has a non trivial finite cyclic subgroup, which has periodical cohomology (Chap. 2.3.9)).

Theorem 3.17. *Let H be an open subgroup of a torsion free pro-p-group G. Then $cd(H) = cd(G)$.* ⊠ (Serre (1965)).

Proposition 3.18. *Let G be a profinite group with $cd_p(G) = n < \infty$ and $\mathbb{Z}$ the G-module with trivial action of G. Then $scd_p(G) = n$ if and only if the p-component of $H^{n+1}(H, \mathbb{Z})$ vanishes for all open subgroups H of G.* ⊠

Proposition 3.19. *Let H be a closed normal subgroup of G. Then*

$$cd_p(G) \leqslant cd_p(H) + cd_p(G/H). \boxtimes$$

See also Proposition 3.25.

1.11. The Dualizing Module. Let n be a natural number and G a profinite group such that $cd(G) \leqslant n$ and for every finite G-module A the group $H^n(G, A)$ is finite.

Proposition 3.20. *There exists a G-module I which is a torsion group and a map $\imath: H^n(G, I) \to \mathbb{Q}/\mathbb{Z}$ such that the induced homomorphism φ of $\mathrm{Hom}_G(A, I)$ into $H^n(G, A)^*$ given by*

$$\varphi(f)(\alpha) = \imath f_*(\alpha) \qquad \text{for } f \in \mathrm{Hom}_G(A, I),\ \alpha \in H^n(G, A)$$

is an isomorphism for every finite G-module A. ⊠

The pair I, ι is uniquely determined up to isomorphism and is called the *dualizing module* of G for the dimension n.

Proposition 3.21. *The dualizing module of G is also the dualizing module for every open subgroup of G.* ⊠

Proposition 3.22. $scd_p(G) = n + 1$ *if and only if there exists an open subgroup H of G such that I^G contains a subgroup isomorphic to $\mathbb{Q}_p/\mathbb{Z}_p$.* ⊠

Example 16. Let $G = \hat{\mathbb{Z}}$ and $n = 1$. Furthermore let σ be the automorphism of the finite G-module A induced by $1 \in \mathbb{Z}$. Then $H^1(G, A)$ can be identified with $A/(\sigma - 1)A$ by means of the map which associates to a cocycle $h: \hat{\mathbb{Z}} \to A$ its value $h(1)$. This shows that the dualizing module of $\hat{\mathbb{Z}}$ is $\mathbb{Q}/\mathbb{Z}$ with trivial action of $\hat{\mathbb{Z}}$. □

1.12. Cohomology of Pro-p-Groups. Let G be a pro-p-group and A a simple G-module which is annihilated by p. Then A is finite and there is an open normal subgroup U of G such that U acts trivially on A. Hence A is a G/U-module and it is known from representation theory that A is isomorphic to $\mathbb{Z}/p\mathbb{Z}$ with trivial action of G. Therefore according to Proposition 3.13 we have the following result.

Proposition 3.23. *A pro-p-group G has cohomological dimension $cd(G) \leqslant n$ if and only if $H^{n+1}(G, \mathbb{Z}/p\mathbb{Z}) = \{0\}$.* ⊠

In the following we write $H^i(G) := H^i(G, \mathbb{Z}/p\mathbb{Z})$ for short.

It is easy to see that $cd(G) = 0$ if and only if $G = \{1\}$. The following characterization of free pro-p-groups is a corner stone in the theory of pro-p-groups:

Proposition 3.24. *A pro-p-group G is free if and only if $cd(G) \leqslant 1$.* ⊠

Proposition 3.25. *Let G be a pro-p-group and H a closed normal subgroup of G. Let $n := cd(H)$ and $m := cd(G/H)$ be finite. Then one has $cd(G) = n + m$ in the following two cases:*

1. *$H^n(H)$ is finite.*
2. *H is contained in the centre of G.* ⊠

Example 17. Let $\mathbb{Z}_p^n$ be the n-fold direct product of $\mathbb{Z}_p$. Then $cd(\mathbb{Z}_p^n) = n$. □

If the groups $H^i(G)$ are finite for $i = 0, 1, \ldots, n$, one can define the *partial Euler-Poincaré characteristic*

$$\chi_n(G) := \sum_{i=0}^{n} (-1)^i \dim_{\mathbb{Z}/p\mathbb{Z}} H^i(G).$$

If $H^i(G) = \{0\}$ for $i > n$, then $\chi(G) := \chi_n(G)$ is called the *Euler-Poincaré characteristic of G and we say that G is a pro-p-group with Euler-Poincaré characteristic.*

Proposition 3.26. *Let G be a pro-p-group with Euler-Poincaré characteristic and H an open subgroup of G. Then also H has an Euler-Poincaré characteristic and*

$$\chi(H) = [G : H]\chi(G). \boxtimes$$

Proposition 3.27. *Let G be a pro-p-group and N a closed normal subgroup of G such that N and G/N have an Euler-Poincaré characteristic. Then also G has an Euler-Poincaré characteristic and*

$$\chi(G) = \chi(N)\chi(G/N). \boxtimes$$

Proposition 3.28. *Let G be a pro-p-group with partial Euler-Poincaré characteristic $\chi_n(G)$ and for every open subgroup U let*

$$\chi_n(U) = [G:U]\chi_n(G).$$

Then $cd(G) \leqslant n$. $\boxtimes$

1.13. Presentation of Pro-p-Groups by Means of Generators and Relations. Let G be a pro-p-group and q a power of p. The descending central q-series $\{G^{(n,q)} | n = 1, 2, \dots\}$ is defined recurrently as follows:

$G^{(1,q)} := G$, $G^{(n+1,q)}$ is the closed subgroup of G generated by the elements of the form g^q and $(g, h) := g^{-1}h^{-1}gh$ for $g \in G^{(n,q)}$, $h \in G$.

The *generator rank* $d(G)$ of G is defined as the dimension of the $\mathbb{Z}/p\mathbb{Z}$-vector space $H^1(G)$. This definition is motivated by the following *Burnside's basis theorem*

Proposition 3.29. *Let $\{g_i | i \in I\}$ be a minimal generator system of G* (§ 1.5). *Then*

$$d(G) = |I| = d(G/G^{(2,p)}). \boxtimes$$

Example 18 (theorem of Schreier). *Let F be a free pro-p-group and U an open subgroup of F. Then*

$$d(U) = 1 + [F:U](d(F) - 1) \text{(Proposition 3.24, Proposition 3.26).} \square$$

Let G be a pro-p-group and F a free pro-p-group with free generator system $\{t_i | i \in I\}$. We call an exact sequence

$$\{1\} \to R \to F \underset{\varphi}{\to} G \to \{1\} \tag{3.4}$$

a presentation of G by F. The presentation is called minimal if $\{\varphi t_i | i \in I\}$ is a minimal generator system of G.

A subset E of R is called a (generating) *relation system* of G (for the presentation (3.4)) if 1. R is generated by E as closed normal subgroup of F and 2. every open normal subgroup of R contains almost all elements of E. We say that G is presented by the generator system $\{t_i | i \in I\}$ and the relation system E for short.

E is called *minimal* if no proper subset of E is a relation system of G.

Proposition 3.30. *Let G be a pro-p-group and E a minimal relation system of G. Then*

$$|E| = \dim_{\mathbb{Z}/p\mathbb{Z}} H^2(G). \boxtimes \tag{3.5}$$

The number (3.5) is called the *relation rank* of G.

In our application to number fields (Chap. 3.2.6) we will have a family $\{G_i | i \in I\}$ of pro-p-groups and morphisms $\{\varphi_i\colon G_i \to G\}$ into a pro-p-group G. We study the notion of relation system in this connection. Some complications are due to condition 2. of the definition of relation systems.

For every i let T_i be a normal subgroup of G_i such that G_i/T_i is a free pro-p-group. $\{\varphi_i | i \in I\}$ is called *admissible* with respect to $\{T_i | i \in I\}$ if every open subgroup of G contains almost all $\varphi_i(T_i)$.

Example 19. If I is finite and $T_i = G_i$, then $\{\varphi_i | i \in I\}$ is admissible. □

Proposition 3.31. *Let $\{\varphi_i | i \in I\}$ be admissible with respect to $\{T_i | i \in I\}$. Moreover we assume that we have presentations*

$$\{1\} \to R_i \to F_i \underset{\psi_i}{\to} G_i \to \{1\} \tag{3.6}$$

for every $i \in I$. Then there exist morphisms χ_i of F_i into F with restrictions χ_i' to R_i such that the diagrams

$$\begin{array}{ccccc} R_i & \longrightarrow & F_i & \xrightarrow{\psi_i} & G_i \\ \downarrow{\scriptstyle \chi_i'} & & \downarrow{\scriptstyle \chi_i} & & \downarrow{\scriptstyle \varphi_i} \\ R & \longrightarrow & F & \longrightarrow & G \end{array} \tag{3.7}$$

are commutative and $\{\chi_i' | i \in I\}$ is admissible with respect to $\{R_i | i \in I\}$. □

We call (3.7) in this case an admissible presentation of $\{\varphi_i | i \in I\}$.

For the rest of this section we assume that an admissible presentation (3.7) is given.

Proposition 3.32. *The groups $\chi_i(R_i)$, $i \in I$, generate R as closed normal subgroup of F if and only if the restriction to $H^1(R)^G$ of the map*

$$\prod_{i \in I} \chi_i^* \colon H^1(R) \to \prod_{i \in I} H^1(R_i) \tag{3.8}$$

is injective. □

By our assumptions the image of $\prod_{i \in I} \chi_i^*$ lies already in $\sum_{i \in I} H^1(R_i)^{G_i}$.

Now we come to the main result of this section. A subset E of R is called *supplementary set* of $\{\chi_i | i \in I\}$ if E together with $\bigcup_{i \in I} \chi_i(R_i)$ generates R as closed normal subgroup of F and every open subgroup of R contains almost all elements of E. The supplementary set E is called *minimal* if no proper subset of E is supplementary set of $\{\chi_i | i \in I\}$.

Proposition 3.33. *Let E be a minimal supplementary set of $\{\chi_i | i \in I\}$ and*

$$\varphi^* := \prod_{i \in I} \varphi_i^* \colon H^2(G) \to \prod_{i \in I} H^2(G_i).$$

Then

$$\dim \operatorname{Ker} \varphi^* = |E|. \boxtimes$$

The proof of Proposition 3.33 combines Proposition 3.32 with the consideration of the exact diagram (Proposition 2.43)

$$\begin{array}{ccccccccc}
H^1(G) & \xrightarrow[\text{Inf}]{} & H^1(F) & \xrightarrow[\text{Res}]{} & H^1(R)^G & \xrightarrow[\text{Tra}]{} & H^2(G) & \longrightarrow & \{0\} \\
\downarrow & & \downarrow & & \downarrow & & \downarrow & & \\
\Pi H^1(G_i) & \xrightarrow[\text{Inf}]{} & \Pi H^1(F_i) & \xrightarrow[\text{Res}]{} & \Pi H^1(R_i)^{G_i} & \xrightarrow[\text{Tra}]{} & \Pi H^2(G_i) & \longrightarrow & \{0\}
\end{array}$$

Since (3.4) and (3.6) are minimal presentations, the transgressions are isomorphisms.

Proposition 3.30 is a special case of Proposition 3.33.

Proposition 3.34. *The assumptions being as above, R is generated as closed normal subgroup of F by the subgroups $\chi_i(R_i)$, $i \in I$, if and only if φ^* is injective.* □

1.14. Poincaré Groups. Let G be a pro-p-group of cohomological dimension n. Then G is called a *Poincaré group* of dimension n if the following conditions are satisfied:

1. $H^i(G)$ is finite for $i \geqslant 0$, $H^n(G) \cong \mathbb{Z}/p\mathbb{Z}$.
2. (Poincaré duality) The cup product

$$H^i(G) \times H^{n-i}(G) \to H^n(G)$$

given by the product $\mathbb{Z}/p\mathbb{Z}$ (Chap. 2, example 16) is not degenerated for $0 \leqslant i \leqslant n$.

Example 20 (theorem of Lazard). Any open torsion free subgroup of a compact analytic group over $\mathbb{Q}_p$ of dimension n is a Poincaré group of dimension n (Lazard (1965), Serre (1965)). ⊠

Proposition 3.35. *An open subgroup of a Poincaré group is a Poincaré group of the same dimension.* ⊠

Proposition 3.36. *Let I be the dualizing module of a Poincaré group G of dimension n. Then as an abelian group, I is isomorphic to $\mathbb{Q}_p/\mathbb{Z}_p$.* ⊠

The profinite group $U := \mathbb{Z}_p^\times$ is canonically isomorphic to the automorphism group of $\mathbb{Q}_p/\mathbb{Z}_p$. Hence the action of G on I induces a canonical homomorphism $\chi: G \to U$. Since G is a pro-p-group, the image of χ lies in the group U_1 of principal units in $\mathbb{Q}_p$ (Chap. 1.4.4).

Proposition 3.37. *Let G be a Poincaré group of dimension n and let $\chi: G \to U$ be the corresponding morphism. Then $scd(G) = n + 1$ if and only if* $\operatorname{Im} \chi$ *is finite.* ⊠

$\mathbb{Z}_p^+$ is up to isomorphism the unique Poincaré group of dimension 1. Poincaré groups of dimension 2 are called *Demushkin groups* (Serre (1963)). These groups are completely classified:

Theorem 3.38. *Let G be a Demushkin group of relation rank d. Then $G/[G, G]$ is isomorphic to $\mathbb{Z}_p^{d-1} \times (\mathbb{Z}_p/q\mathbb{Z}_p)$ where $q = q(G)$ is a uniquely determined power of p ($p^\infty = 0$).* ⊠

Beside $d(G)$ and $q(G)$ we need the invariant $\operatorname{Im}\chi \subseteq U_1$ where χ is the morphism of G into U_1 defined above. If $q(G) \neq 2$, then $\operatorname{Im}\chi = 1 + q\mathbb{Z}_p$ and we get nothing new. If $q(G) = 2$ and d odd, then $\operatorname{Im}\chi = \{\pm 1\} \times U_f$ where $f = f(G) \geqslant 2$ or ∞, $U_f = 1 + 2^f\mathbb{Z}_2$. If $q(G) = 2$ and d even, then

$$a(G) := [\operatorname{Im}\chi : (\operatorname{Im}\chi)^2] = 2 \text{ or } 4.$$

If $a(G) = 2$, then $\operatorname{Im}\chi = (-1 + 2^f)$ where $f = f(G) \geqslant 2$ or ∞. If $a(G) = 4$, then $d \geqslant 4$ and $\operatorname{Im}\chi = \{\pm 1\} \times U_f$ where $f = f(G) \geqslant 2$.

If $q = q(G) \neq 2$, we have $a = 2$ and we put $f = \infty$. A quadruple d, q, a, f which satisfies the above conditions is called an admissible set of invariants.

Theorem 3.39. *Let G be a Demushkin group with invariants d, q, a, f. Then G can be presented by generators $s_1, \ldots, s_d$ and the single relation*

$$\rho = s_1^q(s_1, s_2)(s_3, s_4)\ldots(s_{d-1}, s_d) \quad \text{if } q \neq 2,$$

$$\rho = s_1^2 s_2^{2^f}(s_2, s_3)\ldots(s_{d-1}, s_d) \quad \text{if } q = 2,\ d \text{ odd},$$

$$\rho = s_1^{2+2^f}(s_1, s_2)(s_3, s_4)\ldots(s_{d-1}, s_d) \quad \text{if } q = 2,\ d \text{ even},\ a = 2,$$

$$\rho = s_1^2(s_1, s_2)s_3^{2^f}(s_3, s_4)\ldots(s_{d-1}, s_d) \quad \text{if } q = 2,\ d \text{ even},\ a = 4.$$

On the other hand, if d, q, a, f is an admissible set of invariants and F the free pro-p-group generated by $s_1, \ldots, s_d$, then $F/(\rho)$ with ρ given as above is a Demushkin group with invariants d, q, a, f. ⊠ (Labute (1967))

1.15. The Structure of the Relations and the Cup Product. In the proof of Theorem 3.39 one uses the connection between the structure of the relations of a pro-p-group and the cup product. Let G be a pro-p-group of finite generator rank d and let $q = p^h$, $1 \leqslant h \leqslant \infty$, be a divisor of the exponent of the torsion group of $G/[G, G]$. Moreover let

$$\{1\} \to R \to F \to G \to \{1\}$$

be a minimal presentation of G with generators $s_1, \ldots, s_d$ of F and relations ρ_i, $i \in I$. By our assumption about q, $R \subseteq F^{(2,q)}$ (§ 1.13) and $\operatorname{Inf}: H^1(G, \mathbb{Z}/q\mathbb{Z}) \to H^1(F, \mathbb{Z}/q\mathbb{Z})$ is an isomorphism. Hence $\operatorname{Tra}: H^1(R, \mathbb{Z}/q\mathbb{Z})^G \to H^2(G, \mathbb{Z}/q\mathbb{Z})$ is an isomorphism (Proposition 2.43). For every $i \in I$ we define a homomorphism

$$\varphi_i : H^2(G, \mathbb{Z}/q\mathbb{Z}) \to \mathbb{Z}/q\mathbb{Z}$$

by means of

$$\varphi_i(\alpha) = (\operatorname{Tra}^{-1}\alpha)(\rho_i) \qquad \text{for } \alpha \in H^2(G, \mathbb{Z}/q\mathbb{Z}).$$

Proposition 3.40. *For $i \in I$, $\kappa, \nu, \mu \in \{1, \ldots, d\}$, $\nu < \mu$ there are uniquely determined integers $a_{i\kappa}, a_{i\nu\mu} \in \{0, 1, \ldots, q-1\}$ such that*

$$\rho_i = \prod_{\kappa=1}^{d} s^{a_{i\kappa}q} \prod_{\nu<\mu} (s_\nu, s_\mu)^{a_{i\nu\mu}} \rho_i' \qquad \text{with } \rho_i' \in F^{(3,q)}.$$

Let $\chi_i, \ldots, \chi_d$ be the basis of $H^1(G, \mathbb{Z}/q\mathbb{Z})$ with

$$\chi_\nu(\bar{s}_\mu) = \delta_{\nu\mu} \qquad \textit{for } \nu, \mu \in \{1, \ldots, d\}.$$

Then

$$\varphi_i(\chi_\nu \cup \chi_\mu) = \begin{cases} -\bar{a}_{i\nu\mu} & \textit{if } \nu < \mu \\ -\binom{q}{2}\bar{a}_{i\nu} & \textit{if } \nu = \mu. \ \square \end{cases}$$

The numbers $a_{i\kappa}$ are determined by the formula above only if $q = 2$. For the determination of $a_{i\kappa}$ in general, we introduce the *Bockstein operator* B: $H^1(G, \mathbb{Z}/q\mathbb{Z}) \to H^2(G, \mathbb{Z}/q\mathbb{Z})$ as the connecting homomorphism induced from the exact sequence

$$\{0\} \to \mathbb{Z}/q\mathbb{Z} \underset{q}{\to} \mathbb{Z}/q^2\mathbb{Z} \to \mathbb{Z}/qZ \to \{0\}.$$

Proposition 3.41. $\varphi_i(B\chi_\kappa) = -\bar{a}_{i\kappa}$. ⊠

There is also a connection between higher commutators and Massey products (Folklore).

Example 21. Let G be a pro-p-group as above with $\chi \cup \chi' = 0$ for all χ, $\chi' \in H^1(G, \mathbb{Z}/p\mathbb{Z})$ and let $\psi_{\nu\mu}(g)$ such that

$$\chi_\nu(g_1)\chi_\mu(g_2) = \psi_{\nu\mu}(g_1) + \psi_{\nu\mu}(g_2) - \psi_{\nu\mu}(g_1 g_2) \qquad \text{for } g_1, g_2 \in G.$$

Then

$$\chi_\nu(g_1)\psi_{\kappa\lambda}(g_2) - \psi_{\nu\kappa}(g_1)\psi_\lambda(g_2)$$

is a cocycle whose class is independent of the choice of $\psi_{\kappa\lambda}$ and $\psi_{\nu\kappa}$ and is connected with the 3-commutators in the relations of G. □

Now we make some remarks about the proof of Theorem 3.39. We restrict to the case $p \neq 2$. Let G be a Demushkin group with invariants d, q. Since the cup product $H^1(G) \times H^1(G) \to H^2(G)$ is not degenerated Proposition 3.40 shows that we can write the single relation of G in the form

$$\rho = s_1^q(s_1, s_2)(s_3, s_4)\ldots(s_{d-1}, s_d)\rho', \ \rho' \in F^{(3,q)}.$$

If ρ' contains e.g. a commutator $(s_1, (s_i, s_j))$, we go over to the new generator $s_2' = s_2(s_i, s_j)$ and get

$$\rho = s_1^q(s_1, s_2')(s_3, s_4)\ldots(s_{d-1}, s_d)\rho''$$

where $\rho''(s_1, s_2, s_3, \ldots, s_d)$ already does not contain $(s_1, (s_i, s_j))$. By means of this method of Demushkin one is able to kill all higher commutators and one ends up with the normal form

$$\rho = s_1^q(s_1, s_2)(s_3, s_4)\ldots(s_{d-1}, s_d).$$

1.16. Group Rings and the Theorem of Golod-Shafarevich. Let Λ be a compact ring and G a profinite group. For open normal subgroups $N \supseteq N'$ of G, the natural map $G/N' \to G/N$ can be uniquely extended to an algebra homomorphism

$$\Lambda[G/N'] \to \Lambda[G/N].$$

We get an inverse system of compact rings $\Lambda[G/N]$, $N \in \mathfrak{U}$, where $\mathfrak{U}$ is the set of all open normal subgroups of G. The completed group ring $\Lambda[[G]]$ is the inverse limit of $\{\Lambda[G/N] | N \in \mathfrak{U}\}$. With $\Lambda[G/N]$ also $\Lambda[[G]]$ is compact. We embed G in $\Lambda[[G]]$ by means of the map

$$g \to \prod_{N \in \mathfrak{U}} gN \qquad \text{for } g \in G.$$

$\Lambda[G]$ is dense in $\Lambda[[G]]$.

For the rest of this section we consider the special case that G is a pro-p-group and $\Lambda = \mathbb{Z}/p\mathbb{Z}$.

$\mathbb{Z}/p\mathbb{Z}[[G]]$ has a unique closed maximal ideal $I(G)$ generated by the elements $g - 1$, $g \in G$.

Proposition 3.42. *Let G be a pro-p-group with finite generator rank. Then the closed powers $I^n(G)$, $n = 1, 2, \ldots$, of the ideal $I(G)$ form a full system of neighborhoods of* 0 *in $\mathbb{Z}/p\mathbb{Z}[[G]]$.* □

The filtration $\{I^n(G) | n = 1, 2, \ldots\}$ induces a filtration $\{G_n | n = 1, 2, \ldots\}$ in G, defined by

$$G_n = \{g | g - 1 \in I^n(G)\},$$

called the *Zassenhaus filtration* of G.

Proposition 3.43. *Let G be a pro-p-group with finite generator rank. Then the groups G_n, $n = 1, 2, \ldots$, are open normal subgroups of G and form a full system of neighborhoods of* 1 *in G.* ⊠

The connection of the Zassenhaus filtration with the filtration $\{G^{(n,q)} | n = 1, 2, \ldots\}$ is given by the following proposition.

Proposition 3.44. *Let G be a pro-p-group and $g \in G_n$, $h \in G_m$. Then $g^p \in G_{np}$ and $(h, g) \in G_{n+m}$.*

Proof. More exactly one has $g^p - 1 = (g - 1)^p$ and

$$\begin{aligned}(h, g) - 1 &= h^{-1} g^{-1} hg - 1 \\ &= \sum_{\nu=0}^{\infty} (1 - h)^\nu \sum_{\mu=0}^{\infty} (1 - g)^\mu (1 + (h - 1))(1 + (g - 1)) - 1 \\ &\equiv (h - 1)(g - 1) - (g - 1)(h - 1) (\text{mod } I^{n+m+1}(G)). \ \square\end{aligned}$$

The *Magnus ring* $\mathbb{Z}/p\mathbb{Z}\{\{x_1, \ldots, x_d\}\}$ in d variables $x_1, \ldots, x_d$ is the ring of non commutative power series in $x_1, \ldots, x_d$ with coefficients in $\mathbb{Z}/p\mathbb{Z}$. The topology of the Magnus ring is defined by means of a system of open neighborhoods D^n, $n = 1, 2, \ldots$, of 0 where D^n is the ideal of all power series in $\mathbb{Z}/p\mathbb{Z}\{\{x_1, \ldots, x_d\}\}$ with homogeneous components of degree $\geqslant n$.

Proposition 3.45. *Let F be the free pro-p-group with generators* $s_1, \ldots, s_d$. *Then the map* $\varphi(x_i) = s_i - 1$, $i = 1, \ldots, d$, *extends to a topological isomorphism of* $\mathbb{Z}/p\mathbb{Z}\{\{x_1, \ldots, x_d\}\}$ *onto* $\mathbb{Z}/p\mathbb{Z}[[F]]$. □

In the following we identify $\mathbb{Z}/p\mathbb{Z}\{\{x_1, \ldots, x_d\}\}$ with $\mathbb{Z}/p\mathbb{Z}[[F]]$. Intuitively, it is clear that a pro-p-group can be finite only if its relation rank $r(G)$ is large with respect to its generator rank $d(G)$. This is quantified by the following *theorem of Golod-Shafarevich* (1964):

Theorem 3.46. *Let G be a finite pro-p-group. Then*

$$r(G) > d(G)^2/4. \boxtimes \tag{3.9}$$

One can show that this unequality is the best possible in the sense that

$$\liminf_{d(G)\to\infty} r(G)/d(G)^2 = \frac{1}{4}.$$

(Wisliceny (1981))

The basic idea of the proof of Theorem 3.46 is as follows: Let

$$\{1\} \to R \to F \to G \to \{1\}$$

be a minimal presentation of G with generators $s_1, \ldots, s_d$ and relations $\rho_1, \ldots, \rho_r$. Since

$$\mathbb{Z}/p\mathbb{Z}[G] \cong \mathbb{Z}/p\mathbb{Z}[[F]]/M$$

where M is the closed ideal of $\mathbb{Z}/p\mathbb{Z}[[F]]$ generated by $\rho_1 - 1, \ldots, \rho_r - 1$, we look on M instead on R. Let L be the $\mathbb{Z}/p\mathbb{Z}$-module generated by $\rho_1 - 1, \ldots, \rho_r - 1$. To get the inequality (3.9) one has to build up M from L as economically as possible. ⊠

By means of filtrations of G one can get refinements of Theorem 3.46 (Andozhskij (1975), Koch (1978). We mention here only the following result:

Theorem 3.47. *Let G be a finite pro-p-group with a presentation*

$$\{1\} \to R \to F \to G \to \{1\}$$

such that $R \subseteq F_m$ *where* $\{F_n | n = 1, 2, \ldots\}$ *is the Zassenhaus filtration of* F. *Then*

$$r(G) > d(G)^m (m-1)^{m-1}/m^m. \boxtimes$$

§2. Galois Cohomology of Local and Global Fields

(Main reference: Serre (1964))

2.1. Examples of Galois Cohomology of Arbitrary Fields. Let K be a field and L a normal extension of K (§1.8). *Galois cohomology* is the study of cohomology groups of $G(L/K)$-modules A. Instead of $H^n(G(L/K), A)$ we shall often write $H^n(L/K, A)$.

Important examples of $G(L/K)$-modules are the additive group L^+ and the multiplicative group $L^\times$ of L.

Proposition 3.48. $H^n(L/K, L^+) = \{0\}$ *for* $n > 0$.

Proof. If L/K is a finite extension, the existence of a normal basis for L/K (Appendix 1) means $L^+ \cong M_{G(L/K)}(K^+)$, hence $H^n(L/K, L^+) = \{0\}$ for $n > 0$ by Proposition 2.38. If L/K is infinite, the claim follows from Proposition 3.9. □

Proposition 3.49. $H^1(L/K, L^\times) = \{0\}$.

Proof. This follows from Proposition 2.6 and Proposition 3.9. □

Proposition 3.50. *$H^2(L/K, L^\times)$ is isomorphic to the Brauer group $B(L/K)$. In particular if L is a separable algebraic closure of K, then $H^2(L/K, L^\times)$ is isomorphic to $B(K)$* (Chap. 2.5.4).

Proof. This follows from Proposition 2.79, Theorem 2.81, Proposition 3.9, and the commutativity of the diagram

$$\begin{array}{ccc} H^2(F_1/K, F_1) & \longrightarrow & B(F_1/K) \\ \downarrow & & \downarrow \\ H^2(F_2/K, F_2) & \longrightarrow & B(F_2/K) \end{array}$$

where $F_1 \subseteq F_2$ are finite normal subextensions of L/K. □

2.2. The Algebraic Closure of a Local Field. Let K be a finite extension of $\mathbb{Q}_p$ and $\overline{K}$ the algebraic closure of K. We denote the Galois group of $\overline{K}/K$ by G_K.

In Chap. 2.4.1 we have considered the map

$$\mathrm{inv}\colon H^2(L/K, L^\times) \to \mathbb{Q}/\mathbb{Z}$$

for finite normal extensions L/K. Let L'/K be a normal extension with $L' \subseteq L$. Since the diagram

$$\begin{array}{ccc} H^2(L'/K, L'^\times) & \xrightarrow{\mathrm{inv}} & \\ \downarrow{\scriptstyle \mathrm{Inf}} & & \mathbb{Q}/\mathbb{Z} \\ H^2(L/K, L^\times) & \xrightarrow{\mathrm{inv}} & \end{array}$$

commutes by (2.44), inv induces a homomorphism $\mathrm{inv}_K\colon H^2(G_K, \overline{K}^\times) \to \mathbb{Q}/\mathbb{Z}$.

Theorem 3.51. *inv_K is an isomorphism of $H^2(G_K, \overline{K}^\times)$ onto $\mathbb{Q}/\mathbb{Z}$. If K' is a finite extension of K in $\overline{K}$, then*

$$\mathrm{inv}_{K'}\, \mathrm{Res}_{K \to K'} = [K' : K]\, \mathrm{inv}_K.$$

Proof. This follows from Theorem 2.56 and (2.42). □

An important role in Galois cohomology plays the *Kummer sequence*

$$\{1\} \to \mu_n \to L^\times \underset{n}{\to} L^\times \to \{1\}, \; n(\alpha) = \alpha^n \qquad \text{for } \alpha \in L^\times \tag{3.10}$$

where L is a normal extension of K containing the group μ_n of n-th roots of unity. We demonstrate this role in the simple situation $L = \overline{K}$:

Theorem 3.52. 1. $H^1(G_K, \mu_n) \cong K^\times/K^{\times n}$, 2. $H^2(G_K, \mu_n) \cong \mathbb{Z}/n\mathbb{Z}$.

Proof. The cohomology sequence (2.31) of (3.10) together with Proposition 3.49 and Theorem 3.51 implies the exact sequences

$$K^\times \underset{n}{\to} K^\times \to H^1(G_K, \mu_n) \to \{0\},\ \{0\} \to H^2(G_K, \mu_n) \to \mathbb{Q}/\mathbb{Z} \underset{n}{\to} \mathbb{Q}/\mathbb{Z}. \square$$

Theorem 3.53. *Let L/K be an extension in $\overline{K}$ of degree $\prod_l l^\infty$, i.e. for every natural number n there is a subextension of L/K of degree a multiple of n. Then $H^2(G_L, \overline{K}^\times) = \{0\}$ and $cd(G_L) \leqslant 1$.*

Proof. Let $L = \bigcup_{i=1}^\infty L_i$ where $L_1 \subseteq L_2 \subseteq \cdots$ are finite extensions of K. Then $H^2(G_L, \overline{K}^\times) = \lim_{i\to\infty} H^2(G_{L_i}, \overline{K}^\times)$, hence the first claim follows from Theorem 3.51.

Let l be a prime and G_l the l-Sylow group of G_L. For the proof of the second claim we have to show $H^2(G_l, Z/lZ) = \{0\}$ (Proposition 3.16, Proposition 3.23). The Kummer sequence

$$\{1\} \to \mu_l \to \overline{K}^\times \underset{l}{\to} \overline{K}^\times \to \{1\}$$

together with $H^2(G_L, \overline{K}^\times) = \{0\}$ and Proposition 3.49 implies $H^2(G_l, \mu_l) = \{0\}$. Since G_l operates trivially on μ_l, $H^2(G_l, \mathbb{Z}/l\mathbb{Z}) \cong H^2(G_l, \mu_l) = \{0\}$. $\square$

Theorem 3.54. $cd(G_K) = 2$.

Proof. Let K_{nr} be the maximal unramified subextension of $\overline{K}$. Then $cd(G(K_{nr}/K)) = 1$ and $cd(G(\overline{K}/K_{nr})) = 1$, hence Proposition 3.19 implies $cd(G_K) \leqslant 2$. Moreover $cd(G_K) = 2$ by Theorem 3.52.2. $\square$

Let A be a finite G_K-module. From our results above it follows easily that the groups $H^i(G_K, A)$ are finite. Since $cd(G_K) = 2$, there is a dualizing module of G_K (§ 1.11).

Theorem 3.55. *The dualizing module of G_K is the module μ of all roots of unity in $\overline{K}$.* ⊠

Let $A' := \mathrm{Hom}_{\mathbb{Z}}(A, \mu)$. Then G_K acts on A' according to

$$(gf)(a) = g(f(g^{-1}a)) \qquad \text{for } f \in A',\ g \in G_K,\ a \in A.$$

Theorem 3.56. *The cup product*

$$H^i(G_K, A) \times H^{2-i}(G_K, A') \to H^2(G_K, \mu) \cong \mathbb{Q}/\mathbb{Z},\ i = 0, 1, 2,$$

induced by the natural pairing from $A \times A'$ into μ defines a duality between the finite groups $H^i(G_K, A)$ and $H^{2-i}(G_K, A')$. ⊠

For $i = 0, 2$ the duality in Theorem 3.56 is part of the definition of the dualizing module.

The (multiplicative) Euler Poincaré characteristic $\chi(A)$ is defined by

$$\chi(A) := h^0(A)h^1(A)^{-1}h^2(A), \qquad h^i(A) = |H^i(G_K, A)|.$$

Theorem 3.57. *Let a be the order of the G_K-module A. Then*

$$\chi(A) = |a|_K$$

where $|\ |_K$ denotes the normalized valuation of K (Chap. 1.4.1). ⊠ (Serre (1964), Chap. 2.5.7)

Theorem 3.58. $scd(G_K) = 2$.

Proof. This follows from Proposition 3.22 and Theorem 3.55. □

Let $PGL_n(\mathbb{C}) := GL_n(\mathbb{C})/\mathbb{C}^\times$ denote the projective linear group. Theorem 3.58 implies the following result which is important for the representation theory of G_K.

Theorem 3.59. *Every continuous homomorphism of G_K into $PGL_n(\mathbb{C})$ can be lifted to $GL_n(\mathbb{C})$.*

Proof. The exact sequence

$$\{1\} \to \mathbb{C}^\times \to GL_n(\mathbb{C}) \to PGL_n(\mathbb{C}) \to \{1\}$$

induces the exact sequence

$$\mathrm{Hom}(G_K, GL_n(\mathbb{C})) \to \mathrm{Hom}(G_K, PGL(\mathbb{C})) \to H^2(G_K, \mathbb{C}^\times).$$

Moreover the group $\mathbb{C}^+$ is uniquely divisible by any natural number, hence $H^i(G_K, \mathbb{C}^+) = \{0\}$ for $i > 0$. Therefore the sequence

$$\{0\} \longrightarrow \mathbb{Z} \longrightarrow \mathbb{C}^+ \xrightarrow[\exp]{} \mathbb{C}^\times \longrightarrow \{1\}$$

induces an isomorphism $H^2(G_K, \mathbb{C}^\times) \to H^3(G_K, \mathbb{Z})$ and $H^3(G_K, \mathbb{Z}) = \{0\}$ by Theorem 3.58. □

The following considerations are necessary for the study of the cohomology of global fields.

A G_K-module A is called unramified if $G(\overline{K}/K_{nr})$ operates trivially on A. Such a module can be considered as $\hat{\mathbb{Z}}$-module since $G(K_{nr}/K)$ is canonically isomorphic to $\hat{\mathbb{Z}}$. We define

$$H^i_{nr}(G_K, A) := H^i(\hat{\mathbb{Z}}, A).$$

Then $H^0_{nr}(G_K, A) = H^0(G_K, A)$, $H^1_{nr}(G_K, A)$ can be considered as subgroup of $H^1(G_K, A)$, and $H^2_{nr}(G_K, A) = \{0\}$ since $cd(\hat{\mathbb{Z}}) = 1$.

Proposition 3.60. *Let A be a finite unramified G_K-module of order prime to p. Then A′ has the same properties and in the duality between $H^1(G_K, A)$ and $H^1(G_K, A')$ the subgroups $H^1_{nr}(G_K, A)$ and $H^1_{nr}(G_K, A')$ are orthogonal to each other.* ⊠

2.3. The Maximal *p*-Extension of a Local Field. Let K be a finite extension of $\mathbb{Q}_p$ of degree N and $K(p)$ the maximal p-extension of K, i.e. the union of all normal extension of K of p-power degree. Then $G_K(p) := G(K(p)/K)$ is the maximal factor group of G_K which is a pro-p-group. The main purpose of this section is the determination of the structure of this group.

Theorem 3.61. *Let A be a p-primary $G_K(p)$-module. Then the inflation*

$$H^i(G_K(p), A) \to H^i(G_K, A)$$

is an isomorphism for all $i > 0$.

Proof. We put $I := G(\overline{K}/K(p))$. Analogously to the proof of Theorem 3.53 one shows $cd_p(I) \leqslant 1$. Since $H^1(I, A) = \{0\}$, the claim follows from Proposition 2.43. □

Theorem 3.62. *If K does not contain the p-th roots of unity, $G_K(p)$ is a free pro-p-group of generator rank $N + 1$.*

Proof. We have

$$d(G_K(p)) = \dim_{\mathbb{Z}/p\mathbb{Z}} H^1(G_K(p), \mathbb{Z}/p\mathbb{Z}) = \dim_{\mathbb{Z}/p\mathbb{Z}}(K^\times/K^{\times p}) = N + 1$$

by Proposition 1.73. If we apply this to the finite extension L of K we get

$$\chi_1(G_L(p)) = [L : K]\chi_1(G_K(p)).$$

The claim follows therefore from Proposition 3.28. □

Theorem 3.63. *Let q be the largest power of p such that K contains the q-th roots of unity, $q \neq 1$. Then $G_K(p)$ is a Demushkin group with $d(G_K(p)) = N + 2$ and $q(G_K(p)) = q$. The dualizing module of $G_K(p)$ is the module of all roots of unity of p-power order.*

Proof. One determines $d(G_K(p))$ as in the proof of Theorem 3.62. Moreover $r(G_K(p))$ by Theorem 3.52 and 3.61. Hence $cd(G_K(p)) = 2$ by Proposition 3.28. $G_K(p)$ is a Demushkin group since the cup product

$$H^1(G_K(p), \mathbb{Z}/p\mathbb{Z}) \times H^1(G_K(p), \mathbb{Z}/p\mathbb{Z}) \to H^2(G_K(p), \mathbb{Z}/p\mathbb{Z}) \cong \mathbb{Z}/p\mathbb{Z}$$

induces up to an isomorphism $\mathbb{Z}/p\mathbb{Z} \cong \mu_p$ the Hilbert symbol $K^\times/K^{\times p} \times K^\times/K^{\times p} \to \mu_p$ (Chap. 2.1.5) (Serre (1962), Chap. 14). Finally Theorem 3.55 and 3.61 imply the claim about the dualizing module. □

Theorem 3.63 together with Theorem 3.39 determines the structure of $G_K(p)$ in the case $\mu_p \subset K$ completely.

Example 22. Let $K = \mathbb{Q}_2$. Then $d = 3$, $q = 2$, $f = 2$. Therefore $G_{\mathbb{Q}_2}(2)$ is a pro-2-group with generators s_1, s_2, s_3 and one relation $s_1^2 s_2^4(s_2, s_3)$. □

Example 23. Let $l \neq p$ be a prime and let K be a finite extension of $\mathbb{Q}_l$. Then the maximal p-extension $K(p)$ of K is tamely ramified. Example 14 shows that $G(K(p)/K)$ can be presented as pro-p-group with two generators s, t and one relation $t^{N(\mathfrak{l})}st^{-1}s^{-1} = t^{N(\mathfrak{l})-1}(t^{-1}, s^{-1})$, where $N(\mathfrak{l})$ is the number of elements in the residue class field of K. If $N(\mathfrak{l}) \not\equiv 1 \pmod{p}$, then $K(p)/K$ is unramified and $G(K(p)/K) \cong \mathbb{Z}_p$. □

2.4. The Galois Group of a Local Field (Main references: Jannsen (1982), Jannsen, Wingberg (1982), Wingberg (1982)). In this section we describe the structure of G_K in the case $p \neq 2$. For $p = 2$ only the case $\sqrt{-1} \in K$ is settled (Diekert (1984), (1972)). If $[K : \mathbb{Q}_p]$ is even, such a description by means of generators and relations was first given by Jakovlev (1968), (1978).

A p-closed extension of K is a normal extension which has no proper p-extension. Since the case of p-closed extensions L is not more difficult than the case of the algebraic closure of K, we consider such extensions L.

Let L_1 be the maximal tamely ramified subextension of L/K. Then $G(L_1/K)$ is a factor group of $G(\overline{K}_1/K)$ with $p^\infty | [L_1 : K]$. $G(\overline{K}_1/K)$ is a profinite group with two generators σ, τ and one relation $\sigma\tau\sigma^{-1} = \tau^{N(\mathfrak{p})}$ where $N(\mathfrak{p})$ denotes the number of elements in the residue class field of K. (More precisely this means that there is an exact sequence $\{1\} \to R \to F \underset{\varphi}{\to} G(\overline{K}_1/K) \to \{1\}$ such that F is a free profinite group with generators s, t (§ 1.5), the closed normal subgroup R of F is generated by $sts^{-1}t^{-N(\mathfrak{p})}$ and $\varphi s = \sigma$, $\varphi t = \tau$.) We call L_1/K a p-closed tame extension and consider σ and τ also as generators of $G(L_1/K)$.

The study of $G(L/K)$ will be modelled by that of $G_K(p)$. First we introduce the group theoretical notion of *Demushkin formation*, we classify such objects by means of invariants, and then we show that $G(L/K) \to G(L_1/K)$ is a Demushkin formation.

Let G be the Galois group of a p-closed tame extension of K, let $n, h \geqslant 1$ be integers, and let $\alpha: G \to (\mathbb{Z}/p^h\mathbb{Z})^\times$ be a character of G. Then a pair (X, ϕ) consisting of a profinite group X and a morphism ϕ of X onto G is called a *Demushkin formation* if the following conditions are fulfilled for all open normal subgroups H of G with $H \subseteq \operatorname{Ker} \alpha$:

1. Let X_H be the maximal pro-p-factor group of $\phi^{-1}(H)$. Then X_H is a Demushkin group (§ 1.14) with $d(X_H) = n[G : H] + 2$ and $q(X_H) = p^h$.

2. Let φ_1 (resp. φ_2) be the inflation map from $H^1(X_H, \mathbb{Z}/p\mathbb{Z})$ (resp. $H^1(H, \mathbb{Z}/p\mathbb{Z})$) into $H^1(\phi^{-1}(H), \mathbb{Z}/p\mathbb{Z})$, let $M = \varphi_1^{-1}\varphi_2(H^1(H, \mathbb{Z}/p\mathbb{Z}))$, and let $M^\perp$ be the orthogonal complement of M with respect to the bilinear form in $H^1(X_H, \mathbb{Z}/p\mathbb{Z})$ given by the cup product. Then $M^\perp/M$ is a free $\mathbb{Z}/p\mathbb{Z}[[G/H]]$-module of rank n and decomposes in the direct sum of two totally isotropic submodules.

3. $\gamma x = \alpha(\gamma)x$ for $\gamma \in G$, $x \in H^2(X_H, \mathbb{Z}/p^h\mathbb{Z})$.

In the following we assume that the invariants n, α satisfy the following condition:

4. Let f_0 denote the natural number with $p^{f_0} = N(\mathfrak{p})$. If n is odd, then f_0 is odd too and

$$\alpha(\tau)^{(p-1)/2} \equiv -1 \pmod{p}.$$

Theorem 3.64. *A Demushkin formation over G is uniquely determined up to isomorphism by its invariants n, s, α.* ☒

Theorem 3.65. *Let n, s be natural numbers and* $\alpha: G \to (\mathbb{Z}/p^h\mathbb{Z})$ *a homomorphism satisfying* 4. *Then there exists a Dumushkin formation over G with the invariants n, s, α.* ☒

Theorem 3.66. *Let K be a p-adic number field of degree n over $\mathbb{Q}_p$, $p \neq 2$, let L be a p-closed extension of K, and L_1 the maximal tamely ramified subextension of L/K. Let μ_{p^h} be the group of roots of unity of p-power order in L_1 and α the homomorphism of $G(L_1/K)$ into $\mathbb{Z}/p^h\mathbb{Z}$ given by*

$$\gamma\zeta = \zeta^{\alpha(\gamma)} \qquad \text{for } \gamma \in G(L_1/K),\ \zeta \in \mu_{p^h}.$$

Then $G(L/K)$ is a Demushkin formation over $G(L_1/K)$ with invariants n, s, α.

Proof. Axiom 1 and 3 are verified by means of Theorem 3.63. Axiom 2 can be reformulated by class field theory as an assertion about the $G(T/K)$-modules U_1/Ω, where T is a normal tamely ramified finite extension of K, U_1 is the group of principal units of T, and Ω is the group of p-primary numbers in U_1 (Chap. 2.6.1). In this form the first claim of axiom 2 is proved in Iwasawa (1955), see also Pieper (1972), Jannsen (1982). The second claim of axiom 2 is a consequence of Shafarevich's explicit reciprocity law (Chap. 2.6.1), which has an excellent transformation behavior with respect to the Galois group of tamely ramified extensions (see Koch (1963)). It is easy to see that axiom 4 is fulfilled. ⊠

A Demushkin formation can be presented as profinite group with $n + 3$ generators. We need some preparations for the description of this presentation.

Let Y be any profinite group. $\hat{\mathbb{Z}}$ operates on Y according to

$$y^m = \lim_{i\to\infty} y^{m_i} \qquad \text{for } m_i \in \mathbb{Z},\ m = \lim_{i\to\infty} m_i.$$

If $x, y, z \in Y$ we write

$$x^y := yxy^{-1},\ x^{y+z} := x^y x^z.$$

We identify $\hat{\mathbb{Z}}$ with $\prod_l \mathbb{Z}_l$, where the product runs over all primes l, such that $m = \prod_l m$ if $m \in \mathbb{Z}$ and we identify $\mathbb{Z}_p$ with its image in $\hat{\mathbb{Z}}$ under the map $\varphi_p(\alpha) = \prod_l \alpha\delta_{lp}$. We put $\varphi_p(1) =: \pi_p$.

Now we consider the free profinite group F with generators $s, t, s_0, \ldots, s_n$. We define a morphism ϕ_1 of F onto G by means of

$$\phi_1(s) = \sigma,\ \phi_1(t) = \tau,\ \phi_1(s_i) = 1,\ i = 0, \ldots, n.$$

Let $\beta: F \to G \to \mathbb{Z}_p^\times$ be a continuous lifting of α. We put

$$w := (s_0^{\beta(1)} t s_0^{\beta(t)} t \ldots s_0^{\beta(t^{p-2})})^{\pi_p (p-1)^{-1} \beta(s)^{-1}},$$

$$\tilde{x} := x^{\pi_p},\ \{x, y\} := (x^{\beta(1)} y^2 x^{\beta(y)} y^2 \ldots x^{\beta(y^{p-2})} y^2)^{\pi_p (p-1)^{-1}} \qquad \text{for } x, y \in F,$$

$$y_1 := s_1^{\tilde{t}^{p+1}} \{s_1, \tilde{t}^{p+1}\}^{\tilde{s}} \{\{s_1, \tilde{t}^{p+1}\}, \tilde{s}\}^{\tilde{s}\tilde{t} + \tilde{t}(p+1)/2} \qquad \text{if } \left(\frac{-\alpha(\sigma)}{p}\right) = 1,$$

$$y_1 := s_1^{\tilde{t}^{p+1}} \{s_1, \tilde{t}^{p+1}\}^{\tilde{s}\tilde{t}} \{\{s_1, \tilde{t}^{p+1}\}, \tilde{s}\tilde{t}\}^{\tilde{s} + \tilde{t}(p+1)/2} \qquad \text{if } \left(\frac{-\alpha(\sigma)}{p}\right) = -1.$$

Theorem 3.67. *Let (X, ϕ) be a Demushkin formation with invariants n, s, α satisfying* 4. *Then X has the presentation*

$$\{1\} \to R \to F \to X \to \{1\}$$

and ϕ is induced by ϕ_1. The closed normal subgroup R of F is generated by the smallest normal subgroup N of $\phi_1^{-1}(G)$ such that $\phi_1^{-1}(G)/N$ is a pro-p-group, by the kernel of the restriction of ϕ_1 to the closed subgroup of F generated by s, t, and by

$$s_0^{-s} w s_1^{p^s}(s_1, s_2)(s_3, s_4)\dots(s_{n-1}, s_n) \qquad \text{if } n \equiv 0 \pmod 2,$$

$$s_0^{-s} w s_1^{p^s}(s_1, y_1)(s_2, s_3)\dots(s_{n-1}, s_n) \qquad \text{if } n \equiv 1 \pmod 2. \boxtimes$$

2.5. The Maximal Algebraic Extension with Given Ramification (Main reference: Haberland (1978)). Let K be a finite extension of $\mathbb{Q}$ and S a set of places of K containing all infinite places. K_S denotes the maximal extension of K unramified outside S. We put $G_S := G(K_S/K)$ and $G_v := G(\overline{K}_v/K_v)$ for $v \in S$. For every $v \in S$ we fix an injection $\iota_v: K_S \to \overline{K}_v$. Let A be a finite G_S-module. The induced homomorphism

$$\iota_{v*}: H^n(G_S, A) \to H^n(G_v, A)$$

is independent of the choice of the injection ι_v.

In this section we study $H^n(G_S, A)$ by means of the localizations $H^n(G_v, A)$. Therefore we consider the localization map

$$H^n(G_S, A) \to \prod_{v \in S} H^n(G_v, A). \tag{3.11}$$

If v is infinite we take for $H^0(G_v, A)$ the modified cohomology groups $\hat{H}^0(G_v, A)$ (Chap. 2.3.8), in particular $H^0(G_v, A) = \{0\}$ if v is imaginary.

The image of the map (3.11) lies in a subgroup of $\prod_{v \in S} H^n(G_v, A)$ which is defined as follows: Since A is finite, there is an open subgroup of G_S acting trivially on A. Hence the G_v-module A is unramified at almost all $v \in S$ (§2.2). We define $P^n(G_S, A)$ to be the restricted direct product of the groups $H^n(G_v, A)$ for $v \in S$ with respect to the subgroups $H^n_{nr}(G_v, A)$ i.e. $P^n(G_S, A)$ consists of the elements $\prod_{v \in S} a_v$ of $\prod_{v \in S} H^n(G_v, A)$ such that $a_v \in H^n_{nr}(G_v, A)$ for almost all $v \in S$. We consider $P^n(G_S, A)$ as topological group taking as basis of neighborhoods of 0 the subgroups $\prod_{v \notin T} H^n_{nr}(G_v, A)$, where T runs through the finite subsets of S containing all infinite places and the places for which A is ramified. Then

$$P^0(G_S, A) = \prod_{v \in S} H^0(G_v, A) \text{ is compact,}$$

$$P^1(G_S, A) =: \mathop{\square}_{v \in S} H^1(G_v, A) \text{ is locally compact,}$$

$$P^2(G_S, A) = \sum_{v \in S} H^2(G_v, A) \text{ is discrete,}$$

$$P^n(G_S, A) = \sum_{v \in S} H^n(G_v, A) \text{ is finite for } n \geqslant 3.$$

Let μ_S be the group of roots of unity in K_S and let M be a finite G_S-module such that $v \nmid |M|$ for $v \notin S$. As in §2.2 we put $M' := \operatorname{Hom}_{\mathbb{Z}}(M, \mu_S)$. According to Theorem 3.56 the bilinear form

$$P^n(G_S, M) \times P^{2-n}(G_S, M') \to \mathbb{Q}/\mathbb{Z} \tag{3.12}$$

is not degenerated for $n = 0, 1, 2$. In this section we denote the Pontrjagin dual of a topological abelian group B by B^*.

We denote the map $H^n(G_S, M) \to P^n(G_S, M)$ by α_n. (3.12) induces a homomorphism

$$\beta_n\colon P^n(G_S, M) \longrightarrow P^{2-n}(G_S, M')^* \xrightarrow[\alpha^*_{2-n}]{} H^{2-n}(G_S, M')^*.$$

Moreover we denote the kernel of α_n by $\mathrm{Ker}^n(G_S, M)$.

Theorem 3.68 (Theorem of Poitou-Tate). *Let S be a set of places of K containing all infinite places and let M be a finite G_S-module such that $v \nmid |M|$ for $v \notin S$. Then*

1. $\mathrm{Ker}^1(G_S, M)$ *and* $\mathrm{Ker}^2(G_S, M')$ *are finite and there is a canonical non-degenerate pairing*

$$\mathrm{Ker}^1(G_S, M) \times \mathrm{Ker}^2(G_S, M') \to \mathbb{Q}/\mathbb{Z}. \tag{3.13}$$

2. *Let*

$$\gamma_M\colon H^2(G_S, M')^* \to \mathrm{Ker}^2(G_S, M')^* \to \mathrm{Ker}^1(G_S, M)$$

be the homomorphism induced by (3.13). *Then the sequence*

$$\begin{array}{ccccccccc}
\{0\} & \longrightarrow & H^0(G_S, M) & \xrightarrow[\alpha_0]{} & P^0(G_S, M) & \xrightarrow[\beta_0]{} & H^2(G_S, M')^* & \xrightarrow[\gamma_M]{} & H^1(G_S, M) \\
 & & & & & & & & \Big\downarrow{\scriptstyle \alpha_1} \\
 & & & & & & & & P^1(G_S, M) \\
 & & & & & & & & \Big\downarrow{\scriptstyle \beta_1} \\
\{0\} & \longleftarrow & H^0(G_S, M')^* & \xleftarrow[\beta_2]{} & P^2(G_S, M) & \xleftarrow[\alpha_2]{} & H^2(G_S, M) & \xleftarrow[\gamma^*_{M'}]{} & H^1(G_S, M')^*
\end{array}$$

is exact.

3. *The maps α_n are proper, i.e. the inverse image of every compact set is compact.*
4. *α_n is bijective for $n \geqslant 3$.* ⊠

4. together with Theorem 3.54 implies

Theorem 3.69. *$cd_p(G_S) \leqslant 2$ if $p \neq 2$. $cd_2(G_S) \leqslant 2$ if K has no real place.* □

The strict cohomological dimension of G_S is unknown, but one has the following result about $G_K := G(\overline{K}/K)$:

Theorem 3.70. *$H^{2n+1}(G_K, \mathbb{Z}) = \{0\}$ for $n \geqslant 1$, $H^{2n}(G_K, \mathbb{Z}) \cong (\mathbb{Z}/2\mathbb{Z})^{r_1}$ for $n \geqslant 2$ where r_1 denotes the number of real places of K.* ⊠

Theorem 3.70 implies that Theorem 3.59 is valid also for algebraic number fields K.

Let p be a fixed prime number and S a set of places of K containing the set S_∞ of infinite places and the set S_p of places above p. We denote by $T(K_S, p)$ the statement that $scd_p(G_S) \leqslant 2$, if $p = 2$ it is assumed that K has no real place.

$T(K_S, p)$ was announced as a theorem in Tate (1963) but has not been proved so far. It is called the Tate conjecture. It is connected with the following *conjecture of Leopoldt*:

Let U_v (resp. U_{v1}) be the group of units (resp. principal units) of K_v and E the group of units of K. We put

$$U_p := \prod_{v \in S_p} U_v, \qquad U_{p1} := \prod_{v \in S_p} U_{v1}.$$

Then we have an embedding of E in U_p. The group U_{p1} is a $\mathbb{Z}_p$-module. The Leopoldt conjecture states that $E_1 := E \cap U_{p1}$ generates in U_{p1} a $\mathbb{Z}_p$-module of rank $|S_\infty| - 1$. We denote this statement by $L(K, p)$.

Theorem 3.71. *If $L(L, p)$ holds for all finite extensions L/K in K_S/K, then $T(K_S, p)$. If $T(K_S, p)$ holds for a finite set S, then $L(L, p)$ holds for all finite subextensions L/K of K_S/K.* ☒

We come back to the Leopoldt conjecture in Chap. 4.3.5.

Theorem 3.72. *Let M be as in Theorem 3.68. Moreover let S be finite. Then $H^n(G_S, M)$ is finite for all $n \geqslant 0$. Hence the partial Euler-Poincaré characteristic*

$$\chi_2(M) := h^0(M)h^1(M)^{-1}h^2(M),\ h^i(M) = |H^i(G_S, M)|,$$

is defined. Let m be the degree of K over $\mathbb{Q}$, then

$$\chi_2(M) = |M|^{-m} \prod_{v \in S_\infty} |M^{G_v}|. \boxtimes$$

Theorem 3.72 is an essential step in the proof of Theorem 3.68.

2.6. The Maximal p-Extension with Given Ramification (Main reference: Koch (1970)). Let p be a fixed prime number, let K be a finite extension of $\mathbb{Q}$, let S be a set of places of K, and let $K_S(p)$ be the maximal p-extension of K unramified outside S.

The following places v cannot be ramified in a p-extension:

1. Finite places v with $N(v) \not\equiv 0, 1 \pmod p$ (Proposition 1.52.4),
2. complex places,
3. real places if $p \neq 2$.

In the following we assume that such places are not in S.

We put $G_S(p) := G(K_S(p)/K)$.

Theorem 3.73. *Let S be a set of places of K containing all places above p and all real places if $p = 2$. Then the canonical homomorphism*

$$H^n(G_S(p), \mathbb{Z}/p\mathbb{Z}) \to H^n(G_S, \mathbb{Z}/p\mathbb{Z})$$

is an isomorphism for all $n \geqslant 0$. (Neumann (1975), Haberland (1978), Proposition 22) ☒

By means of Theorem 3.73 one transfers the results about the cohomology of G_S to $G_S(p)$: We put

$$V_S(K) := \{\alpha \in K^\times | \alpha \in U_v K_v^{\times p} \text{ for all finite places } v, \alpha \in K_v^{\times p} \text{ for } v \in S\},$$

$$Б_S(K) := (V_S(K)/K^{\times p}).$$

Theorem 3.75. *Let S be a finite set of places of K containing all places above p and all real places if p = 2.*

1. $\mathrm{Ker}^1(G_S, \mu_p) = V_S/K^{\times p}$.
2. *If K contains the p-th roots of unity, there is an exact sequence*

$$\{0\} \to Б_S \to H^2(G_S(p)) \to \sum_{v \in S} H^2(G_v(p)) \to \mathbb{Z}/p\mathbb{Z} \to 0.$$

3. *If K does not contain the p-th roots of unity, there is an exact sequence*

$$\{0 \to Б_S \to H^2(G_S(p)) \to \sum_{v \in S} H^2(G_v(p)) \to \{0\}.$$

4. *If $p \neq 2$ or K has no real place, then $cd(G_S(p)) \leqslant 2$.*
5. *If S is finite, $\chi_2(G_S(p)) = -r_2$ where r_2 denotes the number of complex places of K (notation as in § 1.12).* ⊠ (Haberland (1978), § 6.3)

For arbitrary S the kernel Ш$_S$ of

$$H^2(G_S(p)) \to \sum_{v \in S} H^2(G_v(p), \mathbb{Z}/p\mathbb{Z})$$

must not be isomorphic to Б$_S$. One has only the following result:

Theorem 3.75. *There is a canonical injection of Ш$_S$ into Б$_S$. If the p-th roots of unity are contained in K and $S \neq \varnothing$; then Ш$_S$ is already the kernel of the map*

$$H^2(G_S(p)) \to \sum_{v \in S - v_0} H^2(G_v(p))$$

where v_0 is an arbitrary place in S. ⊠

For any field L we put $\delta(L) = 0$ if $\mu_p \not\subset L$ and $\delta(L) = 1$ if $\mu_p \subset L$. Moreover we put $\theta(K, S) = 1$ if $\delta(K) = 1$ and $S = \varnothing$, and $\theta(K, S) = 0$ otherwise. Theorem 3.75 together with the results of § 2.3 implies the following *theorem of Shafarevich* (1964):

Theorem 3.76. *Let S be a finite set of places of K. Then*

$$r(G_S(p)) = \dim_{\mathbb{Z}/p\mathbb{Z}} H^2(G_S(p)) \leqslant \sum_{v \in S} \delta(K_v) - \delta(K) + \dim_{\mathbb{Z}/p\mathbb{Z}} Б_S + \theta(K, S). \ \square$$

On the other hand it is not difficult to compute the generator rank $d(G_S(p)) = d(G_S(p)/G_S(p)^{(2,p)})$ by means of class field theory. $G_S(p)/G_S(p)^{(2,p)}$ is the Galois group of the maximal p-elementary extension of K unramified outside S. One finds

$$d(G_S(p)) = \sum_{\substack{v \in S \\ v | p}} [K_v : \mathbb{Q}_p] - \delta(K) - r + 1 + \sum_{v \in S} \delta(K_v) + \dim Б_S \qquad (3.14)$$

where r denotes the number of infinite places of K.

Let S be an arbitrary set of places of K and φ_v: $G_v(p) \to G_S(p)$ the homomorphism corresponding to an embedding of $K_S(p)$ in $K_v(p)$ for $v \in S$. In the sense of § 1.13 we can transfer the known local relation of $G_v(p)$ to a relation of $G_S(p)$. In general there remain unknown relations. But if $Б_S = \{0\}$, then by Theorem 3.75 all relations of $G_S(p)$ come from local relations. In this case one has a full description of $G_S(p)/G_S(p)^{(3,p)}$ and if $\delta(K) = 1$ (and in some other cases) even of $G_S(p)/G_S(p)^{(3)}$.

In particular we have a simple situation in the case that K is the field of rational numbers or an imaginary-quadratic number field with class number prime to p. In both cases $\dim Б_\varnothing = \delta(K)$.

In the case $K = \mathbb{Q}$ and $\infty \in S$ such a description has been given already by Fröhlich (1954), see also Fröhlich (1983a).

In special cases it is possible to determine the full structure of $G_S(p)$ on the basis of Theorem 3.75. We give here only three examples:

Example 24. Let $K = \mathbb{Q}$, $p \neq 2$, and $S = \{p, q\}$, $q \not\equiv 1 \pmod{p^2}$. (By our assumption above, $q \equiv 1 \pmod{p}$.) Then $G_S(p)$ has a presentation with two generators s_q, t_q and one defining relation $t_q^{q-1}(t_q^{-1}, s_q^{-1})$. □

Example 25. Let $K = \mathbb{Q}$, $p = 2$, and $S = \{2, q, \infty\}$, $q \equiv \pm 3 \pmod{8}$. Then $G_S(2)$ has a presentation with three generators s_q, t_q, t_∞ and two defining relations $t_q^{q-1}(t_q^{-1}, s_q^{-1})$ and t_∞^2. □

Example 26. Let $K = \mathbb{Q}(\sqrt{-23})$, $p = 3$, and $S = \{\mathfrak{p}_1, \mathfrak{p}_2, q, \mathfrak{p}\}$, where $\mathfrak{p}_1$, $\mathfrak{p}_2$ are the prime divisors of 3, q is a prime with $q \not\equiv 1 \pmod{9}$ generating a prime ideal in K and $\mathfrak{p}$ is a prime ideal which is not principal. Then $G_S(3)$ has a presentation with four generators s_q, t_q, $s_\mathfrak{p}$, $t_\mathfrak{p}$ and two defining relations $t_q^{q-1}(t_q^{-1}, s_q^{-1})$, $t_\mathfrak{p}^{N(\mathfrak{p})-1}(t_\mathfrak{p}^{-1}, s_\mathfrak{p}^{-1})$. □

Let K be a quadratic number field and $G_\infty(2)$ the Galois group of the maximal 2-extension of K unramified at finite places (hence $G_\infty(2) = G_\varnothing(2)$ if K is imaginary). Using the theory of the inertia group (Chap. 1.3.7) and some group theory, one derives from our knowledge about the Galois group of the maximal 2-extension of $\mathbb{Q}$, ramified only in ∞, and the primes which are ramified in K, information about $G_\infty(2)$:

Theorem 3.77. *Let K be a quadratic number field and let $G_\infty(2)$ be as above. Moreover let $d = q_1^* \ldots q_n^*$ be the decomposition of the discriminant d of K in prime discriminants (Chap. 2.7.3). For distinct primes q, q' we define $[q, q'] \in \{0, 1\}$ by means of the Legendre symbol:*

$$(-1)^{[q,q']} = \left(\frac{q}{q'}\right) = \left(\frac{q'^*}{q}\right) \qquad \text{if } q' \neq 2,$$

$$(-1)^{[q,2]} = \left(\frac{2^*}{q}\right).$$

Then there is a minimal presentation

$$\{1\} \to R \to F \to G_\infty(2) \to \{1\}$$

of $G_\infty(2)$ by a free pro-2-group F with generators $s_1, \ldots, s_{n-1}$ and defining relations

$$\rho_i = \prod_{\substack{j=1\\ j\neq i}}^{n} (s_1^2 s_j^2 (s_i, s_j))^{[q_i, q_j]} \rho_i', \qquad i = 1, \ldots, n,$$

with $\rho_i' \in F^{(3,2)}$ *and* $s_n = 1$. ⊠

Let $CL_0(K)$ be the ideal class group in the narrow sense (Chap. 2.7.3). Via class field theory one can determine the structure of $CL_0(K)/CL_0(K)^4$ (theorem of Redei and Reichardt (1934)) by means of Theorem 3.77.

2.7. The Class Field Tower Problem. Let K be a finite extension of $\mathbb{Q}$. A fundamental problem of algebraic number theory is the question whether it is possible to embed K in a finite extension L such that $h(L) = 1$, i.e. $\mathfrak{O}_L$ is a UFD (Chap. 1.2.1). Then according to Theorem 2.2 L contains the Hilbert class field $H(K)$. The *class field tower* is the sequence

$$K \subseteq H(K) \subseteq H(H(K)) = H^2(K) \subseteq \cdots$$

Let $H^\infty(K) := \bigcup_{n=1}^{\infty} H^n(K)$. Then $H^\infty(K) \subseteq L$. Hence a necessary condition for the existence of a finite extension L of K with $h(L) = 1$ is that the extension $H^\infty(K)/K$ is finite. On the other hand, if $H^\infty(K)/K$ is finite, then $h(H^\infty(K)) = 1$.

Our initial problem reduces therefore to the question whether $H^\infty(K)/K$ is finite or infinite. It was an old conjecture that $H^\infty(K)/K$ is always finite (see Hasse (1970), I, § 11.3), but Golod and Shafarevich (1964) found the first examples of algebraic number fields K with infinite class field tower $H^\infty(K)$.

Theorem 3.78. *Let K be an imaginary quadratic number field with at least* 6 *discriminant prime divisors. Then the maximal unramified* 2*-extension $K_\varnothing(2)$ of K has infinite degree.*

Proof. We put $G = G(K_\varnothing(2)/K)$. Then $d(G) \geqslant 5$ (Theorem 2.114) and $r(G) \leqslant d(G) + 1 < d(G)^2/4$ (Theorem 3.76). Therefore G is infinite by Theorem 3.46. □

More general a number field with sufficiently many ramified prime ideals has infinite class field tower. This was shown by Brumer (1965) for absolutely normal fields using a method of Brumer, Rosen (1963) for the estimation of the class number. Roquette (1967) proved the following sharper result.

Theorem 3.79. *Let K be a normal extension of degree n over $\mathbb{Q}$, let r be the number of infinite places of K, and let p be a prime with exponent $v_p(n) > 0$ in n. Moreover let t_p be the number of primes which are ramified in K with ramification index $e \equiv 0 \pmod p$.*

Then K has infinite class field tower if

$$t_p \geqslant \frac{r-1}{p-1} + v_p(n)\delta_p + 2 + 2\sqrt{r + \delta_p},$$

where $\delta_p = 1$ if the p-th roots of unity are in K and $\delta_p = 0$ if not.

Proof. The method of Golod and Shafarevich implies that the class field tower of K is infinite if

$$d(CL(K)(p)) \geqslant 2 + 2\sqrt{r + \delta_p}.$$

Hence Theorem 3.79 follows from the following estimation of $d(CL(K)(p))$:

$$d(CL(K)(p)) \geqslant t_p - \left(\frac{r-1}{p-1} + v_p(n)\delta_p\right). \boxtimes$$

There is a similar but less sharp result for arbitrary number fields K (Roquette, Zassenhaus (1969)).

Let $p \neq 2$ be a prime number and K an imaginary quadratic number field. Using the existence of a quadratic automorphism of $G_\varnothing(p)$, one shows that for a minimal presentation

$$\{1\} \to R \to F \to G(p) \to \{1\}$$

one has $R \subseteq F_3$ (§ 1.16) (Koch Venkov (1975)). Therefore Theorem 3.47 implies that $G_\varnothing(p)$ is infinite if $d(G_\varnothing(p)) \geqslant 3$. (This assertion is also true if K is a real quadratic number field (Schoof (1986).)

Example 27. $K = \mathbb{Q}\sqrt{-3321607})$ is a field with prime discriminant. The 3-component of $CL(K)$ has generator rank 3 (Diaz y Diaz (1978), Schoof (1983)). Therefore K has infinite class field tower. $\boxtimes$

2.8. Discriminant Estimation from Above. In Chap. 1.6.15 we got estimations from below for the absolute value $|d_K|$ of the discriminant of the algebraic number field K. The existence of infinite class field towers enables us to get estimations from above: Let F be a field with infinite class field tower $F_\varnothing$ and let K be an intermediate field of $F_\varnothing/F$ with finite degree n over $\mathbb{Q}$. Then

$$|d_K|^{1/n} = |d_F|^{1/[F:\mathbb{Q}]}.$$

We know from Theorem 3.79 that $F = \mathbb{Q}(\sqrt{-m})$ with $m = 3 \cdot 5 \cdot 7 \cdot 11 \cdot 19$ has infinite class field tower, hence in this case

$$|d_K|^{1/n} = 2m^{1/2} = 296{,}276\ldots.$$

In general let r_1 and r_2 be two non negative integers such that $n_0 := r_1 + 2r_2$ is positive. Let d_n be the minimum of the values $|d_K|$ for all number fields K with degree $n \equiv 0 \pmod{n_0}$ and such that the quotient $r_1(K)/r_2(K)$ of the number of real places $r_1(K)$ and complex places $r_2(K)$ is equal to r_1/r_2. It follows from Chap. 1.6.15 that $\alpha(r_1, r_2) = \liminf d_n^{1/n}$ satisfies the inequality

$$\alpha(r_1, r_2) \geqslant (60, 8)^{r_1/n_0}(22, 3)^{2r_2/n_0}$$

and under assumption of the generalized Riemann conjecture

$$\alpha(r_1, r_2) \geqslant (215, 3)^{r_1/n_0}(44, 7)^{2r_2/n_0}.$$

The example above shows that

$$\alpha(0, 1) \leqslant 296{,}276\ldots.$$

Using real quadratic fields one sees analogously

$$\alpha(1, 0) \leqslant 5123{,}106\ldots.$$

These estimations can be improved by choosing other number fields F with infinite class field tower. Martinet (1978) shows

$$\alpha(0, 1) < 93, \qquad \alpha(1, 0) < 1059.$$

2.9. Characterization of an Algebraic Number Field by its Galois Group. Let K be a finite extension of $\mathbb{Q}$. One may ask whether K is determined by its Galois group G_K. This was shown by J. Neukirch (1969) for normal extensions K of $\mathbb{Q}$. K. Uchida and M. Ikeda proved independently that for arbitrary finite extensions K_1 and K_2 of $\mathbb{Q}$ the isomorphy of G_{K_1} and G_{K_2} implies that K_1 and K_2 are conjugate fields. This is a consequence of the following theorem of Uchida (1976):

Theorem 3.80. *Let Ω be a normal extension of $\mathbb{Q}$ such that Ω has no abelian extensions and let G be the Galois group of $\Omega/\mathbb{Q}$. Furthermore let G_1 and G_2 be open subgroups of G and let σ: $G_1 \to G_2$ be a topological isomorphism. Then σ can be extended to an inner automorphism of G.* ⊠

The proof of Uchida (1976) is based on the results of Neukirch (1969), (1969a). Neukirch gives a characterization of the decomposition groups G_p for every prime p as subgroups of $G_{\mathbb{Q}}$ which allows to determine the product of the inertia degree and the ramification index of p in normal extensions $K/\mathbb{Q}$. Then the theorem of Bauer (Theorem 1.117) shows that the group G_K determines K.

§ 3. Extensions with Given Galois Groups

Let G be a finite group and K an algebraic number field. In this paragraph we ask for normal extensions L of K with $G(L/K) = G$. The *weak inverse problem* is the question whether there exists such an extension. The *strong inverse problem* consists in a description of the set of all normal extensions L with $G(L/K) = G$.

The strong inverse problem is solved in a very satisfying manner for abelian groups G by class field theory. But already for the symmetric group S_3 there is no description of the extensions of $\mathbb{Q}$ with Galois group S_3 by means of data from $\mathbb{Q}$. Of course by class field theory it is easy to characterize the extensions L of degree 3 over a quadratic number field F such that $L/\mathbb{Q}$ is a normal extension with Galois group S_3.

Our knowledge about the inverse problem is much better for other interesting base fields. The most celebrated case is the function field $\mathbb{C}(t)$ in one variable t over the field $\mathbb{C}$ of complex numbers. The Galois group of the algebraic closure of $\mathbb{C}(t)$ is the free profinite group with continuum many generators. Even more beautiful, the maximal extension of $\mathbb{C}(t)$ with ramification only in s given points of the Riemann sphere is the free profinite group with $s - 1$ generators and one has a similar theorem for finite extensions of $\mathbb{C}(t)$ as base field (Example 13). If the base field L is an algebraic extension of the maximal abelian extension $\mathbb{Q}^{ab}$ of $\mathbb{Q}$, then the Galois group of the maximal solvable extension of L is a free pro-solvable group (Theorem 3.92). This result, the analogy between function fields and number fields, and the results in § 3.6 raise some hope that the Galois group of the algebraic closure of $\mathbb{Q}$ over $\mathbb{Q}^{ab}$ is a free profinite group (*Shafarevich conjecture*).

The known results about the weak inverse problem belong mainly to two entirely different theories. If G is solvable, then one can construct extensions with the group G by means of a chain of abelian extensions (theorem of Shafarevich). This is done by the theory of embedding problems. If G is not solvable, then one knows only special cases based on tools from algebraic geometry.

3.1. Embedding Problems (Main reference: Hoechsmann (1968)). Let K be an arbitrary field, L/K a finite separable normal extension with Galois group G, let G_K be the Galois group of a separable algebraic closure of K and $f: E \to G$ a homomorphism of a finite group E onto G. The *embedding problem* asks for a normal extension M/K and an isomorphism $G(M/K) \to E$ such that the diagram

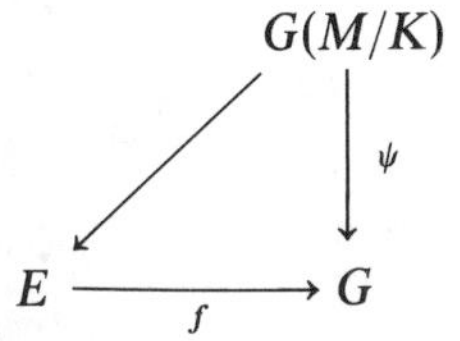

is commutative. (ψ denotes the natural projection.) Ker f is called the *kernel* of the embedding problem.

Obviously this is the same as to ask for a surjective homomorphism $G_K \to E$ such that

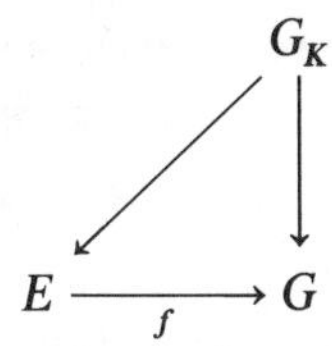

is commutative. Instead of G_K one may be interested to consider also other profinite groups $\mathfrak{G}$ if one is interested in extensions with some given conditions as e.g. restricted ramification.

In this and the following section we restrict ourselves to the case that the kernel A of the homomorphism f is abelian and we define the embedding problem $[\mathfrak{G}, \varphi, \varepsilon]$ for an arbitrary profinite group $\mathfrak{G}$, a surjective homomorphism $\varphi: \mathfrak{G} \to G$, and an $\varepsilon \in H^2(G, A)$ (the class in $H^2(G, A)$ corresponding to f) as the problem to find a morphism $\phi: \mathfrak{G} \to E$ such that $f\,\phi = \varphi$. Later on we consider the question whether ϕ is surjective or not.

Proposition 3.81. *The embedding problem* $[\mathfrak{G}, \varphi, \varepsilon]$ *has a solution if and only if the inflation*

$$\varphi^*: H^2(G, A) \to H^2(\mathfrak{G}, A)$$

maps ε *to* 0.

Proof. We consider the exact and commutative diagram

$$\begin{array}{ccccccccc} \{1\} & \longrightarrow & A & \longrightarrow & E \times_G \mathfrak{G} & \longrightarrow & \mathfrak{G} & \longrightarrow & \{1\} \\ & & \| & & \downarrow & & \downarrow{\scriptstyle\varphi} & & \\ \{1\} & \longrightarrow & A & \longrightarrow & E & \underset{f}{\longrightarrow} & G & \longrightarrow & \{1\} \end{array}$$

where $E \times_G \mathfrak{G} = \{(e, \sigma) \in E \times \mathfrak{G} \mid \varphi(\sigma) = f(e)\}$ is the fibre product. $[\mathfrak{G}, \varphi, \varepsilon]$ has a solution if and only if the upper sequence splits. □

Proposition 3.82. *Let* A *be the direct sum of two G-modules* A_1, A_2 *and* $\varepsilon = \varepsilon_1 + \varepsilon_2$ *with* $\varepsilon_i \in H^2(G, A_i)$, $i = 1, 2$. *Then* $[\mathfrak{G}, \varphi, \varepsilon]$ *has a solution if and only if* $[\mathfrak{G}, \varphi, \varepsilon_1]$ *and* $[\mathfrak{G}, \varphi, \varepsilon_2]$ *have a solution.* □

Proposition 3.83. *Let* G_1 *be a subgroup of* G *such that* $[G : G_1]$ *is prime to* $|A|$. *Then* $[\mathfrak{G}, \varphi, \varepsilon]$ *has a solution if and only if* $[\varphi^{-1}(G_1), \varphi, \mathrm{Res}_{G, G_1}(\varepsilon)]$ *has a solution.*

Proof. We have

$$\varphi^*(\mathrm{Res}_{G, G_1}(\varepsilon)) = \mathrm{Res}_{\mathfrak{G}, \mathfrak{G}_1}(\varphi^*(\varepsilon)) \text{ with } \mathfrak{G}_1 := \varphi^{-1}(G_1)$$

and $[\mathfrak{G} : \mathfrak{G}_1] = [G : G_1]$. Therefore $\mathrm{Res}_{\mathfrak{G}, \mathfrak{G}_1}$ is injective since

$$\mathrm{Cor}_{\mathfrak{G}_1, \mathfrak{G}} \mathrm{Res}_{\mathfrak{G}, \mathfrak{G}_1} = [\mathfrak{G} : \mathfrak{G}_1] \text{ (Chap. 2.3.5).} \ \square$$

In general it is difficult to say something about the existence of a surjective solution $\phi: \mathfrak{G} \to E$, but in the case of a p-group E one has the following remark:

Proposition 3.84. *Let* E *be a p-group such that the generator rank of* E *is equal to the generator rank of* $\mathfrak{G}$. *Then all solutions of* $[\mathfrak{G}, \varphi, \varepsilon]$ *are surjective.*

Proof. Let $\phi(\mathfrak{G}) := E'$. Then $E' \to E/E^{(2,p)} \cong G/G^{(2,p)}$ is surjective, i.e. E' and $E^{(2,p)}$ generate E. But then E is also generated by E', hence $E' = E$. □

Now let $G = G(L/K)$ where K is a field of characteristic 0 and L is contained in a fixed algebraic closure $\overline{K}$ of K. Moreover let $\mathfrak{G} := G(\overline{K}/K)$ and let $\varphi: \mathfrak{G} \to G$ be the natural projection. The dual $\mathfrak{G}$-module A' of A is defined as in § 2.2: $A' := \mathrm{Hom}_{\mathbb{Z}}(A, \overline{K}^\times)$. For $\chi \in A'$ let χ^* be the homomorphism

$$H^2(G, A) \xrightarrow[\mathrm{Inf}]{} H^2(\mathfrak{G}, A) \xrightarrow[\mathrm{Res}]{} H^2(\mathfrak{G}_\chi, A) \xrightarrow[\chi_*]{} H^2(\mathfrak{G}_\chi, \overline{K}^\times) \quad (3.15)$$

where $\mathfrak{G}_\chi$ denotes the stabilizer of χ. (χ is a $\mathfrak{G}_\chi$-module homomorphism by definition of A'.)

Obviously if $[\mathfrak{G}, \varphi, \varepsilon]$ has a solution, then $\chi^*(\varepsilon) = 0$ for every $\chi \in A'$. If on the other hand $\chi^*(\varepsilon) = 0$ for every $\chi \in A'$ implies that $[\mathfrak{G}, \varphi, \varepsilon]$ has a solution, we say that $V(\mathfrak{G}, A, \varepsilon)$ is true. ($V(\mathfrak{G}, A, \varepsilon)$ is equivalent to a conjecture of Hasse (1948), which was disproved by Beyer (1956).)

In the following we look at embedding problems from a slightly different view-point: We start with a finite (discrete) $\mathfrak{G}$-module A. Let $\mathfrak{G}_A$ be the fixed group of A and $\mathfrak{U}$ an open normal subgroup of $\mathfrak{G}$ contained in $\mathfrak{G}_A$. Then A is a $\mathfrak{G}/\mathfrak{U}$-module and for any $\varepsilon \in H^2(\mathfrak{G}/\mathfrak{U}, A)$ we have an embedding problem $[\mathfrak{G}, \varphi_{\mathfrak{U}}, \varepsilon]$ with $\varphi_{\mathfrak{U}}: \mathfrak{G} \to \mathfrak{G}/\mathfrak{U}$. We say that $V(\mathfrak{G}, A)$ is true if $V(\mathfrak{G}/\mathfrak{U}, A, \varepsilon)$ is true for every open normal subgroup $\mathfrak{U}$ of $\mathfrak{G}$ with $\mathfrak{U} \subseteq \mathfrak{G}_A$ and every $\varepsilon \in H^2(\mathfrak{G}/\mathfrak{U}, A)$. By this definition $V(\mathfrak{G}, A)$ is true if and only if the map

$$\prod \chi^*: H^2(G, A) \to \prod_{\chi \in A'} H^2(\mathfrak{G}_\chi, \bar{K}^\times) \tag{3.16}$$

is injective.

Proposition 3.85. *If $\mathfrak{G}$ acts trivially on A', then $V(\mathfrak{G}, A)$ is true.*

Proof. A' and therefore A is a direct sum of cyclic modules. By Proposition 3.82 we can assume that A is cyclic. Let $n = |A|$ and $\chi \in A'$ a character of order n. Then the sequence

$$\{0\} \longrightarrow A \xrightarrow[\chi]{} \bar{K}^\times \xrightarrow[n]{} \bar{K}^\times \longrightarrow \{1\}$$

is an exact sequence of $\mathfrak{G}$-modules. Hence by Proposition 3.49 the induced homomorphism $H^2(\mathfrak{G}, A) \xrightarrow[\chi_*]{} H^2(\mathfrak{G}, \bar{K}^\times)$ is injective. □

More generally one has the following theorem:

Theorem 3.86. *If $\mathfrak{G}$ acts cyclic on A', then $V(\mathfrak{G}, A)$ is true.* ⊠

3.2. Embedding Problems for Local and Global Fields (Main references: Hoechsmann (1968), 5.–6., Demushkin, Shafarevich (1962)).

Theorem 3.87. *Let K be a finite extension of $\mathbb{Q}_p$. Then $V(\mathfrak{G}, A)$ is true for all finite $\mathfrak{G}$-modules A.*

Proof. By Theorem 3.56 for every $\varepsilon \in H^2(\mathfrak{G}, A)$, $\varepsilon \neq 0$, there exists a $\chi \in H^0(G, A')$ such that $\varepsilon \cup \chi = \chi^*(\varepsilon) \neq 0$, hence (3.16) is injective. □

Theorem 3.87 is also true if $K = \mathbb{R}$.

Now let K be an algebraic number field. For every place v of K we choose an extension w_v to K. The decomposition group of w_v will be denoted by $\mathfrak{G}_{w_v}$.

Theorem 3.88 (Local-global-principle for embeddings). *The localization map*

$$H^2(\mathfrak{G}, A) \to \prod_v H^2(\mathfrak{G}_{w_v}, A)$$

is injective if and only if $V(\mathfrak{G}, A)$ is true.

Proof. This follows from Theorem 2.86 and Theorem 3.87 applied to (3.16). □

Theorem 3.89. *Let F be the fixed field of the fixed group of A' and $G := G(F/K)$. Then $V(\mathfrak{G}, A)$ is true if and only if*

$$H^1(G, A') \to \prod_v H^1(G_{w_v}, A')$$

is injective where G_w denotes the decomposition group of $w|_F$.

Proof. This follows easily from Theorem 3.87, Theorem 3.68.1. and Proposition 3.85. □

As a corollary we get the following theorem:

Theorem 3.90. *With the notation as above, $V(\mathfrak{G}, A)$ is true if the g.c.d. of the indices $[G : G_w]$ is equal to 1.* □

By Theorem 2.119 the last condition is equivalent to $H^3(G, F^\times) = \{0\}$.
The following theorem is due to Scholz (1925) and Ikeda (1960).

Theorem 3.91. *Let K be an algebraic number field. Then every embedding problem $[G_K, \varphi, \varepsilon]$ which has a solution has a surjective solution.* ⊠

So far we have considered only finite extensions K of $\mathbb{Q}$. We know already that the Brauer group of an infinite algebraic extension L of $\mathbb{Q}$ is trivial if $[L : \mathbb{Q}]$ is divisible by l^∞ for all primes l (Theorem 3.53). Hence all embedding problems over L are solvable; by Theorem 3.91 they even have a surjective solution. This implies the following theorem of Iwasawa (1953):

Theorem 3.92. *Let L be an algebraic extension of $\mathbb{Q}$ with degree $[L : \mathbb{Q}]$ divisible by l for all primes l and let L^{sol} be the maximal algebraic solvable extension of L. Then $G(L^{\text{sol}}/L)$ is a free prosolvable group with countable many generators.* ⊠

The assumptions of Theorem 3.92 are fulfilled e.g. if L is the maximal abelian extension $\mathbb{Q}^{ab}$ of $\mathbb{Q}$. Uchida (1982) shows that also the maximal unramified solvable extension of $\mathbb{Q}^{ab}$ has a Galois group which is a free prosolvable group with countable many generators.

3.3. Extensions with Prescribed Galois Group of l-Power Order (Main reference: Shafarevich (1954a)). Let K be an algebraic number field, l a fixed prime number, and G an l-group. We want to construct normal extensions L/K with $G(L/K) \cong G$. Since every nontrivial l-group has a non trivial center, this can be done in the following way: We choose a chain of groups

$$G_1 \leftarrow G_2 \leftarrow \cdots \leftarrow G_{s-1} \leftarrow G_s = G$$

such that for all $i = 1, 2, \ldots, s-1$ the homomorphism $G_{i+1} \to G_i$ is surjective with kernel of order l lying in the center of G_{i+1} and $G_1 = G/G^{(2,1)}$. Then we try to solve the embedding problems connected with the homomorphisms $G_{i+1} \to G_i$ one after the other starting with an extension with Galois group G_1.

For this purpose we introduce the notion of *Scholz extension*: A Scholz extension L/K with exponent h is a normal extension with the following properties:

1. $G(L/K)$ is an l-group.
2. The places above l and the real places of K split completely in L.
3. The prime ideals $\mathfrak{p}$ of K which are ramified in L/K have inertia degree 1 and their absolute norm $N(\mathfrak{p})$ satisfies

$$N(\mathfrak{p}) \equiv 1 \pmod{l^h}.$$

Theorem 3.93. *Let L/K be a Scholz extension of exponent h and $[G_K, \varphi, \varepsilon]$ an embedding problem such that φ is the projection $G_K \to G(L/K)$, the extension $E \to G(L/K)$ is central with kernel A of order l, and the ranks of E and $G(L/K)$ are equal. Then $[G_K, \varphi, \varepsilon]$ has a surjective solution if h is sufficiently large.*

Proof. First assume that the l-th roots of unity are in K. Then by Proposition 3.84 and Proposition 3.85 it is sufficient to show $\chi^*(\varepsilon) = 0$ for all $\chi \in A'$. Since $\chi^*(\varepsilon) \in H^2(G_K, \bar{K}^\times)$, by Theorem 2.86 one has to show that the localizations of $\chi^*(\varepsilon)$ vanish. But this is easy to see since by definition of a Scholz extension the decomposition groups G_v of $G(L/K)$ are cyclic. Hence the corresponding local embedding problem $[G_v, \varphi_v, \varepsilon_v]$ with $\varphi_v: E_v \to G_v$ either splits or E_v is cyclic. In the first case we have the trivial (non surjective) solution. In the second case one has the unramified solution if v is unramified in L/K and the ramified solution if v is ramified in L/K provided that h is sufficiently large.

If K does not contain the l-th roots of unity one goes over to the extension $L(\mu_l)/K$ and uses Proposition 3.83. ⊠

Our problem of constructing fields with given Galois group of l-power order is reduced by Theorem 3.93 to the problem to show that among the solutions of an embedding problem for Scholz extensions there is one which is again a Scholz extension. If $\mu_l \not\subset K$ this is not difficult to show (Scholz (1937), Reichardt (1937)). In the case $\mu_l \subset K$ one proceeds as follows.

One defines invariants (χ, X) and $(X)_h$ with values in μ_l for $X \in H^2(L/K, \mu_l)$ and $\chi \in H^1(L/K, \mu_l)$ by means of the Artin symbol. These invariants are multiplicative in X and χ. We fix an isomorphism $\theta: \mu_l \to A := \operatorname{Ker} \varphi$. Then to every embedding problem $[G_K, \varphi, \varepsilon]$ for a Scholz extension L/K there corresponds an $X := \theta^* \varepsilon$.

Theorem 3.94. *Let L/K be a Scholz extension of exponent h and $\mu_l \subset K$. The embedding problem $[G_K, \varphi, \varepsilon]$ of Theorem 3.93 has a solution L'/K which is a Scholz extension if and only if $(\chi, X) = 1$ and $(X)_h = 1$ for all $\chi \in H^1(L/K, \mu_l)$.* ⊠

Example 28 (Shafarevich (1954a), § 2.3). Let $l = 2$, $K = \mathbb{Q}$, $L = \mathbb{Q}(\sqrt{17}, \sqrt{281})$ and let $\varphi: E \to G(L/K)$ be the extension which corresponds to the embedding of $\mathbb{Q}(\sqrt{17})/\mathbb{Q}$ in a cyclic extension of order 4. Furthermore let $\chi \in H^1(L/K, \mu_l)$ be the character with

$$\eta(\sqrt{281}) = \chi(\eta)\sqrt{281} \qquad \text{for } \eta \in G(L/K).$$

Then $(\chi, X) = -1$. ⊠

The example shows that in general it is not possible to embed a Scholz extension in a Scholz extension. The way out of this situation is to use the multiplicativity of the invariants and to show that starting with a bigger embedding problem E' we find among the factor embedding problems E those with trivial invariants. If one chooses E' big enough one can manage that E corresponds to a group extension which is group theoretical isomorphic to $E \to G(L/K)$.

By means of this method one proves the following theorem.

Theorem 3.95. *Let G be an arbitrary l-group and K an algebraic number field. Then there exists a normal extension L/K such that $G(L/K)$ is isomorphic to G.* ⊠

3.4. Extensions with Prescribed Solvable Galois Group (Main references: Shafarevich (1954b), Ishchanov (1976)). We remember that a group extension

$$\{1\} \to N \to E \underset{\varphi}{\to} G \to \{1\} \tag{3.17}$$

is called a semi direct product if it splits, i.e. if φ has a section $s: G \to E$.

Theorem 3.96. *Let K be an algebraic number field, L/K a finite normal extension with Galois group G, and let* (3.17) *be a semi direct product of the nilpotent group N. Then the corresponding embedding problem has a solution M/K.*

Proof. Since a nilpotent group is the direct product of l-groups, one can assume that N is an l-group. Let G_l be an l-Sylow group of G and E_l the corresponding group extension with N. By means of Proposition 3.83 and Theorem 3.91 one shows that the embedding problem for E is solvable if the embedding problem for E_l is solvable. Therefore we can assume that E is an l-group. One defines the notion of a Scholz extension L'/L of exponent h with respect to the extension L/K and generalizes the methods of § 3.3 to this situation. ⊠

Now let G be a finite solvable group. Then G is a factor group of a group G' which is a semidirect product of l-groups (Ore (1939)). Hence as a corollary to Theorem 3.96 one gets the following theorem of Shafarevich:

Theorem 3.97. *Let G be a finite solvable group and K an algebraic number field. Then there exists a normal extension L/K such that $G(L/K)$ is isomorphic to G.* ⊠

3.5. Extensions with Prescribed Local Behavior (Main reference: Neukirch (1979)). The embedding problem can be strengthened in the following manner: As above let K be an algebraic number field and let K_v its completion at the place v. We put $G_v := G_{K_v}$. For every place v we choose a fixed embedding $\overline{K} \to \overline{K}_v$ and obtain an embedding $G_v \to G_K$. Let $\varphi: G_K \to G$ be a morphism of G_K onto a finite group G and $\varphi_v: G_v \to G$ its restriction to G_v. If then $f: E \to G$ is a morphism of an arbitrary profinite group E, we obtain the diagrams

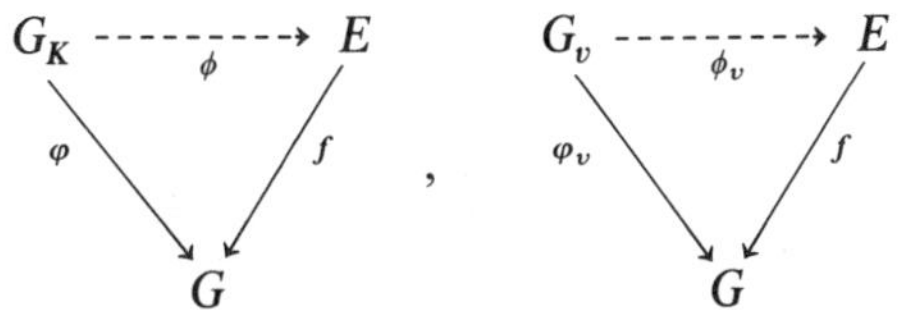

and the corresponding restriction map $\mathrm{Hom}_G(G_K, E) \to \prod_v \mathrm{Hom}_G(G_v, E)$.

The embedding problem with given local behavior asks for morphisms ϕ which induce given local morphisms ϕ_v for certain places v of K.

Theorem 3.98. *Let $\varphi\colon G_K \to G$ be a surjective homomorphism onto the finite group G and let $m(L)$ be the number of roots of unity in the fixed field L of the kernel of φ.*

If $f\colon E \to G$ is a surjective morphism with a pro-solvable separable kernel of finite exponent which is prime to $m(L)$ and if $\mathrm{Hom}_G(G_v, E)$ is not empty for all places v of K then the map

$$\mathrm{Hom}_G(G_K, E)_{\mathrm{sur}} \to \prod_{v \in S} \mathrm{Hom}_G(G_v, E)$$

is surjective for every finite set S of places of K ($\mathrm{Hom}_G(G_K, E)_{\mathrm{sur}}$ denotes the set of surjective morphisms). ⊠

The proof uses ideas of Scholz (1937) and Reichardt (1937) and Galois cohomology of algebraic number fields.

In the following we shall assume that all occuring profinite groups are separable. The following theorems are corollaries to Theorem 3.95. We remember that by the theorem of Feit and Thompson every finite group of odd order is solvable.

Theorem 3.99. *Let K be a finite algebraic number field and E a profinite group of finite odd exponent. Then there exists a normal extension L/K such that $G(L/K)$ is isomorphic to E.*

Proof. One applies Theorem 3.98 to the field $\mathbb{Q}$ of rational numbers and to the set S of primes which ramify in $K/\mathbb{Q}$. We put $G := \{1\}$. If L' is a solution of the corresponding embedding problem, then $L := L'K$ has over K a Galois group isomorphic to E. □

Theorem 3.100. *Let K be a finite algebraic number field, $m(K)$ the number of roots of unity in K, and S a finite set of places of K. Let E be a profinite group of finite exponent prime to $m(K)$ and L_v/K_v, $v \in S$, normal extensions whose Galois groups are embeddable in E.*

Then there exists a normal extension L/K with Galois group isomorphic to E which for $v \in S$ has the given extensions L_v/K_v as completions.

Proof. One applies Theorem 3.98 with $G := \{1\}$. □

Theorem 3.100 is a generalization of the Grunwald-Wang-theorem (Chap. 2.1.12). The assumption that E has finite exponent in the Theorems 3.98–100 can not be dropped since e.g. there is no normal extension $L/\mathbb{Q}$ with $G(L/\mathbb{Q}) \cong \mathbb{Z}_p \times \mathbb{Z}_p$ for a prime p.

3.6. Realization of Extensions with Prescribed Galois Group by Means of Hilbert's Irreducibility Theorem (Main references: Matzat (1987), (1988)). With the methods used up to now we can construct only solvable extensions. In this section we consider the realization of finite groups G as Galois groups of normal extensions L/K by means of Hilbert's irreducibility theorem. This method goes already back to Hilbert who used it in the case of the symmetric and alternating groups but for other groups the method was developed only recently by Matzat (1977), Belyj (1979). Up to now one is able to realize certain groups as Galois groups only over some extension K of $\mathbb{Q}$ as ground field but not over $\mathbb{Q}$ itself. It is convenient to take for K the maximal abelian extension $\mathbb{Q}^{ab}$ of $\mathbb{Q}$. In the case of solvable groups G the theorem of Iwasawa (Theorem 3.92) gives a full insight in the possible extensions L of $\mathbb{Q}^{ab}$ with $G(L/\mathbb{Q}^{ab}) \cong G$.

Let x, t be variables. A field K is called a *Hilbert field* if for any polynomial $f(x, t) \in K[x, t]$ which is irreducible over K there are infinitely many elements t_0 of K such that $f(x, t_0)$ is irreducible over K. Beside the infinite finitely generated fields also the finite extensions of $\mathbb{Q}^{ab}$ are Hilbert fields (Weissauer (1982)). Hilbert showed that $\mathbb{Q}$ is a Hilbert field. This fact and its generalization to other fields is called Hilbert's irreducibility theorem.

Let G be a finite group, K a Hilbert field, and $N/K(t)$ a normal extension with $G(N/K(t)) \cong G$ which is regular, i.e. K is algebraically closed in N. Then there exists a normal extension L/K such that $G(L/K) \cong G$.

Riemann's existence theorem for complex algebraic functions gives a full insight in the possible algebraic extensions of $\mathbb{C}(t)$ and their ramification behavior (Example 13). In particular for any finite group G there is a normal extension $N/\mathbb{C}(t)$ with $G(N/\mathbb{C}(t)) \cong G$. If we can show that there is a subfield K of $\mathbb{C}$ and a normal extension $N_0/K(t)$ such that $N = \mathbb{C}N_0$, then $G(N_0/K(t)) = G$. In this case we say that we can go down from $\mathbb{C}$ to K. If K is a Hilbert field we get a normal extension of K with Galois group isomorphic to G.

By the Lefschetz principle we can go down from $\mathbb{C}$ to $\overline{\mathbb{Q}}$. For certain finite groups G it is possible to go down from $\overline{\mathbb{Q}}$ to $\mathbb{Q}^{ab}$ or even to $\mathbb{Q}$. The background for this procedure is given by the following theorem. For its formulation we need some definitions:

Let G be a finite group of order n and $C_1, \dots, C_s$ some classes of conjugated elements in G. The *class structure* $\mathfrak{C} = (C_1, \dots, C_s)$ is defined by

$$(C_1, \dots, C_s) := \{(g_1, \dots, g_s) | g_j \in C_j\}.$$

For $\kappa \in \mathbb{Z}/n\mathbb{Z}$ we put

$$C_j^\kappa = \{g^\kappa | g \in C_j\}.$$

then

$$\mathfrak{C}^* := \bigcup_{\kappa \in (\mathbb{Z}/n\mathbb{Z})^\times} (C_1^\kappa, \dots, C_s^\kappa)$$

is called the *ramification structure* of G generated by $\mathfrak{C}$.

Moreover for any set $\mathfrak{D}$ of s-tuples in G^s we put

$$[\mathfrak{g}] := \{(h^{-1}g_1h, \ldots, h^{-1}g_sh) | \mathfrak{g} = (g_1, \ldots, g_s), h \in G\},$$

$$\overline{\sum}^i(\mathfrak{D}) := \{[\mathfrak{g}] | \mathfrak{g} \in \mathfrak{D}, g_1 \ldots g_s = 1\},$$

$$\sum^i(\mathfrak{D}) := \{[\mathfrak{g}] \in \overline{\sum}^i(\mathfrak{D}) | (g_1, \ldots, g_s) = G\}.$$

Then $l^i(\mathfrak{C}) := |\sum^i(\mathfrak{C})|$ is a divisor of $l^i(\mathfrak{C}^*) := |\sum^i(\mathfrak{C}^*)|$. Let $e(\mathfrak{C})$ be the natural number with

$$l^i(\mathfrak{C}^*) = e(\mathfrak{C}) l^i(\mathfrak{C}).$$

A normal regular extension $N/K(t)$ has the ramification structure $\mathfrak{C}^*$ if $\overline{K}(t)$ has exactly s places $\mathfrak{p}_1, \ldots, \mathfrak{p}_s$ which are ramified in $N\overline{K}$ and for all $[\mathfrak{g}] \in \sum^i(\mathfrak{C}^*)$ there are generators g_j of the inertia groups $T(\mathfrak{P}_j)$ for places $\mathfrak{P}_j$ of NK with $\mathfrak{P}_j | \mathfrak{p}_j$ such that $\mathfrak{g} = g_1, \ldots, g_s$.

A subgroup U of G has the complement H, where H is a subgroup of G, if every element g of G can be uniquely presented in the form $g = uh$ with $u \in U$, $h \in H$.

Theorem 3.101. *Let G be a finite group whose center $Z(G)$ has a complement and let $\mathfrak{C}$ be a class structure of G with $l^i(\mathfrak{C}) \neq 0$. Then there exists a subfield K of $\overline{\mathbb{Q}}$ and a normal regular extension $N/K(t)$ with $G(N/K(t)) \cong G$ and the ramification structure $\mathfrak{C}^*$. Moreover the constant field K contains an abelian extension K_0 of $\mathbb{Q}$ with $[K : K_0] \leqslant l^i(\mathfrak{C})$ and $[K_0 : \mathbb{Q}] = e(\mathfrak{C})$.* ⊠

The most interesting cases are $l^i(\mathfrak{C}) = 1$, i.e. $K \subseteq \mathbb{Q}^{ab}$ and $l^i(\mathfrak{C}^*) = 1$, i.e. $K = \mathbb{Q}$.

In general it is difficult to decide whether for a given group G there exists a ramification structure with $l^i(\mathfrak{C}) = 1$ or $l^i(\mathfrak{C}^*) = 1$. Belyj (1979), (1983) studied the cases $s = 3$ for subgroups of $GL_m(\mathbb{F}_q)$ and proved the following theorem.

Theorem 3.102. *There are normal regular extensions $N/\mathbb{Q}^{ab}(t)$ with $G(N/\mathbb{Q}^{ab}(t)) \cong G$ for all classical simple groups, i.e. for the simple groups of the form $\mathrm{PSL}_m(\mathbb{F}_q)$, $\mathrm{PSU}_m(\mathbb{F}_{q^2})$, $\mathrm{PSp}_{2m}(\mathbb{F}_q)$, $\mathrm{PSO}_{2m+1}(\mathbb{F}_q)$, and $\mathrm{PSO}^{\pm}_{2m}(\mathbb{F}_q)$.* ⊠

If one knows the character table of G, one can estimate $l^i(\mathfrak{C})$ from above. This leads to the following results about sporadic simple groups (we use the standard notation for these groups).

Theorem 3.103. *There are normal regular extensions $N/\mathbb{Q}^{ab}(t)$ with $G(N/\mathbb{Q}^{ab}(t)) \cong G$ for all sporadic simple groups G with at most the exception of J_4.*

There are normal regular extensions $N/\mathbb{Q}(t)$ with $G(N/\mathbb{Q}(t))$ isomorphic to one of the following 18 sporadic simple groups: M_{11}, M_{12}, M_{22}, J_1, J_2, HS, Sz, ON, Co_3, Co_2, Co_1, Fi_{22}, Fi_{23}, Fi'_{24}, F_5, F_3, F_2, F_1. ⊠

For the proof see Hoyden (1985) (M_{23}, M_{24}, J_1, J_2), Hoyden-Matzat (1986) (J_3, Mc, He, Ru, ON, Ly, F_5), Hunt (1986) (M_{12}, M_{22}, J_2, HS, Sz, Co_3, Co_2, Co_1, Fi_{22}, Fi_{23}, Fi'_{24}, F_3, F_2), Matzat (1984) (M_{12}, M_{22}), Matzat-Zeh (1986) (M_{11}, M_{12}), Thompson (1984) (F_1).

There are also results about finite groups G with certain given simple composition factors (Matzat (1985)). In particular, if K is a finite extension of $\mathbb{Q}^{ab}$, then any embedding problem over K is solvable if its kernel satisfies certain conditions, which are verified for many groups. There is some hope to verify these conditions for all simple groups. By a theorem of Iwasawa (1953) this would prove the Shafarevich conjecture that $G(\overline{\mathbb{Q}}/K)$ is a free profinite group with countable many generators.

Chapter 4
Abelian Fields

(Main reference: Washington (1982))

The finite abelian extensions of $\mathbb{Q}$ are called (absolute) *abelian fields.* So far they appeared as examples for more general theorems. In this chapter we consider further problems about number fields mostly restricted to abelian fields because the theory is much more complete in this restriction as in a more general setting.

Every abelian field M is a subfield of a cyclotomic field (Chap. 2.1.1), i.e. of a field of the form $\mathbb{Q}(\zeta_n)$, where $\zeta_n := \exp(2\pi i/n)$. The smallest n with $M \subseteq \mathbb{Q}(\zeta_n)$ is called the conductor of M.

There is a canonical homomorphism ϕ_n of $(\mathbb{Z}/n\mathbb{Z})^\times$ onto $G(M/\mathbb{Q})$ given by

$$\phi_n(\bar{a})(\zeta_n) = \zeta_n^a \qquad \text{for } \bar{a} \in (\mathbb{Z}/n\mathbb{Z})^\times.$$

The characters of $(\mathbb{Z}/n\mathbb{Z})^\times$ are called *Dirichlet characters* mod n. The abelian field M is uniquely determined by the group $X = X(M)$ of characters of $(\mathbb{Z}/n\mathbb{Z})^\times$ which vanish on the kernel of ϕ_n.

If $n|m$, then every character χ of $(\mathbb{Z}/n\mathbb{Z})^\times$ induces a character χ' of $(\mathbb{Z}/m\mathbb{Z})^\times$ by

$$\chi'(a + \mathbb{Z}/m\mathbb{Z}) := \chi(a + \mathbb{Z}/n\mathbb{Z}).$$

One identifies χ' and χ and calls n and m defining modules of χ. The smallest defining module of χ is called the conductor f_χ of χ. χ considered as character of $(\mathbb{Z}/f_\chi\mathbb{Z})^\times$ is called *primitive character* (compare Chap. 1.5.6–8 where we defined the same notions by means of the corresponding idele character). We denote also by χ the corresponding character of $G(M/\mathbb{Q})$.

In this chapter we use the convention for the definition of Gaussian sums in the theory of abelian fields and define

$$\tau(\chi) := \tau_{1/f}(\chi) = \sum_{x \in (\mathbb{Z}/f\mathbb{Z})^\times} \chi(x)\zeta_f^x \qquad \text{for } \chi \in X(M),$$

where f denotes the (finite part of the) conductor of χ. This is the complex conjugate of the Gaussian sum defined in Chap. 1.6.5 as product of local Gaussian sums (Example 39). Then the discriminant d_M of M is given by

$$d_M = \prod_{\chi \in X} \tau(\chi)^2 = \prod_{\chi \in X} \chi(-1) f_\chi \qquad \text{(Theorem 2.5, 2.27).} \tag{4.1}$$

We mentioned already in the introduction to Chapter 1 the fundamental work of Kummer in the middle of the last century about the arithmetic of cyclotomic fields. This work remained to be the main part of the theory of abelian fields until Leopoldt inspired by the work of Hasse, in particular by his book (1952), in the fifties of our century generalized many of the then existing results on cyclotomic fields to abelian fields and together with Kubota found the p-adic analogue to Dirichlet L-functions. In the same time Iwasawa introduced his theory of Γ-extensions, i.e. normal extension with Galois group isomorphic to $\mathbb{Z}_p$, and interpreted p-adic L-functions in terms of Γ-extensions. This led him to his main conjecture which was proved in 1979 by Mazur and Wiles (4.7).

A part of the theory can be established for abelian extensions of imaginary-quadratic number fields with the use of elliptic functions instead of the exponential function. The study of the class number formula in this setting was initiated by Meyer (1957). For the present state of the theory see Rubin (1987), § 12.

Chapter 4 is organized as follows. § 1 develops Leopoldt's description of the ring of integers of an abelian field M by means of $X(M)$. In § 2 we show how to derive from the analytic class number formula (Theorem 2.25) a purely algebraic expression called the arithmetical class number formula and we consider some other results on class numbers connected with the arithmetical class number formula. § 3 is devoted to p-adic L-functions and § 4 to Iwasawa theory.

§ 1. The Integers of an Abelian Field

(Main reference: Leopoldt (1959))

If L/K is any finite normal extension of algebraic number fields and $\mathfrak{O}_L \supseteq \mathfrak{O}_K$ are the corresponding rings of integers, then one is interested in the structure of $\mathfrak{O}_L$ as an $\mathfrak{O}_K[G(L/K)]$-module. In the case of an abelian extension $M/\mathbb{Q}$ one has a beautiful description of this structure which we explain in the following. For the general case see Chap. 5.

1.1. The Coordinates. Let $\mathbb{Q}(\chi)$ be the extension of $\mathbb{Q}$ generated by the values of the character χ and let $\mathbb{Q}(X)$ be the compositum of the fields $\mathbb{Q}(\chi)$ for $\chi \in X$. If $G := G(M/\mathbb{Q})$ has exponent m, then $\mathbb{Q}(X) = \mathbb{Q}(\zeta_m)$.

We introduce the following coordinates $y_M(\alpha, \chi)$ for $\alpha \in M$:

$$y_M(\alpha, \chi) := \frac{1}{\tau(\chi)} \sum_{g \in G} \chi(g^{-1}) g\alpha.$$

These coordinates have the following properties:

1. $y_M(\alpha, \chi) \in \mathbb{Q}(\chi)$.
2. $y_M(\alpha, \chi^a) = \left(\dfrac{a}{\mathbb{Q}(\chi)/\mathbb{Q}}\right) y_M(\alpha, \chi)$ for $a \in \mathbb{Z}$ which are prime to the order of χ.

3. $\alpha = \frac{1}{[M:\mathbb{Q}]} \sum_{\chi \in X} y_M(\alpha, \chi)\tau(\chi)$.
4. $y_M(\alpha, \chi)$ is $\mathbb{Q}$-linear as function of α.
5. $y_M(g\alpha, \chi) = \chi(g) y_M(\alpha, \chi)$ for $g \in G$.
6. Let M' be a subfield of M and χ' a character of M'. Then

$$y_M(\alpha, \chi') = y_{M'}(\mathrm{tr}_{M/M'}(\alpha), \chi').$$

7. Let $y(\chi)$, $\chi \in X$, be a system of numbers with $y(\chi) \in \mathbb{Q}(\chi)$ and $y(\chi^a) = \left(\frac{a}{\mathbb{Q}(\chi)/\mathbb{Q}}\right) y(\chi)$ for $a \in Z$ prime to the order of χ. Then there exists an $\alpha \in M$ such that $y(\chi) = y_M(\alpha, \chi)$ for $\chi \in X$.

1.2. The Galois Module Structure of the Ring of Integers of an Abelian Field. $\alpha \in M$ belongs to $\mathfrak{O}_M$ if and only if the coordinates $y_M(\alpha, \chi)$ are in $\mathfrak{O}_{\mathbb{Q}(\chi)}$ and satisfy certain congruences mod $[M:\mathbb{Q}]$. Hence for a description of $\mathfrak{O}_M$ as $\mathbb{Z}[G]$-module it is sufficient to describe these congruences. This can be done as follows.

Let θ be an element of M such that $g\theta$, $g \in G$, is a basis of $M/\mathbb{Q}$. Then every $\alpha \in \mathfrak{O}_M$ can be written uniquely in the form $\alpha = w\theta$, where w is an element of a certain subgroup $\mathfrak{W}$ of $\mathbb{Q}[G]$ of $\mathbb{Z}$-rang $[M:\mathbb{Q}]$:

$$\mathfrak{W} := \{w \in \mathbb{Q}[G] | w\theta \in \mathfrak{O}_M\}$$

The order

$$O_M := \{v \in \mathbb{Q}[G] | v\mathfrak{W} \subseteq \mathfrak{W}\}$$

of $\mathfrak{W}$ and the class of $\mathfrak{W}$ as O_M-module are independent of the choice of θ. Any other element θ' of M such that $g\theta'$, $g \in G$, is a basis of $M/\mathbb{Q}$ has the form $e\theta$ with $e \in \mathbb{Q}[G]^\times$. Therefore

$$\mathfrak{W}' := \{w \in \mathbb{Q}[G] | w\theta' \in \mathfrak{O}_M\} = \mathfrak{W}e.$$

We introduce some more notations about characters. Two characters χ and χ' of M are called similar if they have the same wild ramification, i.e. if $v_p(f_\chi) > 1$ or $v_p(f_{\chi'}) > 1$, then $v_p(f_\chi) = v_p(f_{\chi'})$ for all primes p with $p \| G|$. Let ϕ be a class of similar characters, then

$$1_\phi := \frac{1}{|G|} \sum_{\chi \in \phi} \sum_{g \in G} \chi(g^{-1}) g$$

is an idempotent in $\mathbb{Q}[G]$. Furthermore we set

$$f_\phi := \mathrm{l.c.m.}\{f_\chi | \chi \in \phi\},$$

$M_\phi :=$ fixed field of the group $\{g \in G | \chi(g) = 1 \text{ for } \chi \in \phi\}$,

$$\tau(\chi|_\phi) := \sum_{x \in (\mathbb{Z}/f_\phi\mathbb{Z})^\times} \chi(x)\zeta_{f_\phi}^x$$

$$\theta_\phi := \frac{1}{[M_\phi:\mathbb{Q}]} \sum_{\chi \in \phi} \tau(\chi|\phi).$$

Now we are able to formulate the theorems about the structure of $\mathfrak{O}_M$:

Theorem 4.1. 1. *$O_M = \sum \mathbb{Z}[G]1_\phi$ where the sum runs over all classes ϕ of similar characters.*

2. *$\mathfrak{W}$ is equivalent to $\mathfrak{O}_M$ considered as O_M-module.* ⊠

Theorem 4.1 implies that there is a $\theta \in M$ such that $\mathfrak{O}_M = O_M\theta$. Hence for an explicit description of the integers in M it suffices to find such an element θ.

Theorem 4.2. *Let $\theta := \sum \theta_\phi$ where the sum runs over all classes of similar characters. Then $\mathfrak{O}_M = O_M\theta$.* ⊠

As a corollary to Theorem 4.2 one gets the following result.

Theorem 4.3. *$\mathfrak{O}_M$ is the direct sum of the $\mathbb{Z}[G]$-modules $1_\phi\mathfrak{O}_M$. In particular $\mathfrak{O}_M$ has an integral normal basis, i.e. a basis of the form $\{g\omega, g \in G\}$ for some $\omega \in \mathfrak{O}_M$, if and only if $M/\mathbb{Q}$ is tamely ramified.* ⊠

By means of Theorem 4.2 one finds the congruences which characterize the coordinates of the integers in M.

Example 1. If $M/\mathbb{Q}$ is tamely ramified, then $\theta = \theta_\phi$ generates an integral normal basis. □

Example 2. As an illustration of Theorem 4.3 we consider the case $[M : \mathbb{Q}] = 2$. In this case the result is well known (Chap. 1.1.1, Example 2). The two characters χ and χ_0 of G, where χ_0 denotes the unit-character, belong to the same class ϕ of similar characters if and only if $M/\mathbb{Q}$ is tamely ramified. Furthermore

$$f_\chi = d_M,\ f_{\chi_0} = 1,\ \theta_\chi = \pm\tfrac{1}{2}\sqrt{d_M},\ \theta_{\chi_0} = 1$$

if $M/\mathbb{Q}$ is wildly ramified, i.e. $d_M \equiv 0 \pmod 4$.

$$f_\phi = d_M,\ \theta_\phi = \tfrac{1}{2}(\pm 1 \pm \sqrt{d_M})$$

if $M/\mathbb{Q}$ is tamely ramified, i.e. $d_M \equiv 1 \pmod 4$.

§2. The Arithmetical Class Number Formula

We want to reformulate the analytical class number formula (Theorem 2.25) in the case of abelian fields as a purely algebraic expression.

2.1. The Arithmetical Class Number Formula for Complex Abelian Fields. First we consider the relation between a complex abelian field M and its maximal real subfield M^+. In the following proposition we consider more generally *CM-fields*, i.e. totally complex extensions K of $\mathbb{Q}$ which have a totally real subfield K^+ with $[K : K^+] = 2$.

Proposition 4.4. *Let K be a CM-field and K^+ its maximal real subfield, let w be the number of roots of unity in K. Then the class number $h(K^+)$ divides $h(K)$.*

The homomorphism $\varphi: E \to E$, $\varphi(\varepsilon) = \bar{\varepsilon}\varepsilon^{-1}$, *of the unit group* $E = E(K)$ *of* K *has the image* μ_w *or* μ_w^2.

Proof. $h(K^+) | h(K)$ follows from class field theory: since K/K^+ is ramified at infinite places, the Hilbert class field of K^+ does not contain K (Theorem 2.2, 2.112). Since $|\varphi(\varepsilon)| = 1$ the image of φ is contained in μ_w (Chap. 1.1.3). Therefore the claim follows from $[E(K): E(K^+)] \geqslant w/2$. □

$h^- := h(K)/h(K^+)$ is called the relative class number and the index $Q(K) := [E(K): \mu_w E(K^+)]$ is called the unit index. By Proposition 4.4 the unit index is equal to 1 or 2, see Hasse (1952), Chap. 3, for its computation in the case of abelian fields,[2] e.g. $Q(\mathbb{Q}(\zeta_n)) = 1$ if and only if n is a power of a prime.

Theorem 4.5. *Let* M *be an complex abelian field and* M^+ *its maximal real subfield. Then*

$$h(M)/h(M^+) = Q(M/M^+) w \prod \left(-\tfrac{1}{2} B_{1,\psi}\right)$$

where w *is the number of roots of unity in* M, *the product runs over all odd characters* ψ *of* M, *i.e.* $\psi(-1) = -1$, *and* $B_{1,\psi}$ *is defined as*

$$B_{1,\psi} := \frac{1}{f_\psi} \sum_{x=1}^{f_\psi} \psi(x) x$$

(as usually one puts $\psi(x) = 0$ *if* x *is not prime to* f_ψ*).*

Proof. One compares the analytical class number formula for M and M^+. It is easy to see that the regulator $R(M)$ is equal to $2^{n^+ - 1} R(M^+) Q(M/M^+)^{-1}$ with $n^+ = [M^+ : \mathbb{Q}]$. Furthermore we compute $L(1, \chi)$:

Proposition 4.6. *Let* χ *be a Dirichlet character with conductor* f. *Then*

$$L(1, \chi) = \frac{\tau(x)}{f} \sum_{x=1}^{f} \overline{\chi(x)} \log |1 - \zeta_f^x| \qquad \text{if } \chi(-1) = 1 \tag{4.2}$$

and

$$L(1, \chi) = \pi i \frac{\tau(x)}{f^2} \sum_{x=1}^{f} \overline{\chi(x)} x \qquad \text{if } \chi(-1) = -1. \tag{4.3}$$

Proof. We put $\zeta := \zeta_f$. Then

$$L(s, \chi) = \sum_{m=1}^{\infty} \frac{\chi(m)}{m^s} = \sum_{z=0}^{f-1} \chi(z) \sum_{\substack{n \equiv z (\mathrm{mod}\, f) \\ n \geqslant 1}} \frac{1}{n^s}$$

$$= \sum_{z=0}^{f-1} \chi(z) \sum_{n=1}^{\infty} \frac{1}{n^s} \frac{1}{f} \sum_{x=0}^{f-1} \zeta^{(z-n)x} = \frac{1}{f} \sum_{x=1}^{f-1} \sum_{z=0}^{f-1} \chi(z) \zeta^{zx} \sum_{n=1}^{\infty} \frac{\zeta^{-nx}}{n^s}$$

for $s > 1$. Hence by Abel's continuity principle

[2] Warning: Theorem 29 of Hasse (1952) is incorrect.

$$L(1, \chi) = -\frac{1}{f}\left(\sum_{x=1}^{f-1} \tau(\chi)\overline{\chi(x)}) \log(1 - \zeta^{-x})\right)$$

where one has to take the main value of the logarithm, which in our case lies between the lines $-\frac{\pi i}{2}$ and $\frac{\pi i}{2}$.

Moreover

$$\overline{\chi(x)} \log(1 - \zeta^{-x}) + \overline{\chi(-x)} \log(1 - \zeta^{x}) = \overline{\chi(x)} \log|1 - \zeta^{x}|^2 \tag{4.4}$$

for $\chi(-1) = 1$ and

$$\begin{aligned} \log(1 - \zeta^{\pm x}) &= \log 2\sin\frac{\pi x}{f} \pm \pi i\left(\frac{x}{f} - \frac{1}{2}\right), \\ \overline{\chi(x)} \log(1 - \zeta^{-x}) + \overline{\chi(-x)} \log(1 - \zeta^{x}) &= -\overline{\chi(x)} 2\pi i\left(\frac{x}{f} - \frac{1}{2}\right) \end{aligned} \tag{4.5}$$

for $1 \leqslant x \leqslant f - 1$ and $\chi(-1) = -1$.

(4.4) and (4.5) imply (4.2) and (4.3). □

Now, inserting the expressions (4.2), (4.3) for the characters of M and M^+ in the analytic class number formula for M and M^+ (Theorem 2.25) and taking into account

$$\prod_{\chi \in X(M)} \tau(\chi) = \begin{cases} d_M^{1/2} & \text{if } M \text{ is real} \\ i^{n/2} d_M^{1/2} & \text{if } M \text{ is complex}, n = [M : \mathbb{Q}] \end{cases}$$

((1.55), (2.28)), we get the desired formula of Theorem 4.5. □

Example 3. Let $M \neq \mathbb{Q}(\sqrt{-1})$ be an imaginary-quadratic field and ψ its nontrivial character. In this case it is possible to simplify the formula of Theorem 4.5 (see e.g. Borevich, Shafarevich (1985), Chap. 5, § 4):

$$h(M) = \frac{1}{2 - \psi(2)} \sum \psi(x),$$

where the sum runs over all $x \in \mathbb{Z}$ with $0 < x < d_M/2$ and $(x, d_M) = 1$. □

2.2. The Arithmetical Class Number Formula for Real Quadratic Fields. It remains to study the case of a real abelian field. First as an example we consider real quadratic fields $M = \mathbb{Q}(\sqrt{d})$ with discriminant d and fundamental unit $\varepsilon > 1$. Then the regulator of M is $\log \varepsilon$. Let χ be the character corresponding to M. It has conductor d. The class number formula gives

$$h(M) \log \varepsilon = \frac{\sqrt{d}}{2} L(1, \chi) = \frac{1}{2} \sum_{x=1}^{d} \log|1 - \zeta_d^x|^{-\chi(x)}$$

Moreover

$$|1 - \zeta_d^x| = \zeta_{2d}^x - \zeta_{2d}^{-x}$$

implies

$$h(M)\log\varepsilon = \log \prod_{1\leqslant x<d/2} (\zeta_{2d}^{x} - \zeta_{2d}^{-x})^{-\chi(x)}.$$

It follows

$$\varepsilon^{h(M)} = \prod_{x} (\zeta_{2d}^{x} - \zeta_{2d}^{-x})^{-\chi(x)}.$$

The last equation shows that

$$\theta := \prod_{x} (\zeta_{2d}^{x} - \zeta_{2d}^{-x})^{-\chi(x)}$$

is a unit in M. It is called a cyclotomic unit. θ together with -1 generates a subgroup of the unit group of M of index $h(M)$. This is the arithmetical form of the class number formula, which we want to generalize to arbitrary real abelian fields M.

2.3. The Arithmetical Class Number Formula for Real Abelian Fields (Main reference: Sinnott (1980)). Let M be an arbitrary abelian field and M^+ its maximal real subfield. $h^+(M) := h(M^+)$ is often called the second class number factor of M.

For the formulation of the arithmetical class number formula for $h^+(M)$ we introduce some notations and definitions:

Let G be the Galois group of $M/\mathbb{Q}$. To every prime number p we associate the element

$$(p, M) := \frac{1}{|T_p|} F_p^{-1} \sum_{g\in T_p} g$$

in the group ring $\mathbb{Q}[G]$, where T_p denotes the inertia group and $F_p \in G/T_p$ the Frobenius automorphism. For a natural number r we denote by T_r the subgroup of G generated by the inertia groups T_p for $p|r$, in particular $T_1 = \{1\}$. Let m be the product of all primes ramified in M. We denote by U the $\mathbb{Z}[G]$-module in $\mathbb{Q}[G]$ generated by the elements

$$\left(\sum_{g\in T_r} g\right) \prod_{p|m/r} (1 - (p, M)),$$

where r varies over the divisors of m.

Proposition 4.7. *As a $\mathbb{Z}$-module, U is free of rank $[M:\mathbb{Q}]$.* ⊠

Let n be a divisor of the conductor of M and let a be an integer not divisible by n. We put $M_n = M \cap \mathbb{Q}(\zeta_n)$. Then the group D of cyclotomic numbers in M is defined as the group generated in M by the roots of unity in M and all elements of the form $N_{\mathbb{Q}(\zeta_n)/M_n}(1 - \zeta_n^a)$. The group C of cyclotomic units of M is defined by $C := E \cap D$.

The following proposition together with the ideal equation

$$(\zeta_{p^h} - 1)^{\varphi(p^h)} = (p)$$

for prime numbers p (Chap. 1.3.9) shows that for given M it is easy to see which elements of D are in C.

Proposition 4.8. *If the natural number n has at least two distinct prime factors, then $1-\zeta_n$ is a unit in $\mathbb{Z}[\zeta_n]$ and*

$$\prod_{j\in(\mathbb{Z}/n\mathbb{Z})^\times}(1-\zeta_n^j)=1.$$

Proof. $x^{n-1}+x^{n-2}+\cdots+x+1=\prod_{j=1}^{n-1}(x-\zeta_n^j)$ implies

$$n=\prod_{j=1}^{n-1}(1-\zeta_n^j).$$

Moreover

$$n=\prod_{p|n}p^a=\prod_{p|n}\prod_{i=1}^{p^a-1}(1-\zeta_{p^a}^i),$$

hence

$$\prod_{j\in(\mathbb{Z}/n\mathbb{Z})^\times}(1-\zeta_n^j)=1. \square$$

In the case $M=\mathbb{Q}(\zeta_{p^m})$ we have a simple description of the group of cyclotomic units by means of independent generators:

Proposition 4.9. The cyclotomic units of $\mathbb{Q}(\zeta_{p^m})^+$ are generated by -1 and the units

$$\zeta_{p^m}^{(1-a)/2}\frac{1-\zeta_{p^m}^a}{1-\zeta_{p^m}},\qquad 1<a<\tfrac{1}{2}p^m,\ (a,p)=1.$$

The cyclotomic units of $\mathbb{Q}(\zeta_{p^m})$ are generated by ζ_{p^m} and the cyclotomic units of $\mathbb{Q}(\zeta_{p^m})^+$. $\square$

Now we are able to formulate the main result (Sinnott (1980)):

Theorem 4.10. *Let M be a real abelian field. Then, with the notation above,*

$$[E:C]=h(M)2^{[M:\mathbb{Q}]-1}[M:\mathbb{Q}]^{-1}\left(\prod_{p|m}[M_{p^e}:\mathbb{Q}]\right)(\mathbb{Z}[G]:U),$$

where p^e denotes the largest power of p dividing the conductor of M. The index $(\mathbb{Z}[G]:U)$ is defined by

$$(\mathbb{Z}[G]:U):=\left[\frac{1}{h}\mathbb{Z}[G]:U\right]\left[\frac{1}{h}\mathbb{Z}[G]:\mathbb{Z}[G]\right]^{-1}$$

with a natural number h such that $U\subseteq\dfrac{1}{h}\mathbb{Z}[G]$. ⊠

One can say something more about the index $(\mathbb{Z}[G]:U)$:

Theorem 4.11. *$(\mathbb{Z}[G]:U)$ is an integer divisible only by the primes dividing $|G|$. If at most two primes ramify in M, or if G is the direct product of its inertia groups, then $(\mathbb{Z}[G]:U)=1$.* ⊠

Theorem 4.12. *Let $n \not\equiv 2 \pmod 4$, let M be the maximal real subfield of $\mathbb{Q}(\zeta_n)$, let g be the number of distinct prime factors of n, and let $b = 0$ if $g = 1$, $b = 2^{g-2} + 1 - g$ if $g \geqslant 2$. Then*

$$[E:C] = 2^b h(M). \boxtimes$$

Further information about the class group of M is contained in the following theorem which was conjectured by G. Gras (1977):

Let $p > 2$ be a prime which is not a divisor of $[M:\mathbb{Q}]$. For any cyclic subextension $F/\mathbb{Q}$ of $M/\mathbb{Q}$ let f be the minimal integer such that $F \subseteq \mathbb{Q}(\zeta_f)$, i.e. f is the conductor of F. Let H_M be the subgroup of M generated by

$$N_{\mathbb{Q}(\zeta_f)/F}(\zeta_f - 1)$$

and its conjugates for all cyclic subfields F of M. We define the group of *circular units* of M to be $C_M = H_M \cap E$. Let $A(M)$ be the p-component of the ideal class group of M and let $B(M)$ be the p-component of E/C_M.

Theorem 4.13. *The $\mathbb{Z}_p[G(M/\mathbb{Q})]$-modules $A(M)$ and $B(M)$ have isomorphic Jordan-Hölder series.* $\square$

Greenberg (1977) showed that Theorem 4.13 is a consequence of the "main conjecture" of Iwasawa (§ 4.7), which was proved by Mazur and Wiles (1984).

2.4. The Stickelberger Ideal (Main reference: Washington (1982), Chap. 6). In this section we study the ideal class group of an abelian field M as a $\mathbb{Z}[G]$-module, $G := G(M/\mathbb{Q})$. We define an ideal in $\mathbb{Z}[G]$, the Stickelberger ideal, which annihilates the class group. In § 2.3 we have connected the second class number factor of M with a group index. The same is possible for the relative class number factor $h^-(M)$. In the case of cyclotomic fields we connect $h^-(M)$ with the Stickelberger ideal. For the general case of complex abelian fields see Sinnott (1980).

For the study of the Stickelberger ideal we need Gauss sums over finite fields, which are also interesting in their own right.

Let $\mathbb{F}_q$ be the finite field with q elements, q being a power of the prime p. Let ζ_p be a fixed primitive p-th root of unity and let tr be the trace from $\mathbb{F}_q$ to $\mathbb{Z}/p\mathbb{Z}$. The function $\psi: \mathbb{F}_p \to \mathbb{C}^\times$ with $\psi(x) = \zeta_p^{\mathrm{tr}(x)}$ for $x \in \mathbb{F}_q$ is a character of $\mathbb{F}_q^+$. Let $\chi: \mathbb{F}_q^\times \to \mathbb{C}^\times$ be any character of $\mathbb{F}_q^\times$. We set $\chi(0) := 0$ and define the Gauss sum $\tau(\chi)$ by

$$\tau(\chi) = -\sum_{a \in \mathbb{F}_q} \chi(a)\psi(a).$$

The following facts are easy to prove:

1. If χ has order m, then $\tau(\chi) \in \mathbb{Q}(\zeta_{mp})$, $\tau(\chi)^m \in \mathbb{Q}(\zeta_m)$.
2. $\tau(\bar{\chi}) = \chi(-1)\overline{\tau(\chi)}$, $\tau(\chi^p) = \tau(\chi)$.
3. If $\chi \neq 1$, then $\tau(\chi)\tau(\bar{\chi}) = \chi(-1)q$.
4. If χ_1, χ_2 have orders dividing m, then $\tau(\chi_1)\tau(\chi_2)\tau(\chi_1\chi_2)^{-1}$ is an algebraic integer in $\mathbb{Q}(\zeta_m)$.
5. $J(\chi_1, \chi_2) = -\sum_{a \in \mathbb{F}_q} \chi_1(a)\chi_2(1-a)$ is called Jacobi sum of the characters χ_1

and χ_2. If $\chi_1 \neq 1$, $\chi_2 \neq 1$, and $\chi_1\chi_2 \neq 1$, then

$$J(\chi_1, \chi_2) = \tau(\chi_1)\tau(\chi_2)\tau(\chi_1\chi_2)^{-1}.$$

6. Let m be prime to p and let σ_b be the automorphism of $\mathbb{Q}(\zeta_m, \zeta_p)$ with $\sigma_b(\zeta_p) = \zeta_p$, $\sigma_b(\zeta_m) = \zeta_m^b$. If χ has order m, then

$$\tau(\chi)^b/\sigma_b\tau(\chi) := \tau(\chi)^{b-\sigma_b} \in \mathbb{Q}(\zeta_m).$$

Let M be an abelian field with conductor f. We identify $G(M/\mathbb{Q})$ with the corresponding quotient of $(\mathbb{Z}/f\mathbb{Z})^\times$. The element in $G := G(M/\mathbb{Q})$ which corresponds to $\bar{a} \in (\mathbb{Z}/f\mathbb{Z})^\times$ will be denoted by σ_a. The *Stickelberger element* $\theta = \theta(M)$ is defined by

$$\theta := \sum_{a \in (\mathbb{Z}/f\mathbb{Z})^\times} \left\{\frac{a}{f}\right\} \sigma_a^{-1} \in \mathbb{Q}[G],$$

where $\{x\}$ denotes for a real number x, the unique real number x' with $0 \leqslant x' < 1$ and $x - x' \in \mathbb{Z}$. The *Stickelberger ideal* $I = I(M)$ is defined by

$$I := \mathbb{Z}[G] \cap \mathbb{Z}[G]\theta.$$

It is an ideal in the group ring $\mathbb{Z}[G]$.

Theorem 4.14 (Stickelberger's theorem). *Let $\mathfrak{A}$ be a fractional ideal of M and $\beta \in I$. Then $\mathfrak{A}^\beta$ is a principal ideal, i.e. the Stickelberger ideal annihilates the ideal class group of M.* □

The proof of Theorem 4.14 is based on the study of Gaussian sums introduced above. Let p be a prime which does not divide the conductor f of M. If h is the order of p (mod f), we set $q = p^h$. Let $\mathfrak{p}_0$ (resp. $\mathfrak{p}$) be a prime ideal of $\mathbb{Q}(\zeta_f)$ (resp. $\mathbb{Q}(\zeta_{q-1})$) with $\mathfrak{p} | \mathfrak{p}_0 | p$. Since $\mathbb{Z}[\zeta_{q-1}]/\mathfrak{p}$ is the finite field with q elements, there is an isomorphism $\omega\colon \mathbb{F}_q^\times \to (\zeta_{q-1})$ satisfying $\omega(\bar{\zeta}_{q-1}) = \zeta_{q-1}$. Then the character $\chi = \omega^{-d}$ with $d = (q-1)/f$ has order f. Hence $\tau(\chi) \in \mathbb{Q}(\zeta_{fp})$ (§3.1.1). One shows

$$(\tau(\chi)^f) = \mathfrak{p}_0^{f\theta}.$$

Therefore

$$(\tau(\chi)^\beta) = \mathfrak{p}_0^{\beta\theta} \qquad \text{if } \beta \in \mathbb{Z}[G],\ \beta\theta \in I.$$

Let $\mathfrak{A}$ be a fractional ideal of M, prime to f. We represent $\mathfrak{A}$ as a product of prime ideals in $\mathbb{Q}(\zeta_f)$. Then $\mathfrak{A}^{\beta\theta} = (\gamma^\beta)$, where γ is a product of Gauss sums $\tau(\chi)$. Finally one shows $\gamma^\beta \in M$. □

Example 4. Let M be a real abelian field. Then $\sigma_a = \sigma_{-a}$ and

$$\theta(M) = \frac{1}{2} \sum_{a \in (\mathbb{Z}/f\mathbb{Z})^\times} \sigma_a = \frac{\varphi(f)}{2[M:\mathbb{Q}]} N_{M/\mathbb{Q}},$$

where $N_{M/\mathbb{Q}}$ denotes the ideal norm. Therefore in this case the theorem is trivial. □

Example 5. Let M be imaginary quadratic. Then $f\theta(M) = \sum a\sigma_a^{-1} \in \mathbb{Z}[G]$ and $f\theta(M)$ acts on the ideal class group as $\sum a\chi(a)$, where χ is the quadratic character of M. Since the class number formula (§ 2.1) implies that this is $-fh(M)$ (if $f > 4$), we have a weak form of the class number formula. □

Now we confine to cyclotomic fields $M = \mathbb{Q}(\zeta_n)$, $n \not\equiv 2 \pmod 4$. We denote the complex conjugation by ι and set $R := \mathbb{Z}[G]$,

$$R^- := \{x \in R \mid \iota x = -x\} = (1 - \iota)R,$$

$$I^- := I \cap R^- = R\theta \cap R^-.$$

Theorem 4.15. *Let g be the number of distinct prime divisors of n and let $b = 0$ if $g = 1$, $b = 2^{g-2} - 1$ if $g \geqslant 2$. Then*

$$[R^- : I^-] = 2^b h^-(\mathbb{Q}(\zeta_n)). \boxtimes$$

(Sinnott (1980), for the case $n = p^m$ see Washington (1982), § 6.4)

The proof of Theorem 4.15 consists in a transformation of the class formula for $h^-(\mathbb{Q}(\zeta_n))$ (Theorem 4.5). This can be done for the prime components of $[R^- : I^-]$ separately. We show the principle of the proof in the easier case of primes q with $q \nmid 2n$. We put

$$R_q := R \otimes_{\mathbb{Z}} \mathbb{Z}_q,\ I_q := I \otimes_{\mathbb{Z}} \mathbb{Z}_q,\ R_q^- := R^- \otimes_{\mathbb{Z}} \mathbb{Z}_q,\ I_q^- := I^- \otimes_{\mathbb{Z}} \mathbb{Z}_q.$$

Then $I_q^- = R_q^- \theta$ and the q-component of $[R^- : I^-]$ is $[R_q^- : I_q^-]$. Consider the linear map $A: R_q^- \to R_q^-$ given by $x \to x\theta$. By the theory of elementary divisors we know that $[R_q^- : R_q^- \theta]$ is the q-part of $\det(A)$. We compute $\det(A)$ by working in the vector space $R^- \otimes_{\mathbb{Z}} \overline{\mathbb{Q}}_q$, which has the basis $\{\varepsilon_\psi | \psi \text{ odd}\}$, where

$$\varepsilon_\psi := \frac{1}{n} \sum_{a=1}^{n} \psi(a) \sigma_a^{-1}.$$

In this basis A becomes a diagonal matrix,

$$\varepsilon_\psi \theta = B_{1,\overline{\psi}} \varepsilon_\psi$$

and

$$\det(A) = \prod_\psi B_{1,\overline{\psi}}.$$

In the case $q | 2n$ one has to take into account the denominators. For the special case $n = p^m$, $q = p$ compare § 2.5.

In the case $M = \mathbb{Q}(\zeta_{p^m})$ one could hope that R^-/I^- is isomorphic to the factor group $CL(M)/CL(M^+)$ since both groups have the same order. In general there is not such an isomorphism as is shown by the following example:

Example 6. Let $p = 4027$, $n = 1$. We set $CL := CL(M)$, $CL^+ := CL(M^+)$. The class group CL_1 of $Q(\sqrt{-4027}) \subseteq M$ is $\mathbb{Z}/3\mathbb{Z} + \mathbb{Z}/3\mathbb{Z}$. Suppose $R^-/I^- \cong CL/CL^+$. Since R^-/I^- is generated by $1 - \iota$ as R-module, there exists $c \in CL$ such that $Rc \cdot CL^+ = CL$. Therefore $CL_1 = N_{M/\mathbb{Q}(\sqrt{-p})}(CL)$ is generated by $c_1 := N_{M/\mathbb{Q}(\sqrt{-p})}(c)$ over R. Since $\sigma c_1 = c_1^{\pm 1}$ for $\sigma \in G(M/\mathbb{Q})$, it follows that Rc_1 is the cyclic group generated by c_1. This is a contradiction since CL_1 is not cyclic. □

2.5. On the p-Component of the Class Group of $\mathbb{Q}(\zeta_{p^m})$ (Main reference: Washington (1982), § 10.3). Let $p \neq 2$ be a prime. In this section we study the p-component $A(m)$ of the class group $CL(\mathbb{Q}(\zeta_{p^m}))$ as a $\mathbb{Z}_p[G]$-module. This is the starting point of Iwasawa theory (§ 3). We keep the notations of § 2.4. For any $\mathbb{Z}_p[G]$-module X we put $X^+ := (1 + \iota)X$, $X^- := (1 - \iota)X$. Moreover we put $R(m) = R_p(\mathbb{Q}(\zeta_{p^m}))$, $I(m) = I_p(\mathbb{Q}(\zeta_{p^m}))$. By Theorem 2.112 the group $A(m)^+$ is the p-component of $CL(\mathbb{Q}(\zeta_{p^m})^+)$.

First of all we consider the Stickelberger ideal $I(m)$ of the group ring $R(m) = \mathbb{Z}_p[G]$. We decompose $G = (\mathbb{Z}/p^m\mathbb{Z})^\times$ in its p-component and its p-prime-component H. The group H is cyclic of order $p - 1$. Let ω be the character $\omega: H \to \mathbb{Z}_p^\times$ given by $\omega(\bar{a}) \equiv a \pmod p$, $\bar{a} \in H$ (for $a \in \mathbb{Z}$, $(a, p) = 1$, there is one and only one $p - 1$-th root of unity $\zeta \in \mathbb{Z}_p^\times$ with $\zeta \equiv a \pmod p$). ω generates the character group of H. We denote by $\varepsilon_i \in \mathbb{Z}_p[H]$ the idempotent corresponding to ω^i:

$$\varepsilon_i := \frac{1}{p-1} \sum_{a \in H} \omega^i(a)\sigma_a^{-1}, \; i = 0, 1, \ldots, p-2.$$

Theorem 4.16. *$I(m)$ is a principal ideal in $R(m)$ with generator*

$$\sum_{i=2}^{p-1} \theta\varepsilon_i + (\sigma_{1+p} - 1 - p)\theta\varepsilon_1 = \theta + (\sigma_{1+p} - 2 - p)\theta\varepsilon_1 . \; \square$$

The proof of Theorem 4.16 (Shafarevich (1969), Chap. 3.8) consists in two steps. First one shows

$$I(m) = \mathbb{Z}_p p^m \theta + \bigoplus_{\substack{1<a<p \\ (a,p)=1}} \mathbb{Z}_p(\sigma_a - a)\theta$$

and then one considers the components $\varepsilon_i R(m)$ separately. If $\zeta^{p-1} = 1$, then $(\gamma(\zeta) - \zeta)\theta \in R(m)$, where γ denotes the inverse of ω. Since $\gamma(\zeta)\varepsilon_i = \omega^i\gamma(\zeta)\varepsilon_i$, we get

$$(\gamma(\zeta) - \zeta)\theta\varepsilon_i = (\omega^i\gamma(\zeta) - \zeta)\theta\varepsilon_i.$$

If $i \neq 1$, then $\omega^i\gamma(\zeta) - \zeta$ is a unit in $\mathbb{Z}_p$, hence $\varepsilon_i I(m) = \theta\varepsilon_i R(m)$. This shows the special role of the character ω. If $i = 1$ one shows $(\sigma_{1+p} - 1 - p)\theta\varepsilon_1 \in R(m)$. ⊠

Kummer (1975), p. 85, conjectured that the class number of $\mathbb{Q}(\zeta_p)^+$ is not divisible by p. This conjecture is known as *Vandiver's conjecture*. It has been verified for all $p < 125000$ (Wagstaff (1978)). By Theorem 2.113 $p \nmid h(\mathbb{Q}(\zeta_p)^+)$ implies $p \nmid h(\mathbb{Q}(\zeta_{p^m})^+)$ for all $m = 1, 2, \ldots$ and therefore $A(m) = A(m)^-$.

Theorem 4.17. *Let $p \neq 2$ be a prime with $p \nmid h(\mathbb{Q}(\zeta_p)^+)$. Then the $\mathbb{Z}_p[G]$-modules $R(m)^-/I(m)^-$ and $A(m)^-$ are isomorphic for all $m = 1, 2, \ldots$.* $\square$

$p \nmid h(\mathbb{Q}(\zeta_{p^m})^+$ implies that the index of the group of cyclotomic units in the group of all units of $\mathbb{Q}(\zeta_{p^m})^+$ is prime to p (Theorem 4.12). This together with Kummer theory (Chap. 1.4.7) shows that $A(m)^-$ is cyclic as a $\mathbb{Z}_p[G]$-module. Therefore any generator a_0 of $A(m)^-$ leads to an isomorphism $r \to ra_0$ of $R(m)^-/I(m)^-$ onto $A(m)^-$.

Now we confine ourselves to the cyclotomic fields $\mathbb{Q}(\zeta_p)$. Already Kummer proved the following theorem:

Theorem 4.18. *Let p be an odd prime and let $h_p = h_p^- h_p^+$ be the class number of $\mathbb{Q}(\zeta_p)$. Then the following conditions are equivalent:*

1. $p | h_p^-$,
2. $p | h_p$,
3. *p divides the numerator of the Bernoulli number B_j for some $j = 2, 4, \ldots, p-3$.*

Proof. Theorem 4.5 together with Appendix 4 implies

$$h_p^- = 2p \prod_{\substack{j=1 \\ j \text{ odd}}}^{p-2} \left(-\frac{1}{2} B_{1,\omega^j}\right), \tag{4.6}$$

where $\omega: (\mathbb{Z}/p\mathbb{Z})^\times \to \mathbb{Z}_p^\times$ is the Teichmüller character, i.e. the character with $\omega(\bar{a}) \equiv a \pmod p$. Furthermore the numbers B_{1,ω^n} and $B_{n+1}/(n+1)$ are in $\mathbb{Z}_p$ and

$$B_{1,\omega^n} \equiv B_{n+1}/(n+1) \pmod p$$

if $n \not\equiv -1 \pmod{p-1}$ (Appendix 4),

$$pB_{1,\omega^{p-2}} = pB_{1,\omega^{-1}} = \sum_{a=1}^{p-1} a\omega^{-1}(a) \equiv -1 \pmod p. \tag{4.7}$$

Therefore

$$h_p^- \equiv \prod_{\substack{j=1 \\ j \text{ odd}}}^{p-4} \left(-\frac{1}{2} \frac{B_{j+1}}{j+1}\right) \pmod p.$$

This shows 1.$\Leftrightarrow$3., 1.$\Leftrightarrow$2. follows from Theorem 2.113. □

Let $A := A(1)$ be the p-component of $CL(\mathbb{Q}(\zeta_p))$. We put $A_i := \varepsilon_i A$. Then

$$A = \bigoplus_{i=0}^{p-2} A_i.$$

We have

$$A^+ = \bigoplus_{\substack{i=0 \\ i \text{ even}}}^{p-2} A_i, \qquad A^- = \bigoplus_{\substack{i=1 \\ i \text{ odd}}}^{p-2} A_i.$$

The following theorems are refinements of a part of Theorem 4.18.

Theorem 4.19 (theorem of Herbrand (1932) $A_0 = A_1 = \{0\}$. *Let i be odd and $3 \leqslant i \leqslant p-2$. If $A_i \neq \{0\}$, then $p | B_{p-i}$.*

Proof. $A_0 = \{0\}$ is obvious since $\varepsilon_0 = N_{\mathbb{Q}(\zeta_p)/\mathbb{Q}}/(p-1)$. Let $c \in \mathbb{Z}$, $c \not\equiv 0 \pmod p$. By Stickelberger's theorem $(c - \sigma_c)\theta$ annihilates A, hence A_i. It follows that $(c - \omega^i(c))B_{1,\omega^{-i}}$ annihilates A_i. For $i = 1$ and $c = p + 1$ we have

$$(c - \omega(c))B_{1,\omega^{-1}} = pB_{1,\omega^{-1}} = \sum_{a=1}^{p-1} a\omega^{-1}(a) \equiv p - 1 \not\equiv 0 \pmod p,$$

hence $A_1 = 0$. For $i \neq 1$ and i odd we choose a c with $c \not\equiv c^i \equiv \omega^i(c) \pmod p$. Then

$$B_{1,\omega^{-i}} A_i = \{0\}. \tag{4.8}$$

Since $B_{1,\omega^{-i}} \equiv B_{p-i}/(p-i) \pmod p$, we have $p | B_{p-i}$ if $A_i \neq \{0\}$. □

Theorem 4.20 (theorem of Mazur and Wiles (1984)). *If i is odd, $3 \leqslant i \leqslant p-2$, then*

$$v_p|A_i| = v_p(B_{1,\omega^{-i}}). \boxtimes$$

This is a consequence of the proof of the "main conjecture" by Mazur and Wiles (see § 4.7). If $p \nmid h_p^+$, i.e. if we assume Vandiver's conjecture then the groups A_i are cyclic by Theorem 4.18. Hence (4.8) implies $v_p|A_i| \leqslant v_p(B_{1,\omega^{-i}})$ and the claim of Theorem 4.20 follows from (4.6).

The following *theorem of Ribet* (1976) is a consequence of Theorem 4.20.

Theorem 4.21. *Let $i \in \mathbb{Z}$ be odd, $3 \leqslant i \leqslant p-2$. If $p | B_{p-i}$, then $A_i \neq \{0\}$.* ⊠

Theorem 4.20 generalizes to odd characters of arbitrary complex abelian extensions K of $\mathbb{Q}$ with $p \nmid [K:\mathbb{Q}]$. This follows again from the main conjecture by a method due to Greenberg (1973), (1974), (1977). See Mazur, Wiles (1984), p. 216.

2.6. Application to Fermat's Last Theorem III (Main reference: Borevich, Shafarevich (1985) Chap. 5). In Chap. 1.3.10 and Chap. 2.6.6 we have considered the first case of Fermat's last theorem. If p is a regular prime, i.e. $p \nmid h_p$, Kummer settled the second case, too.

Theorem 4.22. *The second case of Fermat's last theorem holds for regular primes.* ⊠

The proof is much more difficult than that for the first case. It is based on Theorem 4.18 and the following Lemma.

Lemma 4.23 (Kummer). *Let p be a regular prime and let ε be a unit in $\mathbb{Q}(\zeta_p)$ with $\varepsilon \equiv a \pmod p$, $a \in \mathbb{Z}$. Then ε is a p-th power in $\mathbb{Q}(\zeta_p)$.* ⊠

Further progress in the proof of the second case is mainly due to Vandiver who gave a series of sufficient conditions. The most appropriate condition for computations with high speed calculation mashines is the following (Lehmer D.H., Lehmer E., Vandiver (1954))

Theorem 4.24. *Let l be a prime, $l = kp + 1 < p^2 + p$ and let t be a natural number such that $t^k \not\equiv 1 \pmod l$. We put*

$$d = \sum_{j=1}^{(p-1)/2} j^{p-2a}, \qquad Q_{2a} = t^{-kd/2} \prod_{i=1}^{(p-1)/2} (t^{ki} - 1)^{i^{p-1-2a}}.$$

If $Q_{2a}^k \not\equiv 1 \pmod l$ for all $a \in \mathbb{Z}$ with $2 \leqslant 2a \leqslant p-3$, $p | B_{2a}$, then the second case of Fermat's last theorem holds for p and $p \nmid h_p^+$. ⊠

Wagstaff (1978) verified that the above conditions are fulfilled for $p < 125000$. In all cases it was possible to use the smallest prime $l = kp + 1$ with $2^k \equiv 1 \pmod{l}$, $t = 2$.

Since the first case holds at least for all $p < 3 \cdot 10^9$, Fermat's last theorem is true for all $p < 125000$.

§3. Iwasawa's Theory of Γ-Extensions

(Main reference: Washington (1982), Chap. 13)

Let p be a prime. A *Γ-extension* of an algebraic number field K is a normal extension $K^{(\infty)}$ with Galois group isomorphic to $\Gamma := \mathbb{Z}_p^+$ (Chap. 3.1.8). For every p-power p^n, $n = 0, 1, \ldots$, there is one and only one extension $K^{(n)}$ with $K \subseteq K^{(n)} \subset K^{(\infty)}$ and $G(K^{(n)}/K) \cong \mathbb{Z}/p^n\mathbb{Z}$. The main purpose of Iwasawa's theory of Γ-extensions is the study of the p-class group of the sequence of fields

$$K = K^{(0)} \subset K^{(1)} \subset K^{(2)} \subset \cdots \subset K^{(\infty)} = \bigcup_n K^{(n)}.$$

By class field theory (Chap. 2.1.7) the p-class groups X_n of $K^{(n)}$, $n = 1, 2, \ldots,$ form an inverse system (Chap. 3.1.1) with the norm maps as morphisms. Iwasawa theory studies the inverse limit $X = \varprojlim X_n$ as a $\mathbb{Z}_p[[\Gamma]]$-module (Chap. 3.1.16). Results about the structure of X imply corresponding results about X_n.

The most interesting case of a Γ-extension and in fact the paradigma of the whole theory is the case of cyclotomic fields, $K^{(n)} = \mathbb{Q}(\zeta_{p^{n+1}})$, already studied in §2.3.

3.1. Class Field Theory of Γ-Extensions. We begin with some facts about Γ-extensions which are easy to prove on the basis of class field theory (Chap. 2.1).

Theorem 4.25. *Let $K^{(\infty)}/K$ be a Γ-extension and let v be a place of K which does not lie above p. Then $K^{(\infty)}/K$ is unramified at v, i.e. a Γ-extension is unramified outside p.*

Proof. The inertia group T_v of the local extension $K^{(\infty)}K_v/K_v$ corresponds by local class field theory to a factor group of the unit group of K_v. Since v is tamely ramified, T_v is finite. On the other hand T_v is a closed subgroup of $\Gamma \cong \mathbb{Z}_p^+$. But the unique finite subgroup of $\mathbb{Z}_p^+$ is $\{0\}$. □

Over $\mathbb{Q}$ one has a unique Γ-extension $\mathbb{Q}^{(\infty)} = \mathbb{Q}^{(\infty)}(p)$. The corresponding extensions $\mathbb{Q}^{(n)}$, $n = 1, 2, \ldots$, are constructed as follows.

For $p \neq 2$ we consider the extension $\mathbb{Q}(\zeta_{p^{n+1}})/\mathbb{Q}$. Its Galois group is $(\mathbb{Z}/p^{n+1}\mathbb{Z})^\times \cong \mathbb{Z}/p^n\mathbb{Z} \times \mathbb{Z}/(p-1)\mathbb{Z}$. Hence $\mathbb{Q}(\zeta_{p^{n+1}})$ contains one and only one subfield $\mathbb{Q}^{(n)}$ with $G(\mathbb{Q}^{(n)}/\mathbb{Q}) \cong \mathbb{Z}/p^n\mathbb{Z}$. We have

$$\mathbb{Q} = \mathbb{Q}^{(0)} \subset \mathbb{Q}^{(1)} \subset \cdots \subset \mathbb{Q}^{(n)}$$

and we set $\mathbb{Q}^{(\infty)} = \bigcup_n \mathbb{Q}^{(n)}$.

For $p = 2$ we consider the extension $\mathbb{Q}(\zeta_{2^{n+2}})/\mathbb{Q}$. Its Galois group is isomorphic to $(\mathbb{Z}/2^{n+2}\mathbb{Z})^\times \cong \mathbb{Z}/2^n\mathbb{Z} \times \mathbb{Z}/2\mathbb{Z}$. The Galois group of the maximal real subfield $\mathbb{Q}^{(n)}$ of $\mathbb{Q}(\zeta_{2^{n+2}})$ over $\mathbb{Q}$ is isomorphic to $\mathbb{Z}/2^n\mathbb{Z}$. We set $\mathbb{Q}^{(\infty)} = \bigcup_n \mathbb{Q}^{(n)}$.

For an arbitrary number field K the extension $K\mathbb{Q}^{(\infty)}/K$ is a Γ-extension. It is called the *cyclotomic Γ-extension.* In general there are other Γ-extensions of K beside $K\mathbb{Q}^{(\infty)}$.

Let $r_1 + r_2 - 1 - \delta$ be the $\mathbb{Z}_p$-rank of the group $\bar{E}_1$ considered in Chap. 3.2.5 in connection with Leopoldt's conjecture $\delta = 0$.

Theorem 4.26. *Let F be the maximal abelian p-extension of K unramified outside p. Then the $\mathbb{Z}_p$-rank ρ of $G(F/K)$ is $r_2 + 1 + \delta$.*

Proof. $G(F/K)$ is isomorphic to the p-completion C of the group $J(K)/K^\times \prod_{v \nmid p} U_v$ (Chap. 2.1.1). Since $J(K)/K^\times \prod_v U_v$ is isomorphic to the ideal class group of K, which is finite, the $\mathbb{Z}_p$-rank of C is equal to the $\mathbb{Z}_p$-rank of the p-completion of

$$K^\times \prod_v U_v / K^\times \prod_{v \nmid p} U_v.$$

The p-completion of this group is $\prod_{v|p} U_{v1}/\bar{E}_1$. Therefore

$$\rho = \sum_{v|p} \operatorname{rank}_{\mathbb{Z}_p} U_{v1} - \operatorname{rank}_{\mathbb{Z}_p} \bar{E}_1 = [K:\mathbb{Q}] - (r_1 + r_2 - 1 - \delta) = r_2 + 1 + \delta.$$

□

Let $t := r_2 + 1 + \delta$. By Theorem 4.26 the group $G(F/K)$ is isomorphic to $\mathbb{Z}_p^t \oplus H$ for some finite group H. Let $K^{(\infty)}$ be a Γ-extension of K. Then $K^{(\infty)} \subset F$ by Theorem 4.25. Hence to $K^{(\infty)}$ there corresponds a subgroup of $\mathbb{Z}_p^t \oplus H$ containing H. The extension of K corresponding to H contains all Γ-extensions of K and is the compositum of t "independent" Γ-extensions.

Let $\bar{E}$ be the closure of the unit group E of K, embedded in $\prod_{v|p} U_v$ diagonally. The proof of Theorem 4.26 shows also the following theorems:

Theorem 4.27. *Let H be the Hilbert class field of K* (Chap. 2.1.2) *and let F be the maximal abelian extension of K unramified outside p. Then*

$$G(F/K) \cong \prod_{v|p} U_v/\bar{E}. \ \square$$

Theorem 4.28. *Let $K^{(\infty)}/K$ be a Γ-extension. Then there is a finite subextension M/K such that all prime divisors of p in M are fully ramified in $K^{(\infty)}$.* □

3.2. The Structure of Λ-Modules. We denote the ring $\mathbb{Z}_p[[\Gamma]]$ in the following by Λ. As mentioned in the introduction to § 3, the main subject of Iwasawa theory is the limit group X considered as Λ-module. We show in § 3.4 that X is a finitely generated Λ-torsion module. In this section we study finitely generated Λ-modules in general.

If γ is a topological generator of Γ, we set $T := \gamma - 1$. Then the group rings $\mathbb{Z}_p[[\Gamma]]$ and $\mathbb{Z}_p[[T]]$ are isomorphic and will be identified (Chap. 3.1.16).

The *preparation theorem of Weierstrass* is valid for the ring Λ:

Proposition 4.29. *Let $f(T) \in \Lambda$ be a nonzero power series. Then there is one and only one representation*

$$f(T) = p^m P(T) U(T),$$

where m is a non negative integer, $U(T)$ is a unit in Λ, and $P(T)$ is a distinguished polynomial, i.e.

$$P(T) = T^n + a_{n-1} T^{n-1} + \cdots + a_0$$

with $p | a_i$ for $i = 0, 1, \ldots, n-1$.

Proof. The proof is based on the following proposition:

Proposition 4.30. *Let $f, g \in \Lambda$ and assume $f = a_0 + a_1 T + \cdots$ with $p | a_i$ for $i = 0, 1, \ldots, n-1$ and $a_n \in \mathbb{Z}_p^\times$. Then there is one and only one representation $g = qf + r$, where $q \in \Lambda$ and $r \in \mathbb{Z}_p[T]$ is a polynomial of degree at most $n-1$.* ⊠ □

It follows from Proposition 4.26 that Λ is unique factorization domain, whose irreducible elements are p and the irreducible distinguished polynomials. The units are the power series with constant term in $\mathbb{Z}_p^\times$.

Let $\mathfrak{M}$ be the category of finitely generated Λ-modules and $\mathfrak{C}$ the category of finite Λ-modules. We are going to classify the modules in $\mathfrak{M}$ modulo $\mathfrak{C}$ and we call a morphism of $\mathfrak{M}$ a $\mathfrak{C}$-isomorphism if it has finite kernel and cokernel. If M_1 and M_2 are Λ-modules which are $\mathfrak{C}$-isomorphic we write $M_1 \sim M_2$. Modulo $\mathfrak{C}$-isomorphism one operates with Λ-modules as if Λ were a principal ideal domain. This leads to the following result:

Theorem 4.31. *Let M be a finitely generated Λ-module. Then*

$$M \sim \Lambda^r \oplus \bigoplus_{i=1}^{s} \Lambda/(p^{n_i}) \oplus \bigoplus_{j=1}^{t} \Lambda/(f_j(T))^{m_j},$$

where r, s, t, n_i, m_j are natural numbers and f_j is a distinguished and irreducible polynomial in Λ. ⊠ (Serre (1958))

3.3. The p-Class Group of a Γ-Extension. Let $K^{(\infty)}/K$ be a Γ-extension and $L^{(\infty)}$ the maximal unramified abelian p-extension of $K^{(\infty)}$. Moreover let $L^{(n)}$, $n = 0, 1, \ldots$, be the maximal unramified abelian p-extension of $K^{(n)}$. Then $L^{(\infty)} = \bigcup_{n=1}^{\infty} L^{(n)}$. We have the exact sequence

$$\{1\} \to G(L^{(\infty)}/K^{(\infty)}) \to G(L^{(\infty)}/K) \to \Gamma \to \{1\}$$

and $G(L^{(\infty)}/K^{(\infty)})$ is a Γ-module. On the other hand since $G(L^{(\infty)}/K^{(\infty)})$ is a pro-p-group, it is a $\mathbb{Z}_p$-module, hence a $\mathbb{Z}_p[[\Gamma]]$-module.

Iwasawa's idea consists in the study of $X := G(L^{(\infty)}/K^{(\infty)})$ as a $\mathbb{Z}_p[[\Gamma]]$-module and in the transfer of information about this module to the group $X_n := G(L^{(n)}/K^{(n)})$, which is isomorphic to the ideal class group of $K^{(n)}$ by class field theory.

Since we are interested mainly in results which are valid for all sufficiently large n, we may assume by Theorem 4.25 that all places $\mathfrak{p}_i$ of K with $\mathfrak{p}_i|p$ are fully ramified in $K^{(\infty)}/K$, $i = 1, \ldots, s$. For every i we fix a place $\tilde{\mathfrak{p}}_i$ of $L^{(\infty)}$. Let T_i be the inertia group of $\tilde{\mathfrak{p}}_i$ with respect to K. Then

$$G(L^{(\infty)}/K) = T_i X, \qquad i = 1, \ldots, s.$$

Let γ be a generator of Γ and let $g_1, \ldots, g_s$ be preimages of γ in T_i. Then g_i is a generator of T_i. We put $T := \gamma - 1$ and consider X as an additively written $\mathbb{Z}_p[[T]]$-module (§ 3.2).

Proposition 4.32. *Let $K^{(\infty)}/K$ be fully ramified with respect to places above p, let N_n be the normal subgroup of $G(L^{(\infty)}/K)$ generated by $g_1^{p^n}, \ldots, g_s^{p^n}$, and let $Y_n := X \cap N_n$.*

1. $X_n \cong X/Y_n$.
2. *Y_0 is generated by $a_i := g_i g_1^{-1}$, $i = 2, \ldots, s$, and by TX.*
3. *Let*

$$\nu_n = ((1+T)^{p^n} - 1)T^{-1} = 1 + \gamma + \cdots + \gamma^{p^n-1},$$

then $Y_n = \nu_n Y_0$.

Proof. By definition we have $X_n = XN_n/N_n \cong X/X \cap N_n$. 2. is easy and 3. follows from

$$g_i^{p^n} = (a_i g_1)^{p^n} = a_i g_1 a_i g_1^{-1} g_1^2 a_i g_i^{-2} \ldots g_1^{p^n-1} a_i g_1^{-p^n+1} g_1^{p^n} = (\nu_n a_i) g_1^{p^n}. \ \square$$

3.4. Iwasawa's Theorem. Now we apply the results of section 2 to the Λ-module X.

Theorem 4.33. *X is a finitely generated Λ-torsion module.*

Proof. First we show that X is finitely generated. Since $X \sim Y_0$ it is sufficient to show that Y_0 is finitely generated. ν_1 lies in the maximal ideal (p, T) of Λ and $Y_0/\nu_1 Y_0 = Y_0/Y_1$ is contained in the finite module X/Y_1. Hence Y_0 is finitely generated by Nakayama's lemma.

Now we can apply Theorem 4.31. Since $\Lambda/\nu_1\Lambda$ is infinite and $Y_0/\nu_1 Y_0$ is finite, Y_0 and hence X has no component of the form Λ. $\square$

Theorem 4.34 (Iwasawa's theorem). *Let M be a finitely generated Λ-torsion module and suppose that for all non negative integers n, $M/\nu_n M$ is a finite group of order p^{e_n}. Then there are non negative integers m, l and an integer c such that for n sufficiently large*

$$e_n = mp^n + ln + c.$$

Proof. First of all one shows that the question is invariant by $\mathfrak{C}$-isomorphisms. Therefore we can assume that M is of the form

$$M = \bigoplus_{i=1}^{h} \Lambda/(f_i),$$

where $f_i = p^{n_i}$ or f_i is a distinguished polynomial (Theorem 4.31). We consider the direct summands $M_i := \Lambda/(f_i)$.

ν_n is a distinguished polynomial of degree $p^n - 1$. Therefore the division algorithm shows that every element of $\Lambda/(p^{n_i}, \nu_n)$ is represented uniquely by a polynomial of degree less than $p^n - 1$. Hence if $f_i = p^{n_i}$, we have

$$\log_p[M_i : \nu_n M_i] = n_i(p^n - 1).$$

If f_i is a distinguished polynomial of degree m_i, then M_i is a free $\mathbb{Z}_p$-module of rank m_i. The Λ-module structure is given by the automorphism $\gamma\colon M_i \to M_i$, i.e. by a $m_i \times m_i$ matrix A with coefficients in $\mathbb{Z}_p$ which is unipotent mod p. This implies $\nu_n M_i = p\nu_{n-1} M_i$ for sufficiently large n, say for $n \geqslant n_0$. We set $L_i := \nu_{n_0} M_i$. Then

$$\log_p[M_i : \nu_n M_i] = \log_p[M_i : L_i] + \log_p[L_i : p^{n-n_0} L_i] = c_i + m_i(n - n_0). \ \square$$

Theorem 4.33 and Theorem 4.34 imply the final result:

Theorem 4.35 (Iwasawa's theorem). *Let $K^{(\infty)}/K$ be a Γ-extension and let p^{e_n} be the order of the p-class group of $K^{(n)}$. Then there are non negative integers μ, λ and an integer ν such that*

$$e_n = \mu p^n + \lambda n + \nu$$

for sufficiently large n. $\square$

Now suppose that every $K^{(n)}$ is a CM-field. Then $K^{(\infty)+}/K^+$ is a Γ-extension (cyclotomic if Leopoldt's conjecture is true, see Theorem 4.26). We remember that K^+ denotes the maximal real subfield of K. Let p be odd and let ι denote the complex conjugation, $\Delta := \{1, \iota\}$. Then for any $\mathbb{Z}_p[\Delta]$-module M we have the decomposition

$$M = M^+ + M^-$$

with

$$M^+ := \{m + \iota m | m \in M\}, M^- := \{m - \iota m | m \in M\}.$$

We put $e_n^\pm = \log_p|X_n^\pm|$. Then $e_n = e_n^+ + e_n^-$ and

$$e_n^\pm = \mu^\pm p^n + \lambda^\pm n + \nu^\pm \tag{4.9}$$

for n sufficiently large with

$$\lambda = \lambda^+ + \lambda^-, \qquad \mu = \mu^+ + \mu^-, \qquad \nu = \nu^+ + \nu^-. \tag{4.10}$$

If $p = 2$, we set

$$X_n^- = \{x \in X_n | \iota x = -x\}, \qquad X_n^+ = X_n(K_n^+).$$

Again one obtains

$$e_n^\pm = \mu^\pm 2^n + \lambda^\pm n + \nu^\pm \tag{4.11}$$

and the exact sequence

$$\{0\} \to X_n^- \to X_n \to X_n(K_n^+) \to \{0\}$$

implies (4.10).

Moreover with the techniques developed above it is easy to show that for arbitrary primes p one has $\mu^\pm = 0$ if and only if the p-rank of $X_n^\pm$ is bounded. By the Spiegelungssatz (Chap. 2.7.4) we know that $\mu^- = 0$ implies $\mu^+ = 0$ if $\mu_p \subset K$.

Theorem 4.36 (theorem of Ferrero-Washington (1979)). *Let K be an abelian number field and let $K^{(\infty)}/K$ be the cyclotomic Γ-extension. Moreover let μ be as in Theorem* 4.35. *Then* $\mu = 0$. ⊠

One first shows that it is sufficient to prove the theorem for the case $K = \mathbb{Q}(\zeta_{dp})$. Then $\mu = \mu^+ + \mu^-$ and $\mu^- = 0$ implies $\mu^+ = 0$. Therefore one has to show $\mu^- = 0$. This is the most difficult part of the proof and is done by using the uniform distribution of certain sequences of numbers mod 1. For examples of Γ-extensions with $\mu \neq 0$ see Iwasawa (1973) and Kuz'min (1979).

For the non p-part of the class number of the cyclotomic Γ-extension of an abelian field one has the following result of Washington (1979). The proof is similar to the proof of $\mu = 0$ for the p-part of the class number.

Theorem 4.37. *Let l be a prime distinct from p and let $K^{(\infty)}$ be the cyclotomic Γ-extension of an abelian number field K. Then $v_l(h(K^{(n)}))$ is constant for large n.* ⊠

§ 4. p-adic L-Functions

(Main reference: Washington (1982), Chap. 5)

In this paragraph we shall construct p-adic analogues of Dirichlet L-functions (Chap. 1.6, introduction). Since the usual series for these functions do not converge p-adically, we must resort to another procedure. The values of $L(s, \chi)$ at negative integers are algebraic: $L(1-n, \chi) = -B_{n,\chi}/n$ (§ 4.1) (see Appendix 4 for the definition of the generalized Bernoulli numbers $B_{n,\chi}$.). We denote by $\mathbb{C}_p$ the completion of $\overline{\mathbb{Q}}_p$. The field $\mathbb{C}_p$ is algebraically closed (Proposition 1.88). We fix in the whole § 4 an embedding of $\overline{\mathbb{Q}}$ in $\mathbb{C}_p$ and consider $B_{n,\chi}$ as an element of $\overline{\mathbb{Q}}_p$. We look for a p-adic function which agrees with $L(s, \chi)$ at negative integers.

First let $\chi = \chi_0$ the trivial character and $L(s, \chi_0) = \zeta(s)$ Riemann's ζ-function. Then $\zeta(1-n) = -B_n/n$ (1.44). Already Kummer knew the congruences

$$(1 - p^{m-1})\frac{B_m}{m} \equiv (1 - p^{n-1})\frac{B_n}{n} \pmod{p^{a+1}} \qquad (4.12)$$

for even positive integers with $m \equiv n \pmod{(p-1)p^a}$, $n \not\equiv 0 \pmod{p-1}$, and arbitrary non negative integers a. (4.12) can be understood as a p-adic continuity property for the function $(1 - p^{-s})\zeta(s)$. But we have to consider the classes mod $p - 1$ separately. To get a continuous p-adic function for all integers we have to intertwin various L-functions (Theorem 4.40). In fact we construct p-adic

L-functions independent of (4.12) as analytic p-adic functions (for notions and results in p-adic analysis see Chap. 1.4.8) and prove (4.12) as a corollary. $(1 - p^{-s})^{-1}$ is the p-factor in the product representation of $\zeta(s)$. It is intuitively understandable that this factor has to be removed since the series $\zeta(s) = \sum_n n^{-s}$ has p-adically arbitrary large terms, while the terms of

$$(1 - p^{-s})\zeta(s) = \prod_{p \nmid n} n^{-s}$$

are at least bounded for $s \in \mathbb{Z}$.

The initial motivation of Kubota and Leopoldt (1964) for the introduction of p-adic L-functions was a better understanding of the p-adic class number formula. Iwasawa found the connection of p-adic L-functions with his theory of Γ-extensions (§ 4.6–7) giving a new insight into the relation between properties of p-adic L-functions and the structure of the class group of abelian number fields.

In this paragraph we consider only primitive characters (Chap. 1.5 introduction). The product of two Dirichlet characters χ_1 and χ_2 with conductors f_1 and f_2 is defined as follows. We consider the characters χ_1' and χ_2' mod g.c.m. (f_1, f_2) and define $\chi_1\chi_2$ as the primitive character of $\chi_1'\chi_2'$.

4.1. The Hurwitz Zeta Function (Main reference: Washington (1982), Chap. 4). Let χ be a Dirichlet character with conductor f. Beside the Dirichlet L-series $L(s, \chi)$ we consider the *Hurwitz zeta function*

$$\zeta(s, b) = \sum_{n=0}^{\infty} (b + n)^{-s} \qquad \text{for } \operatorname{Re}(s) > 1, 0 < b \leqslant 1.$$

Then

$$L(s, \chi) = \sum_{a=1}^{f} \chi(a) f^{-s} \zeta(a, a/f). \tag{4.13}$$

Theorem 4.38. *$\zeta(s, b)$ is a meromorphic function in the whole complex plane with a single simple pole at $s = 1$. Moreover*

$$\zeta(1 - n, b) = -B_n(b)/n \qquad \textit{for } n \geqslant 1 \textit{ and } 0 < b \leqslant 1$$

and therefore

$$L(1 - n, \chi) = -B_{n,\chi}/n \qquad \textit{for } n \geqslant 1 \tag{4.14}$$

(see Appendix 4 for the definition of $B_n(b)$ and $B_{n,\chi}$).

Proof. One uses the method of Riemann's first proof for his functional equation for $\zeta(s) = \zeta(s, 1)$ (Riemann (1859)): Let

$$F(t) := \frac{te^{(1-b)t}}{e^t - 1} = \sum_{n=0}^{\infty} B_n(1 - b) \frac{t^n}{n!}$$

and $H(s) := \int F(z) z^{s-2}\, dz$, where the integral is to be taken over the following path

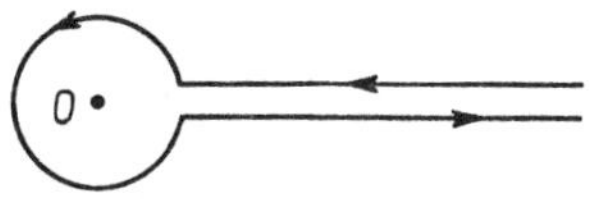

which consists of the positive real axis (top side), a Circle C_ε around 0 of radius ε, and the positive real axis (bottom side). z^s means $\exp(s \log z)$ with the logarithm defined as $\log t$ at the top side and $\log t + 2\pi i$ at the bottom side. Then $H(s)$ is analytic for $s \in \mathbb{C}$. One shows $H(s) = (e^{2\pi i s} - 1)\Gamma(s)\zeta(s, b)$ for $\mathrm{Re}(s) > 1$ and

$$H(1-n) = (2\pi i)\frac{B_n(1-b)}{n!} = (2\pi i)(-1)^n\frac{B_n(b)}{n!}. \square$$

4.2. *p*-adic *L*-Functions. We use the following notations: p is an arbitrary prime. $|x|$, $x \in \mathbb{C}_p$, denotes the valuation of $\mathbb{C}_p$ with $|p| = 1/p$. For convenience we put

$$q := \begin{cases} p & \text{if } p \neq 2 \\ 4 & \text{if } p = 2. \end{cases}$$

Given $a \in \mathbb{Z}_p$, $p \nmid a$, there exists a unique $\varphi(q)$-th root of unity $\omega(a) \in \mathbb{Z}_p$ such that $a \equiv \omega(a) \pmod q$. To ω there corresponds a Dirichlet character mod q, called *Teichmüller character*, which will be denoted also by ω. Let

$$H(s, a, F) = \sum_{\substack{m \equiv a(F) \\ m > 0}} m^{-s} = F^{-s}\zeta(s, a/F),$$

where s is a complex variable and a, F are integers with $0 < a < F$.

Theorem 4.39. *Suppose $q|F$ and $p \nmid a$. Then there exists a p-adic meromorphic function $H_p(s, a, F)$ on*

$$D := \{s \in \mathbb{C}_p \,|\, |s| < qp^{-1/(p-1)}\}$$

such that

$$H_p(1-n, a, F) = \omega^{-n}(a) H(1-n, a, F) \text{ for } n \geqslant 1.$$

In particular, if $n \equiv 0 \pmod{\varphi(q)}$, then

$$H_p(1-n, a, F) = H(1-n, a, F).$$

H_p is analytic except for a simple pole at $s = 1$ with residue $1/F$.

Remark. Since the function $H_p(s, a, F)$ has accumulation points in the set $\{1 - n | n \equiv 0 \pmod{\varphi(q)}\}$, it is already uniquely determined by its values in this set. Therefore the factor $\omega^{-n}(a)$ is not avoidable.

Proof of Theorem 4.39. *One puts*

$$H_p(s, a, F) := \frac{1}{s-1}\frac{1}{F}\left(\frac{a}{\omega(a)}\right)^{1-s}\sum_{j=0}^{\infty}\binom{1-s}{j}B_j(F^j/a^j)$$

and uses Theorem 1.71 to prove that H_p is analytically in D. For $s = 1 - n$ one has

$$H_p(1-n, a, F) = -\frac{1}{nF}\left(\frac{a}{\omega(a)}\right)^n \sum_{j=0}^{n} \binom{n}{j} B_j(F^j/a^j) = -\frac{F^{n-1}\omega^{-n}(a)}{n} B_n\left(\frac{a}{F}\right)$$

$$= \omega^{-n}(a) H(1-n, a, F). \square$$

By means of (4.13) we go over to $L(s, \chi)$:

Theorem 4.40. *Let χ be a Dirichlet character of conductor f and let F be a multiple of q and f. Then*

$$L_p(s, \chi) := \frac{1}{F}\frac{1}{(s-1)} \sum_{\substack{a=1 \\ p \nmid a}}^{F} \chi(a) \left(\frac{a}{\omega(a)}\right)^{1-s} \sum_{j=0}^{\infty} \binom{1-s}{j} B_j(F^j/a^j)$$

is a p-adic meromorphic function on $\{s \in \mathbb{C}_p \,|\, |s| < qp^{-1/(p-1)}\}$ *with*

$$L_p(1-n, \chi) = -(1 - \chi\omega^{-n}(p)p^{n-1}) \frac{B_{n, \chi\omega^{-n}}}{n} = (1 - \chi\omega^{-n}(p)p^{n-1}) L(1-n, \chi\omega^{-n})$$

for $n \geqslant 1$. *$L_p(s, \chi)$ is analytic if $\chi \neq 1$ and has a single pole at $s = 1$ with residue* $(1 - 1/p)$ if $\chi = 1$. ⊠

$L_p(s, \chi)$ is called *p-adic L-function.*

Remark 1. $(1 - \chi\omega^{-n}(p)p^{n-1})^{-1}$ is connected with the Euler factor at p of

$$L(s, \chi\omega^{-n}) = \prod_{l} (1 - \chi\omega^{-n}(l) l^{-s})^{-1}.$$

It is a general principle that to obtain p-adic analogues of complex functions, the p-part must be removed.

Remark 2. If χ is an odd character ($\chi(-1) = -1$), then $B_{n, \chi\omega^{-n}} = 0$. Therefore $L_p(s, \chi)$ is identically zero for odd characters.

4.3. Congruences for Bernoulli Numbers

Theorem 4.41. *Let χ be a non trivial even character of conductor f and $pq \nmid f$. Then*

$$L_p(s, \chi) = a_0 + a_1(s-1) + a_2(s-1)^2 + \cdots$$

with $|a_0| \leqslant 1$ *and* $|a_i| < 1$ *for* $i = 1, 2, \ldots$. ⊠

Theorem 4.41 can be used to prove congruences for Bernoulli numbers:

Theorem 4.42. *Let χ be as above and $m, n \in \mathbb{Z}$. Then*

$$L_p(m, \chi) \equiv L_p(n, \chi) \pmod{p}$$

and both numbers are p-integral.

Proof. Both sides are congruent to a_0. □

Theorem 4.43 (Kummer's congruences). *Let $m \equiv n \not\equiv 0 \pmod{p-1}$ be positive even integers. Then*

$$\frac{B_m}{m} \equiv \frac{B_n}{n} \pmod{p}. \ \square$$

More generally one has the congruence (4.12).

Theorem 4.44. *Let n be a positive odd integer, $n \not\equiv -1 \pmod{p-1}$. Then*

$$B_{1,\omega^n} \equiv \frac{B_{n+1}}{n+1} \pmod{p}. \ \square$$

4.4. Generalization to Totally Real Number Fields. Let K be a totally real number field and M a real abelian extension of K with conductor $\mathfrak{f}$. Moreover let χ be a character of $G(M/K)$, which we consider also as character of the ray class group mod $\mathfrak{f}$ (Chap. 2.1). Then we have the L-series

$$L(s, \chi) = \sum_{\mathfrak{a}} \chi(\mathfrak{a}) N(\mathfrak{a})^{-s},$$

where the sum runs over the integral ideals of K which are prime to $\mathfrak{f}$. $L(s, \chi)$ defines a meromorphic function in the complex plane with at most one simple pole at $s = 1$ (Chap. 1.6.4).

We write $L(s, \chi)$ in the form

$$L(s, \chi) = \sum_{\sigma \in G(M/K)} \chi(\sigma) \zeta_M(s, \sigma),$$

where $\zeta_M(s, \sigma)$ is the partial zeta function

$$\zeta_M(s, \sigma) := \sum_{\left(\frac{\mathfrak{a}}{M/K}\right) = \sigma} N(\mathfrak{a})^{-s}.$$

Here the sum runs over the integral ideals $\mathfrak{a}$ of K with $(\mathfrak{a}, \mathfrak{f}) = 1$ and $\left(\frac{\mathfrak{a}}{M/K}\right) = \sigma$.

Klingen (1962) and Siegel (1969) have shown that $\zeta_M(1-k, \sigma) \in \mathbb{Q}$ for $k \in \mathbb{Z}$, $k > 0$. They use the fact that $\zeta_M(1-k, \sigma)$ appears as constant term of a Hilbert modular form with known higher Fourier coefficients (Eisenstein series). So one may ask whether $L(1-k, \chi)$ has a p-adic interpolation generalizing Theorem 4.40. This was shown by Deligne, Ribet (1980) on the basis of a theory of p-adic Hilbert modular forms which is a consequence of a construction over $\mathbb{Z}$ of certain Hilbert-Blumenthal moduli schemes. On the other hand Barsky (1977) and Cassou-Noguès (1979) got the interpolation of $L(1-k, \chi)$ on the basis of explicit formulas of Shintani. See Gras (1986) for a study of the values of the resulting p-adic L-function in the case that K satisfies the Leopoldt conjecture (§4.5).

4.5. The p-adic Class Number Formula. In this section we consider the p-adic analogue to the analytic class number formula (Theorem 2.17).

First of all we need the notion of the p-adic regulator of an algebraic number field K. We fix an embedding of the normal closure L of K in $\mathbb{C}_p$. Let $g_1, \ldots, g_{r_1}$ be the real embeddings and $g_{r_1+1}, \bar{g}_{r_1+1}, \ldots, g_r, \bar{g}_r$ the complex embeddings of K in $L \subset \mathbb{C}$. Let $l_i = 1$ if g_i is real and $l_i = 2$ if g_i is complex. Moreover let $\varepsilon_1, \ldots, \varepsilon_{r-1}$ be a fundamental set of units of K (Chap. 1.1.3). Then

$$R_p(\varepsilon_1, \ldots, \varepsilon_{r-1}) = \det(l_i \log_p(g_i \varepsilon_j))_{i,j}$$

is up to a change in sign independent of the choice of $\varepsilon_1, \ldots, \varepsilon_{r-1}$ and $g_1, \ldots, g_r$. It is called the *p-adic regulator* of K and it is denoted by $R_p(K)$. In general the definition of $R_p(K)$ depends on the choice of the embedding of L in $\mathbb{C}_p$.

Example 7. Let $K = \mathbb{Q}(\sqrt[3]{2})$. Then the three embeddings of K in $\mathbb{C}_p$ give rise to three different regulators $R_p(K)$. ⊠ (Washington (1982), Exercise 5.12).

If K is totally real or a CM-field, then $R_p(K)$ is independent of the embedding of L in $\mathbb{C}_p$ (Washington (1982), Exercise 5.13).

Theorem 4.45. *Let K be a totally real abelian field of degree n corresponding to a group X of Dirichlet characters. Let $h(K)$ be the class number and $d(K)$ the discriminant of K. Then*

$$\frac{2^{n-1} h(K) R_p(K)}{d(K)} = \prod_{\substack{\chi \in X \\ \chi \neq 1}} \left(1 - \frac{\chi(p)}{p}\right)^{-1} L_p(1, \chi). \boxtimes \tag{4.15}$$

Remark. In (4.15) equality means that the signs of $R_p(K)$ and $d(K)$ can be chosen such that one has equality. □

The proof of Theorem 4.45 is based on the formula

$$L_p(1, \chi) = -\left(1 - \frac{\chi(p)}{p}\right) \frac{\tau(\chi)}{f} \sum_{a=1}^{f} \chi(a)^{-1} \log_p(1 - \zeta_f^a),$$

where $\tau(\chi) = \sum_{a=1}^{f} \chi(a) \zeta_f^a$ is the Gauss sum, and the study of cyclotomic units in analogy to § 2.

Leopoldt conjectured $R_p(K) \neq 0$ for all algebraic number fields K.

Theorem 4.46. *Let K be an abelian field. Then $R_p(K) \neq 0$.* ⊠ (Brumer (1967))

Kuz'min (1981) proves Leopoldt's conjecture in some non abelian cases.

In Chap. 3.2.5 we have considered another form of Leopoldt's conjecture, denoted by $\mathfrak{L}(K, p)$.

Theorem 4.47. *Let K be totally real. Then $R_p(K) \neq 0$ if and only if $\mathfrak{L}(K, p)$.* ⊠

4.6. Iwasawa's Construction of p-adic L-Functions (Main reference: Washington (1982), Chap. 7). In this section we want to connect the theory of Γ-extensions with p-adic L-functions. We consider the cyclotomic Γ-extension for the ground field $K = \mathbb{Q}(\mu_{qd})$, where $q = p$ for $p \neq 2$ and $q = 4$ for $p = 2$ and d is a natural number with $d \not\equiv 0 \pmod p$, $d \not\equiv 2 \pmod 4$. The Galois group $G(K^{(n)}/\mathbb{Q})$

is the direct product of its subgroups $\Gamma_n = G(K^{(n)}/K)$ and $\Delta = G(K^{(n)}/\mathbb{Q}^{(n)}) \cong G(K/\mathbb{Q})$, $n = 0, 1, \ldots$.

Let χ be a Dirichlet character with conductor dp^j for some $j \geqslant 0$. Regarding χ as a character of $G(K^{(\infty)}/\mathbb{Q})$, we see that there is a unique representation $\chi = \tau\psi$, where τ is a character of Δ and ψ a character of Γ. We call τ a character of the first kind and ψ a character of the second kind. ψ is always even since it corresponds to a real field. Let $q_n := qp^n d$ and let ω be as in §2.5. Corresponding to the decomposition $G(K^{(\infty)}/\mathbb{Q}) \cong \Delta \times \Gamma_n$ we write the Stickelberger element (§2.4) of $K^{(n)}$ in the form

$$\theta(K^{(n)}) := \frac{1}{q_n} \sum_a a\delta(a)^{-1}\gamma_n(a)^{-1}, \qquad \delta(a) \in \Delta,\ \gamma_n(a) \in \Gamma_n,$$

with $\sigma_a = \delta(a)\gamma_n(a)$, where the sum runs over all a with $0 < a < q_n$ and $(a, q_n) = 1$. We set $\xi_n := -\theta(K^{(n)})$ and

$$\xi_n(\tau) := -\frac{1}{q_n} \sum_a a\tau\omega^{-1}(a)\gamma_n(a)^{-1},$$

$$\eta_n(\tau) := (1 - (1 + q_0)\gamma_n(1 + q_0)^{-1})\xi_n(\tau).$$

Let $K_\tau/\mathbb{Q}$ be the field extension generated by the values of the character τ and let $\mathfrak{O}_\tau$ be the ring of integers of K_τ. Then $\xi_n(\tau), \eta_n(\tau) \in K_\tau[\Gamma_n]$.

Proposition 4.48. 1. $\frac{1}{2}\eta_n(\tau) \in \mathfrak{O}_\tau[\Gamma_n]$. 2. *If* $\tau \neq 1$, *then* $\frac{1}{2}\xi_n(\tau) \in \mathfrak{O}_\tau[\Gamma_n]$.
3. *If* $m \geqslant n$, *then* $\xi_m(\tau)$ *is mapped onto* $\xi_n(\tau)$ *by the projection from* $K_\tau[\Gamma_m]$ *to* $K_\tau[\Gamma_n]$. ⊠

$\gamma := \varprojlim_n \gamma_n(1 + q_0)$ is a topological generator of Γ. With $T := \gamma - 1$ we get a power series

$$f(T, \tau) = \varprojlim_n \xi_n(\tau) \qquad \text{for } \tau \neq 1$$

in $\mathfrak{O}_\tau[[T]] = \mathfrak{O}_\tau[[\Gamma]]$.

For $\tau = 1$ we set

$$f(T, 1) = \frac{1}{h(T)} \varprojlim_n \eta_n(1)$$

with

$$h(T) := 1 - (1 + q_0)(1 + T)^{-1}.$$

Iwasawa's construction of p-adic L-series is given by the following theorem.

Theorem 4.49. *Let* $\chi = \tau\psi$ *be an even Dirichlet character and let* $\zeta_\psi := \psi(1 + q_0)^{-1} = \chi(1 + q_0)^{-1}$. *Then*

$$L_p(s, \chi) = f(\zeta_\psi(1 + q_0)^s - 1, \tau). \ ⊠$$

Theorem 4.49 can be generalized: Let F be a totally real field, let $K := F(\zeta_p)$, and let $K^{(\infty)}$ be the cyclotomic Γ-extension. The characters of $\Delta := G(K/F)$ can be viewed as Dirichlet characters mod p. Let χ an odd character of Δ. Barsky (1977), Cassou-Noguès (1979) and Deligne, Ribet (1980) have shown that there exists a power series $f_\chi \in \Lambda$ such that

$$L_p(s, \omega\chi^{-1}) = f_\chi((1+p)^s - 1)$$

(§ 4.4).

Comparing Theorem 4.49 with the class number formula of Theorem 4.5 one finds the following result:

Theorem 4.50. *Let* $h_n^- := h(\mathbb{Q}(\zeta_{q_n}))/h(\mathbb{Q}(\zeta_{q_n})^+) = h^-(\mathbb{Q}(\zeta_{q_n}))$. *Then*

$$v_p(h_n^-/h_0^-) = \sum_{\substack{\tau \neq 1 \\ \tau \text{ even}}} \sum_{\substack{\zeta \in \mu_{p^n} \\ \zeta \neq 1}} v_p(\tfrac{1}{2} f(\zeta - 1, \tau)). \square$$

The last formula allows to give a new proof for the minus part of (4.9) in our special situation: We put

$$A(T) := \sum_{\substack{\tau \neq 1 \\ \tau \text{ even}}} \tfrac{1}{2} f(T, \tau).$$

Then $A(T) \in \mathbb{Z}_p[[T]]$ by Proposition 4.48.2. Hence $A(T)$ has the form

$$A(T) = p^m P(T) U(T), \tag{4.16}$$

where $m \geqslant 0$, $P(T)$ is a distinguished polynomial and $U(T)$ is a unit in $\mathbb{Z}_p[[T]]$ (Proposition 4.29). Let l be the degree of $P(T)$. For a primitive p^n-th root of unity and sufficiently large n one has

$$v_p(P(\zeta - 1)) = v_p((\zeta - 1)^l).$$

It follows

$$v_p\left(\sum_{\substack{\zeta \in \mu_{p^n} \\ \zeta \neq 1}} P(\zeta - 1)\right) = ln + c$$

for sufficiently large n and some constant c. Hence

$$v_p(h_n^-) = v_p(h_0^-) + (p^n - 1)m + ln + c.$$

Comparison with (4.9) shows $\mu^- = m$. This interpretation of the invariant μ^- plays an important role in the proof of the theorem of Ferrero-Washington (Theorem 4.36).

4.7. The Main Conjecture. Let p be an odd prime. First we assume $p \nmid h(\mathbb{Q}(\delta_p)^+)$ (Vandiver's conjecture). Let $K = \mathbb{Q}(\zeta_p)$ and let $K^{(\infty)}$ be the cyclotomic Γ-extension. It follows from Theorem 4.15 and from Iwasawa's construction of p-adic L-series that

$$\varepsilon_i X \cong \mathbb{Z}_p[[T]]/(f(T, \omega^{1-i})) \qquad \text{for } i = 3, 5, \ldots, p-2,$$

where $f(T, \omega^{1-i})$ is the power series satisfying

$$f((1+p)^s - 1, \omega^{1-i}) = L_p(s, \omega^{1-i})$$

and

$$\varepsilon_i = \frac{1}{p-1} \sum_{a=1}^{p-1} \omega^i(a)\sigma_a^{-1}.$$

Here $\sigma_a \in G(\mathbb{Q}(\zeta_p)/\mathbb{Q})$ is identified with its image by the injection $G(\mathbb{Q}(\zeta_p)/\mathbb{Q}) \to G(K^{(\infty)}/\mathbb{Q})$ considered in § 4.5.

Moreover since $\mu = 0$, we have $\varepsilon_i X \cong \mathbb{Z}_p[[T]]/(g_i(T))$, where $g_i(T)$ is the distinguished polynomial of $f(T, \omega^{1-i})$.

Now we come back to the general situation considered at the end of § 4.6, i.e. let F be a totally real field, let $K := F(\zeta_p)$, and let $K^{(\infty)}$ be the cyclotomic Γ-extension. We put

$$\varepsilon_\chi = \frac{1}{|\Delta|} \sum_{\delta \in \Delta} \chi^{-1}(\delta)\delta \qquad \text{for a character } \chi \text{ of } \Delta.$$

By § 3.4 we have

$$\varepsilon_\chi X \sim \bigoplus_i \Lambda/(p^{k_i^\chi}) \oplus \bigoplus_j \Lambda/(g_j^\chi(T)).$$

Let

$$\mu_\chi := \sum_i k_i^\chi, \qquad g^\chi(T) := p^{\mu_\chi} \prod_j g_j^\chi(T).$$

Iwasawa announced the following

Main conjecture. *Let $\chi \neq \omega$ be an odd character of Δ. Then*

$$f_\chi(T) = g^\chi(T) U_\chi(T)$$

with $U_\chi(T) \in \Lambda^\times$.

The main conjecture has been proved by Mazur-Wiles (1984) for $F = \mathbb{Q}$, $K = \mathbb{Q}(\zeta_p)$. The most important part of their proof consists in an extension of Ribet's method to prove the converse of Herbrand's theorem (§ 2.5) by means of p-adic representations associated to certain modular forms. K. Rubin (1990) gave a very much simpler proof for the main conjecture using ideas of V.A. Kolyvagin.

Chapter 5
Artin L-Functions and Galois Module Structure

In Chap. 2 we explained class field theory as a theory connecting the abelian extensions of an algebraic number field K with the closed subgroups of finite index of the idele class group of K. A direct generalization of class field theory should consist of a topological group $\mathfrak{G}(K)$, generalizing the idele class group, defined in terms of K with functorial properties with respect to field homomorphisms $K \to L$ and a canonical homomorphism $\varphi_K\colon \mathfrak{G}(K) \to G(\overline{K}/K) := G_K$ such that φ_K respects functorial behavior of $\mathfrak{G}(K)$ and G_K in the sense of Example 12 of Chap. 3 and such that $U \to \varphi_K(U)$ is a one to one correspondence between closed subgroups of finite index in $\mathfrak{G}(K)$ and closed subgroups of finite index in G_K. This last property can also be expressed saying that the induced homomorphism $\hat{\varphi}_K$ of the total completion (Chap. 3.1.1) of $\mathfrak{G}(K)$ into G_K is an isomorphism onto G_K.

Such a generalization of class field theory is unknown so far and perhaps doesn't exist. But there is an obvious reformulation of class field theory in terms of characters, which Langlands (1970), partially relying on earlier research of Gelfand and his school and of Weil, recognized as part of a vaste theory of adelic representations of linear groups including the generalization of Artin's reciprocity law (Chap. 2.1.6). This theory, called Langlands theory, consists mostly of conjectures so far. It postulates a correspondence between adelic representations of a fixed reductive group, e.g. the general linear group GL_n, with certain properties and representations of G_K with certain properties. This correspondence is based on a local correspondence for all completions of K. Adelic representations with certain conditions, called modular representations, can be considered as generalization of modular forms. This is the main motive for their intensive study by many mathematicians with mathematical background mainly in representation theory of Lie groups and functional analysis. Representations of Galois groups appear as classifying parameters. On the contrary for the purposes of algebraic number theory representations of reductive groups appear as classifying parameters for representations of Galois groups.

In this introduction we want to explain the basic idea of Langlands' generalization of class field theory and begin with the reformulation of class field theory in terms of characters of abelian groups: In the following we denote the character group of a compact abelian group G by $X(G)$ (Appendix 3).

Let A_K be the class module of the local or global field K (Chap. 2.1.6) i.e. $A_K := K^\times$ if K is a local nonarchimedean field, $A_K :=$ the idele class group $\mathfrak{C}(K)$ if K is a global field. Furthermore let $X_f(A_K)$ be the group of finite characters χ of A_K, i.e. χ is a continuous homomorphism of A_K into $\mathbb{C}^\times$ with finite image. The norm symbol $(\ , K)$ (Example 3.12) induces an isomorphism ϕ_K of the group $X(G_K)$ onto the group $X_f(A_K)$:

$$(\phi_K\chi)(\alpha) = \chi((\alpha, K)) \qquad \text{for } \chi \in X(G_K),\ \alpha \in A_K.$$

(Since G_K is a profinite group, all characters of G_K are finite (Chap. 3.1.2)).

If K' is a finite extension of K, then the injection $\iota\colon A_K \to A_{K'}$ induces a homomorphism $\iota^*\colon X_f(A_{K'}) \to X_f(A_K)$ and the norm map $N_{K'/K}\colon A_{K'} \to A_K$ induces a homomorphism $N^*_{K'/K}\colon X_f(A_K) \to X_f(A_{K'})$. For any automorphism g of K we have an action $g\colon X_f(A_K) \to X_f(gA_K)$ defined by

$$(g\chi)(\alpha) := \chi(g^{-1}\alpha) \qquad \text{for } \chi \in X_f(A_K),\ \alpha \in gA_K.$$

Furthermore we have induced homomorphisms $\mathrm{Ver}^*\colon X(G_{K'}) \to X(G_K)$ of the transfer from G_K into $G_{K'}$ (Chap. 3, Example 12), $\kappa^*\colon X(G_K) \to X(G_{K'})$ of the inclusion $\kappa\colon G_{K'} \to G_K$ and $g\colon X(G_K) \to X(G_{gK})$ defined by

$$g\chi(h) = \chi(g^{-1}hg) \qquad \text{for } \chi \in X(G_K),\ h \in G_{gK}.$$

Corresponding to Example 12 of Chap. 3 one has the following functorial properties of ϕ_K:

$$\begin{CD} X(G_{K'}) @>{\phi_{K'}}>> X_f(A_{K'}) \\ @V{\mathrm{Ver}}VV @VV{\iota^*}V \\ X(G_K) @>{\phi_K}>> X_f(A_K), \end{CD} \tag{5.1}$$

$$\begin{CD} X(G_K) @>{\phi_K}>> X_f(A_K) \\ @V{\kappa^*}VV @VV{N^*_{K'/K}}V \\ X(G_{K'}) @>{\phi_{K'}}>> X_f(A_{K'}), \end{CD} \tag{5.2}$$

$$\begin{CD} X(G_K) @>{\phi_K}>> X_f(A_K) \\ @V{g}VV @VV{g}V \\ X(G_{gK}) @>{\phi_{gK}}>> X_f(A_{gK}). \end{CD} \tag{5.3}$$

The first step in the generalization of class field theory by means of representations of G_K was done by Artin (1923) who defined a new sort of L-functions attached to representations of the Galois group of normal extensions of algebraic number fields, which in the case of abelian groups led him to the conjecture of the reciprocity law of class field theory (Chap. 2.1.6). In general these L-functions play a similar role for the conjectural reciprocity law of Langlands (see Borel, Jacquet (1979)).

Characters of G_K, as considered above, are one-dimensional representations of G_K. More general let ρ be a continuous n-dimensional irreducible representation of G_K (§ 1). The kernel of ρ is an open normal subgroup U of G_K. Let K_ρ be the fixed field of U. Then K_ρ/K is a finite normal extension. Let $\mathfrak{p}$ be a prime ideal of K unramified in K_ρ/K and let $\mathfrak{P}$ be a prime ideal of K above $\mathfrak{p}$. The conjugacy class of the Frobenius automorphism $F_{\mathfrak{P}}$ (Proposition 1.52.7) depends only on $\mathfrak{p}$. The function

$$L_{\mathfrak{p}}(s, \rho) := \det(I - \rho(F_{\mathfrak{P}})N(\mathfrak{p})^{-s})^{-1}$$

is therefore independent of the choice of $\mathfrak{P}$. It is called the local L-function of ρ for $\mathfrak{p}$. Since $\rho(F_{\mathfrak{P}})$ is semisimple, $L_{\mathfrak{p}}(s, \rho)$ contains all information about the decomposition behavior of $\mathfrak{p}$ in the extension K_ρ/K: The order of $\rho(F_{\mathfrak{P}})$ equals the inertia degree of $\mathfrak{P}$ over $\mathfrak{p}$.

The product

$$L(s, \rho) := \prod_{\mathfrak{p}} L_{\mathfrak{p}}(s, \rho)$$

converges for Re $s > 1$ and has a meromorphic continuation to the whole complex plane. $L(s, \rho)$ is called Artin L-function. If $K = \mathbb{Q}$, the local L-functions are uniquely determined by $L(s, \rho)$. Hence in this case the decomposition behavior of unramified primes is determined by $L(s, \rho)$.

In the case of one-dimensional representations ρ the correspondence ϕ_K associates to ρ the character $\phi_K \rho \in X_f(\mathfrak{C}(K))$ and $L(s, \rho)$ is equal to the L-function $L(s, \phi_K \rho)$ studied in Chap. 1.6. More exactly we have equality of the local L-functions $L_\mathfrak{p}(s, \rho) = L(s, \phi_{K_\mathfrak{p}}(\rho_p))$, where ρ_p is the character of $G_{K_\mathfrak{p}}$ induced by ρ.

Generalizing this picture, the conjectural Langlands correspondence consists in its simplest form in a local correspondence associating to every n-dimensional representation $\rho_\mathfrak{p}$ of G_{K_p} a representation $\rho'_\mathfrak{p}$ of $GL_n(K_\mathfrak{p})$ such that L-functions of $\rho_\mathfrak{p}$ and $\rho'_\mathfrak{p}$ coincide and some other properties, partially generalizing the functorial properties (5.1)–(5.3), and a global correspondence associating to every n-dimensional representation ρ of G_K the representation ρ' of $GL_n(A_K)$ which is the tensor product of the local representations corresponding to the localizations of ρ. The representation ρ' should be a modular representation with associated L-function being a generalization of the L-function associated by Hecke to modular forms. The equality of the L-function of ρ and ρ' is the desired generalization of the reciprocity law of class field theory.

In Chapter 5 we restrict to the theory of Artin L-functions, their application to class number questions, and their connections with Galois module problems.

§ 1. Artin L-Functions

We want to treat Artin L-functions in the spirit of Chapter 2, i.e. we want to avoid infinite field extensions. Hence we associate L-functions to representations of finite Galois groups $G(L/K)$ (§ 1.2). This is almost the same as to consider representations of G_K, since every continuous homomorphism of G_K into $GL_n(\mathbb{C})$ factors through a finite factor group of G_K (compare Chap. 3.1.2). In § 1.3–4 we consider abelian fields with small class number. The results are partially based on the Brauer-Siegel-theorem (Theorem 5.2), which is one of the most interesting applications of Artin L-functions. § 1.5 is mainly devoted to the study of the Artin conductor, which plays for representations of Galois groups a similar role as the conductor for a Hecke character. In § 1.6 we consider the functional equation for Artin L-functions. Finally in § 1.7 we consider Artin L-functions at $s = 0$. This is part of a general philosophy that the values or more general the leading term in the power series development at integral arguments of L-functions contain interesting informations about the objects for which the L-functions are defined (compare the introduction to Chap. 1.6). The most general conjecture in this context is the Beilinson conjecture.

1.1. Representations of Finite Groups (Main reference: Serre (1967a)). Let G be a finite group and let V be a vector space over $\mathbb{C}$ of finite dimension n.

A *representation* $\rho = (\rho, V)$ of G on V of degree n is a homomorphism $\rho: G \to GL(V)$. The *representation module* of ρ is the G-module V defined by

$$gv := \rho(g)v \qquad \text{for } g \in G, v \in V.$$

If ρ' is a representation of G on a vector space V', then ρ and ρ' are called *equivalent* if the corresponding representation modules are isomorphic.

The character χ of ρ is the function $\chi: G \to \mathbb{C}$ such that

$$\chi(g) = \operatorname{tr} \rho(g) \qquad \text{for } g \in G.$$

χ depends only on the equivalence class of ρ and determines this class uniquely. χ is also called a *character* of G. If V is one dimensional, then χ is a homomorphism of G in $\mathbb{C}^\times$, i.e. a character in the sense used in the previous chapters. Characters of degree one are also called *linear characters*. The unit representation 1 is the linear representation χ with $\chi(g) = 1$ for $g \in G$.

A representation ρ is called *irreducible* if the corresponding representation module $V(\rho)$ is simple, i.e. $V(\rho)$ contains no submodule distinct from $V(\rho)$ or $\{0\}$. By the *theorem of Maschke* every representation module is the direct sum of simple modules.

One transfers all notions about representations to the corresponding characters. So one speaks about *irreducible characters* and so on.

Let H be a subgroup of G. A representation ρ of G is called induced if its representation module has the form $M_H^G(V(\sigma))$, where σ is a representation of H (Chap. 2.3.4). ρ is called *monomial* if there is a subgroup H of G and a representation σ of H of degree 1 such that $V(\rho) = M_H^G(V(\sigma))$. The following *theorem of Brauer* states that in general representations are not too far from monomial representations:

Theorem 5.1. *Every character of G is a linear combination with integral coefficients of characters induced from cyclic subgroups of G.* ⊠

For characters χ and φ of G we define

$$\langle \chi, \varphi \rangle_G = \langle \chi, \varphi \rangle := |G|^{-1} \sum_{g \in G} \chi(g)\overline{\varphi(g)}.$$

χ is irreducible if and only if $\langle \chi, \chi \rangle = 1$. If χ and χ' are irreducible and distinct, then $\langle \chi, \chi' \rangle = 0$.

The following formula is called *Frobenius reciprocity*: Let H be a subgroup of G, let φ be a character of H and ψ a character of G. We denote the restriction of ψ to H by $\operatorname{Res} \psi$ and the induced character of φ by $\operatorname{ind} \varphi$. Then

$$\langle \varphi, \operatorname{Res} \psi \rangle_H = \langle \operatorname{ind} \varphi, \psi \rangle_G. \ \square$$

If ρ is a representation of G, then $\det \rho$ with $(\det \rho)(g) = \det(\rho(g))$, $g \in G$, is a linear character of G called the *determinant character*.

Let V be the representation module of the representation ρ. Then $V^* := \operatorname{Hom}_{\mathbb{C}}(V, \mathbb{C})$ is defined as G-module by means of

$$(gv^*)(v) := v^*(g^{-1}v) \qquad \text{for } v^* \in V^*, v \in V, g \in G.$$

The representation ρ^* of V^* is called the *contragredient representation* of ρ. The character χ^* of V^* is the complex conjugate of the character χ of V:

$$\chi^*(g) = \overline{\chi(g)} = \chi(g^{-1}) \qquad \text{for } g \in G.$$

1.2. Artin L-Functions. Let L/K be a finite normal extension of algebraic number fields and (ρ, V) a representation of $G := G(L/K)$ with character χ. For every prime ideal $\mathfrak{p}$ of K we choose a prime divisor $\mathfrak{P}$ in L. Let $\mathfrak{Z}_{\mathfrak{P}}$ be the decomposition group and $\mathfrak{T}_{\mathfrak{P}}$ the inertia group of $\mathfrak{P}$. Then

$$V_{\mathfrak{P}} := \{v \in V \mid gv = v \text{ for } g \in \mathfrak{T}_{\mathfrak{P}}\}$$

is a $\mathfrak{Z}_{\mathfrak{P}}/\mathfrak{T}_{\mathfrak{P}}$-module. For almost all $\mathfrak{p}$ we have $V_{\mathfrak{P}} = V$ (Theorem 1.46). Let $\sigma_{\mathfrak{P}}$ be the Frobenius automorphism, i.e. the generator of $\mathfrak{Z}_{\mathfrak{P}}/\mathfrak{T}_{\mathfrak{P}}$ which induces on the residue class field extension the automorphism

$$\bar{\sigma}_{\mathfrak{P}}: x \to x^q, \qquad x \in \mathfrak{O}_L/\mathfrak{P}, \qquad q = |\mathfrak{O}_K/\mathfrak{p}| \text{ (Proposition 1.52)}.$$

We denote the unit element of $\operatorname{Aut}(V_{\mathfrak{P}})$ by 1. Then $1 - N(\mathfrak{p})^{-s}\sigma_{\mathfrak{p}}$ operates on $V_{\mathfrak{P}}$ and

$$L_{\mathfrak{p}}(s, \chi)^{-1} := \det_{V_{\mathfrak{P}}}(1 - N(\mathfrak{p})^{-s}\sigma_{\mathfrak{P}})$$

is a polynomial in $N(\mathfrak{p})^{-s}$ which does not depend on the choice of $\mathfrak{P}$. The *Artin L-function* of χ is defined by

$$L(s, \chi) := \prod_{\mathfrak{p}} L_{\mathfrak{p}}(s, \chi),$$

where the product runs over all prime ideals of K. One writes also $L(s, \chi) = L(s, \rho) = L(s, V)$. The product $L(s, \chi)$ is convergent for $\operatorname{Re} s > 1$ and represents for this halfplane a holomorphic function with the following basic properties:

1. Let χ_1 and χ_2 be characters of G, then

$$L(s, \chi_1 + \chi_2) = L(s, \chi_1)L(s, \chi_2).$$

2. Let H be a normal subgroup of G and χ a character of G/H. We denote the lifting of χ to G by χ'. Then

$$L(s, \chi') = L(s, \chi).$$

3. Let H be a subgroup of G and χ a character of H. We denote the induced character on G by $\operatorname{ind}_H^G \chi$. Then

$$L(s, \operatorname{ind}_H^G \chi) = L(s, \chi).$$

4. Assume that G is abelian. Let χ be a character of G of degree 1 and let χ^* be the character of the idele class group corresponding to χ by class field theory (Chap. 2.1.5). Then

$$L(s, \chi) = L(s, \chi^*)$$

(Chap. 1.6.3).

Since $L(s, \chi^*)$ has a continuation to the whole complex plane as meromorphic function with at most a simple pole at $s = 1$, the same is true for $L(s, \chi)$.

By the theorem of Brauer (Theorem 5.1) a character χ of an arbitrary finite group G is of the form

$$\chi = n_i \sum_{i=1}^{m} \operatorname{ind}_{H_i}^{G} \chi_i, \qquad n_i \in \mathbb{Z},$$

with cyclic subgroups H_i of G. Then by 1.3, and 4.

$$L(s, \chi) = \prod_{i=1}^{m} L(s, \chi_i^*)^{n_i}. \tag{5.4}$$

Therefore $L(s, \chi)$ is a meromorphic function in the whole complex plane.

Artin conjectured that $L(s, \chi)$ is holomorphic in the whole complex plane if χ does not contain the unit character. This is now called the *Artin conjecture*. From 3. and 4. it follows that the conjecture is true for characters which are induced from one dimensional characters.

The Artin conjecture plays a prominent role in the Langlands correspondence. On the one hand the Artin conjecture follows from the Langlands conjecture and on the other hand in special cases the Artin conjecture implies the Langlands conjecture.

Since

$$\operatorname{ind}_{\{1\}}^{G} 1 = \sum_{\chi} \chi(1)\chi$$

where the sum runs over all irreducible representations χ of G, we have

$$\zeta_L(s) = L(s, \operatorname{ind}_{\{1\}}^{G} 1) = \prod_{\chi} L(s, \chi)^{\chi(1)}$$

If H is a subgroup of G with fixed field M, the Artin conjecture implies that $\zeta_M(s)/\zeta_K(s)$ is holomorphic in the whole plane. In fact, we have $\zeta_M(s) = L(s, \operatorname{ind}_H^G 1_H)$ and $\operatorname{ind}_H^G 1_H - 1_G$ is a character of G which does not contain the unit character.

The holomorphy of $\zeta_M(s)/\zeta_K(s)$ in the whole complex plane was proved by representation theory of groups in the following cases:

1. M/K is normal.
2. G is a Frobenius group to H (i.e. $H \cap gHg^{-1} = \{1\}$ for $g \in G - H$).
3. G is solvable.

See van der Waall (1977) for proofs and further information about this question.

For normal extensions L/K formula (5.4) implies

$$\kappa(L)/\kappa(K) = \lim_{s \to 1} \zeta_L(s)/\zeta_K(s) = \prod_{i=1}^{h} L(1, \chi_i^*)^{n_i}, \tag{5.5}$$

where χ_i^* are non trivial characters of $\mathfrak{C}(L)$ and $\kappa(L)$ (resp. $\kappa(K)$) is the residue of $\zeta_L(s)$ (resp. $\zeta_K(s)$) at $s = 1$.

Let $K = \mathbb{Q}$. The expression

$$\kappa(L) = \frac{2^{r_1}(2\pi)^{r_2} R(L) h(L)}{w\sqrt{|d_{L/\mathbb{Q}}|}}$$

in Proposition 1.99 together with estimations of the right side of (5.5) lead to the proof of the following *theorem of Brauer and Siegel* (Lang, (1970), Chap. 9).

Theorem 5.2. 1. *Let $L_1, L_2, \ldots$ be a sequence of normal extensions of $\mathbb{Q}$ such that*

$$\lim_{n\to\infty} [L_n : \mathbb{Q}]/\log|d_{L_n/\mathbb{Q}}| = 0,$$

then

$$\lim_{n\to\infty} \frac{\log R(L_n)h(L_n)}{\log|d_{L_n/\mathbb{Q}}|} = \frac{1}{2}.$$

2. Let $M_1, M_2, \ldots$ *be a sequence of extensions of $\mathbb{Q}$ of fixed degree, then*

$$\lim_{n\to\infty} \frac{\log R(M_n)h(M_n)}{\log|d_{M_n/\mathbb{Q}}|} = \frac{1}{2}. \boxtimes$$

Example 1. Let $\mathbb{Q}(\sqrt{d})$ be the imaginary-quadratic field with discriminant $d < 0$. Then $R(\mathbb{Q}(\sqrt{d})) = 1$ and Theorem 5.2 shows

$$\lim_{-d\to\infty} h(\mathbb{Q}(\sqrt{d})) = \infty.$$

Hence there are only finitely many imaginary-quadratic fields with given class number. For more information about the class-number of imaginary-quadratic fields see § 1.4. □

1.3. Cyclotomic Fields with Class Number 1 (Main reference: Washington (1982) Chap. 11). Using the Brauer-Siegel theorem (Theorem 5.2) for $\mathbb{Q}(\zeta_n)$ and $\mathbb{Q}(\zeta_n)^+$ one proves the following estimation for h_n^-, which shows that h_n grows rapidly with n:

Theorem 5.3.

$$\lim_{n\to\infty} \log h_n^- /(\tfrac{1}{4}\varphi(n)\log n) = 1,$$

where $\varphi(n)$ denotes Euler's function. ⊠

It follows that there are only finitely many cyclotomic fields with restricted class number, but Theorem 5.3 is not effective in the sense that it does not allow us to compute a constant $n(h)$ such that $h_n^- > h$ if $n \geqslant n(h)$. On the basis of the arithmetical class number formula for h_n^- (Theorem 4.5) one proves an effective estimation:

Theorem 5.4. *Let n be a natural number with $\varphi(n) \geqslant 220$. If n is a prime power, then*

$$\log h_n^- \geqslant \frac{1}{4} d_n - (1,08)\varphi(n).$$

If n is arbitrary, then

$$\log h_n^- \geqslant \frac{1}{4} d_n - (1,08)\varphi(n) - \frac{1}{2}\varphi(n)\sum_{p|n}\frac{1}{(2p^2)}.$$

Here d_n denotes the absolute value of the discriminant of $\mathbb{Q}(\zeta_n)$. ⊠

Now we want to find all cyclotomic fields with class number one. From Theorem 5.4 it follows that $h_{p^a}^- > 1$ if $\varphi(p^a) \geqslant 220$. On the other hand h_n^- is known at least for $\varphi(n) \leqslant 256$ (Schrutka von Rechtenstamm (1964), Washington (1982), Tables, § 3). For a prime power p^a one has $h_{p^a}^- = 1$ if and only if $a = 1$, $p \leqslant 19$ or $p^a = 4, 8, 9, 16, 25, 27, 32$.

Since $h_m | h_n$ if $m | n$ by Theorem 2.112 one has $h_n^- = 1$ only if a p-component of n is of the form p^a above. These are finitely many such cases which can be checked individually using Theorem 5.4. One finds the following numbers $n \not\equiv 2 \pmod 4$ with $h_n^- = 1$:

$$\begin{gathered}1, 3, 4, 5, 7, 8, 9, 11, 12, 13, 15, 16, 17, 19, 20, 21, 24, 25, 27,\\ 28, 32, 33, 35, 36, 40, 44, 45, 48, 60, 84.\end{gathered} \tag{5.6}$$

For all these n one has $h_n^+ = 1$. This can be proved using the estimates of Odlyzko (Chap. 1.6.10) for the Hilbert class field of $\mathbb{Q}(\zeta_n)^+$.

Theorem 5.5 (Masley). *Let n be a natural number with $n \not\equiv 2 \pmod 4$. Then $\mathbb{Q}(\zeta_n)$ has class number one if and only if n is a number of the form* (5.6). ⊠

See Masley (1976) for further information about cyclotomic fields with small class number.

1.4. Imaginary-Quadratic Fields with Small Class Number. Now let $\mathbb{Q}(\sqrt{d})$ be an imaginary-quadratic field with discriminant d and class number $h(d)$. The Brauer-Siegel theorem (Theorem 5.2) has the form

$$\lim_{|d|\to\infty} \frac{\log h(d)}{\log|d|} = \frac{1}{2}.$$

It implies that there are only finitely many imaginary-quadratic fields with restricted class number.

The estimation of $h(d)$ can be made effectively (Baker (1966), Gross-Zagier (1986)): So far one has the complete list of imaginary-quadratic fields with $h(d) = 1, 2$ and 3.

Theorem 5.6 (theorem of Baker (1966), (1971) and Stark (1967), (1975)). *Let $h(d)$ be the class number of the imaginary-quadratic number field with discriminant d. Then $h(d) = 1$ if and only if*

$$d = -3, -4, -7, -8, -11, -19, -43, -67, -163.$$

$h(d) = 2$ if and only if

$$\begin{gathered}d = -15, -20, -24, -35, -40, -51, -52, -88, -91, -115, -123,\\ -148, -187, -232, -235, -267, -403, -427.\end{gathered}$$ ⊠

Example 2. Let $d \equiv 1 \pmod 4$ be the discriminant of an imaginary-quadratic number field. Then $x^2 - x + (1 - d)/4$ is a prime number for $x = 1, 2, \ldots, -(d + 3)/4$ if and only if $\mathbb{Q}(\sqrt{d})$ has class number one. □

In particular $x^2 - x + 41$ is a prime for $x = 1, 2, \ldots, 40$ (Euler 1772).

Heegner (1952) (see also Deuring (1968a) and Stark (1969) developed a method to connect the imaginary-quadratic fields of class number one with the solutions of a Diophantine equation. Let $h(d) = 1$. If 2 splits in $\mathbb{Q}(\sqrt{d})$, then it is easy to see that $d = -4, -7$ or -8. Hence we assume that (2) is a prime ideal in $\mathbb{Q}(\sqrt{d})$. Then $d \equiv 1 \pmod 4$ and $\omega := (\sqrt{d} + 1)/2$, 1 is a basis over $\mathbb{Z}$ for the ring of integers of $\mathbb{Q}(\sqrt{d})$. Therefore $j(\omega)$ generates the Hilbert class field of $\mathbb{Q}(\sqrt{d})$ (Chap. 2.2.2), hence $j(\omega)$ is a rational number and even an integer. On the other hand the value $j(\omega)$ of the modular function $j(z)$ determines ω up to modular equivalence, hence determines the field $\mathbb{Q}(\sqrt{d})$. Beside $j(z)$ one considers other functions in $\mathbb{Q}[j(z)]$ and by means of the algebraic dependence of this functions one finds an equation $F(x, y) = 0$ with the property that to any imaginary quadratic field $\mathbb{Q}(\sqrt{d})$ with $d \notin \{-4, -7, -8\}$ there corresponds one and only one integral solution of the equation. After some computations one ends up with the equation $y^2 = 2x(x^3 + 1)$ which has the same property. The solution of this equation, which is well known in the arithmetical theory of elliptic curves, are the following:

$$(0, 0), (1, 2), (1, -2), (-1, 0), (2, 6), (2, -6).$$

Hence there are only nine imaginary-quadratic fields with class number 1.

We see that the imaginary-quadratic number fields with class number 1 and prime ideal (2) are parametrized by the solutions of the Diophantine equation $y^2 = 2x(x^3 + 1)$. Something similar is known only for imaginary-quadratic fields of class number 2 and even discriminant (Kenku (1971), Abrashkin (1974)).

If the imaginary-quadratic field $\mathbb{Q}(\sqrt{d})$ has class number 1, then $\mathbb{Q}(\sqrt{d})$ has no nontrivial abelian extension which is unramified at all places (Theorem 2.2). More general let K be any finite normal extension of $\mathbb{Q}(\sqrt{d})$ which is unramified at all places. If $K \neq \mathbb{Q}(\sqrt{d})$, then $G(K/\mathbb{Q}(\sqrt{d}))$ is not solvable, hence $[K : \mathbb{Q}] \geqslant 120$. Using Odlyzko's estimate for d_K (Chap. 1.6.11) one proves that $K = \mathbb{Q}(\sqrt{d})$.

1.5. The Artin Representation and the Artin Conductor (Main reference: Serre (1962), Chap. 6). Let L be a finite normal extension of a $\mathfrak{p}$-adic number field K with Galois group G. For $g \in G$, $g \neq 1$, we define $i_G(g)$ as the largest natural number $i + 1$ such that g lies in the i-th ramification group G_i of L/K (Chap. 1.3.7). If π is a prime element of L, then

$$i_G(g) = v_L(g\pi - \pi) \qquad \text{for } g \in G_0.$$

We set $i_G(1) = \infty$.

Theorem 5.7. *Let f be the inertia degree of L/K and*

$$a_G(g) = -f i_G(g) \qquad \text{if } g \neq 1,$$

$$a_G(1) = f \sum_{\substack{g \in G \\ g \neq 1}} i_G(g).$$

Then the function a_G is the character of a representation of G, called the Artin representation. ⊠

We have defined the Artin representation by means of its character. A direct construction of this representation is unknown so far. But it is known that the Artin representation is not rational in general (Serre (1960)). Therefore it is unlikely that a simple construction exists. The Artin representation plays a role in the theory of arithmetical algebraic curves (see Serre (1962), Chap. 6.4).

For any character χ of G we set

$$f(\chi) := |G|^{-1} \sum_{g \in G} \chi(g) a_G(g) = \langle \chi, a_G \rangle.$$

$f(\chi)$ is called the *Artin conductor* of χ. In terms of the representation ρ with character χ it is described as follows:

Theorem 5.8. *Let V be the representation module of ρ. Then*

$$f(\chi) = \sum_{i=0}^{\infty} \frac{|G_i|}{|G_0|} (\dim V - \dim V^{G_i}). \ \square$$

From the last formula it is clear that the Artin conductor does not change if we restrict the representation to the inertia group G_0. Hence for the investigation of the Artin conductor it is sufficient to consider full ramified extensions.

The Artin conductor has the following basic properties:

1. Let χ_1 and χ_2 be characters of G. Then

$$f(\chi_1 + \chi_2) = f(\chi_1) + f(\chi_2).$$

2. Let H be a normal subgroup of G and χ a character of G/H. We denote the lifting of χ to G by χ'. Then

$$f(\chi') = f(\chi).$$

3. Let H be a subgroup of G and χ a character of H. Moreover let M be the fixed field of H and $\mathfrak{d}_{M/K}$ (resp. $f_{M/K}$) the discriminant (resp. the inertia degree) of M/K. Then

$$f(\operatorname{ind}_H^G \chi) = v_K(\mathfrak{d}_{M/K})\chi(1) + f_{M/K} f(\chi). \tag{5.7}$$

4. Let χ be a character of degree 1 of G and let c_χ be the largest integer such that the restriction of χ to the ramification group G_{c_χ} is not the unit character (for $\chi = 1$ we set $c = -1$). Then

$$f(\chi) = \varphi_{L/K}(c_\chi) + 1.$$

(For the definition of $\varphi_{L/K}$ see Chap. 1.3.7.) □

4. and local class field theory (Chap. 2.1.3) shows that for a degree one character χ, $f(\chi)$ is the exponent of the conductor of χ viewed as a character of $K^\times$. The conductor of an arbitrary irreducible character plays a similar role in the local Langlands correspondence.

We apply 1., 3. to the case $H = \{1\}$, $\chi = 1$. Then

$$\nu_K(\mathfrak{d}_{L/K}) = f(\operatorname{ind}_{\{1\}}^G 1) = \sum_{\chi} \chi(1) f(\chi), \tag{5.8}$$

where the sum runs over all irreducible characters χ of G. This is Artin's generalization of the Führerdiskriminantenproduktformel (Theorem 2.5).

Now let L be a finite normal extension of an algebraic number field K and χ a character of the Galois group G of L/K. For all prime divisors $\mathfrak{P}$ in L of a prime ideal $\mathfrak{p}$ of K one has the same Artin conductor $f(\chi, \mathfrak{p})$ of χ restricted to the decomposition group $\mathfrak{Z}_{\mathfrak{p}} = G(L_{\mathfrak{P}}/K_{\mathfrak{p}})$. We define the Artin conductor $\mathfrak{f}_\chi$ of χ by

$$\mathfrak{f}_\chi = \prod_{\mathfrak{p}} \mathfrak{p}^{f(\chi, \mathfrak{p})}.$$

The basic properties of $\mathfrak{f}_\chi$ correspond to the properties 1.–4. of the local Artin conductors. To (5.8) corresponds

$$\mathfrak{d}_{L/K} = \prod_{\chi} \mathfrak{f}_\chi^{\chi(1)}. \tag{5.9}$$

1.6. The Functional Equation for Artin L-Functions. Let L/K be a normal extension of number fields. If χ is a one dimensional character of $G(L/K)$ and χ^* the corresponding character of the idele class group, then $L(s, \chi) = L(s, \chi^*)$ (§1.2). Since $L(s, \chi^*)$ satisfies a functional equation (Theorem 1.104), the same is true for $L(s, \chi)$. We want to generalize this functional equation to arbitrary representations ρ of $G(L/K)$. Let χ denote the character of ρ. As in the case of the functional equation for Hecke characters we have to define local L-factors $L_v(s, \chi)$ at infinite places v of K. We put $\gamma(s) := \pi^{-s/2}\Gamma(s/2)$. If v is complex, then $L_v(s, \chi) := (\gamma(s)\gamma(s+1))^{\chi(1)}$. If v is real, let w be a place of L above v. For w real we put $L_v(s, \chi) := \gamma(s)^{\chi(1)}$. For w complex let σ_v be the generator of $G(L_w/K_v)$ and $V(\rho) = V^+ + V^-$ the decomposition of $V(\rho)$ corresponding to the eigenvalues $+1$ and -1 of $\rho(\sigma_v)$. We put $L_v(s, \chi) := \gamma(s)^{\dim V^+}\gamma(s+1)^{\dim V^-}$. Obviously this definition is independent of the choice of w.

The product $\Lambda_\infty(s, \chi)$ over the L-factors $L_v(s, \chi)$ for infinite places v has the properties 1.–4. of §1.2. Hence it follows from Theorem 5.1 and the functional equation for Hecke L-functions (Theorem 1.104) that the enlarged Artin L-function

$$\Lambda(s, \chi) := \Lambda_\infty(s, \chi) L(s, \chi)$$

satisfies a corresponding functional equation:

Theorem 5.9. *Let L/K be a normal extension of number fields and χ a character of $G(L/K)$. Then there is a complex number $\varepsilon_0(\chi)$ with $|\varepsilon_0(\chi)| = 1$, called Artin root number, such that*

$$\Lambda(s, \chi) = \varepsilon_0(\chi) A(\chi)^{1/2-s} \Lambda(1 - s, \overline{\chi}),$$

where

$$A(\chi) := |d_K|^{\chi(1)} N_{K/\mathbb{Q}}\mathfrak{f}(\chi)$$

with d_K the discriminant of K and $\mathfrak{f}(\chi)$ the Artin conductor, and $\bar{\chi}$ is the complex conjugate character of χ (being the character of the contragredient representation of ρ). □

An n-dimensional representation ρ of $G(L/K)$ lifts to a representation ρ' of G_K:

$$G_K \to G(L/K) \underset{\rho}{\to} GL_n(\mathbb{C}),$$

i.e. a continuous homomorphism of G_K into $GL_n(\mathbb{C})$. On the other hand any continuous homomorphism of G_K into $GL_n(\mathbb{C})$ factors through a finite group since its kernel is an open subgroup.

The invariance of the L-function, the enlarged L-function, and the Artin conductor with respect to liftings (§1.2, §1.5, property 2) permits to define these notions and the Artin root number for ρ':

$$L(s, \rho') = L(s, \rho) = L(s, \chi), \qquad \Lambda(s, \rho') := \Lambda(s, \chi),$$

$$f(\rho') := f(\chi), \qquad \varepsilon_0(\rho') := \varepsilon_0(\chi).$$

The reformulation of §1.2, §1.5 and §1.6 in terms of representations of G_K is left to the reader.

The Artin L-function defined in the introduction to Chap. 5 differs from $L(s, \rho')$ by the finitely many local factors associated to ramified prime ideals $\mathfrak{p}$. One needs these factors for the functional equation.

1.7. The Conjectures of Stark about Artin L-Functions at $s = 0$ (Main reference: Tate (1984)). Let L be a finite normal extension of the algebraic number field K and let χ be the character of a representation (ρ, V) of $G(L/K) =: G$. In this section we consider Artin L-functions $L(s, \chi)$ at $s = 0$. More general let S be a finite set of places of K containing the set S_∞ of infinite places. We remove the local factors of $L(s, \chi)$ for places $v \in S - S_\infty$:

$$L_S(s, \chi) := \prod_{\mathfrak{p} \notin S} L_{\mathfrak{p}}(s, \chi)$$

(§1.2).

The functional equation for $L(s, \chi)$ shows that $L_S(s, \chi)$ is regular at $s = 0$. Let $c(\chi) \in \mathbb{C}^\times$ and $r(\chi) \in \mathbb{Z}$ be defined by

$$L_S(s, \chi) = c(\chi) s^{r(\chi)} + O(s^{r(\chi)+1})$$

in the neighborhood of $s = 0$. Then again the functional equation together with $L(1, \chi) \neq 0$ for one dimensional characters χ (Theorem 2.25) and some character theory shows

$$r(\chi) = \left(\sum_{s \in S} \dim V^{G_w}\right) - \dim V^G = \dim_{\mathbb{C}} \operatorname{Hom}_G(V^*, \mathbb{C} \otimes_{\mathbb{Z}} X),$$

where w is a fixed place of L above v, $G_w := G(L_w/L_v)$, and X is the $\mathbb{Z}[G]$-module

$$X = \left\{ \sum_{w \in S(L)} n_w w \,\middle|\, n_w \in \mathbb{Z}, \sum_{w \in S(L)} n_w = 0 \right\}$$

with $S(L)$ the set of places of L above places in S.

If $\chi = \chi_0$ is the trivial character, then $r(\chi_0) = |S| - 1$, and the functional equation delivers also the structure of $c(\chi_0)$:

$$c(\chi_0) = -\frac{h_S R_S}{e},$$

where h_s denotes the class number of

$$\mathfrak{O}_S := \{\alpha \in K \,|\, v_\mathfrak{p}(\alpha) \geqslant 0 \text{ for } \mathfrak{p} \notin S\},$$

R_S the S-Regulator (Chap. 1.4.1), and e the number of roots of unity in K. The main conjecture of Stark (in the form of Tate (1984)) concerns the structure of $c(\chi)$ for $\chi \neq \chi_0$. (Stark (1975a))

First of all we define the Stark regulator $R(\chi, f)$. By Dirichlet's unit theorem for the group U of $S(L)$-units in L we have an isomorphism λ of $R[G]$-modules $R \otimes_\mathbb{Z} U$ and $R \otimes_\mathbb{Z} X$ defined by

$$\lambda(u) := \sum_{w \in S(L)} \log|u|_w w, \qquad u \in U,$$

where $|u|_w$ is the normalized valuation connected with w (Chap. 1.4.1). This implies that $R \otimes_\mathbb{Z} U$ and $R \otimes_\mathbb{Z} X$ have equal characters. Therefore already the rational representation modules $\mathbb{Q} \otimes_\mathbb{Z} U$ and $\mathbb{Q} \otimes_\mathbb{Z} X$ are isomorphic. Let $f\colon \mathbb{Q} \otimes_\mathbb{Z} X \to \mathbb{Q} \otimes_\mathbb{Z} U$ be an isomorphism of $\mathbb{Q}[G]$-modules. The automorphism λ of $\mathbb{C}X := \mathbb{C} \otimes_\mathbb{Z} X = \mathbb{C} \otimes_R (R \otimes_\mathbb{Z} X)$ induces a linear map

$$\mathrm{Hom}_G(V^*, \mathbb{C}X) \to \mathrm{Hom}_G(V^*, \mathbb{C}X).$$

Let $R(\lambda, f)$ be its determinant. We call $R(\chi, f)$ the Stark regulator.

Example 3. Let f be an injection of X in U. Then

$$R(\chi_0, f) = \pm \frac{R_S[U : f(X)]}{e}. \square$$

Now we can formulate the *main conjecture of Stark*:

Let $A(\chi, f) := R(\chi, f)/c(\chi)$. Then $A(\chi, f)$ lies in the value field $\mathbb{Q}(\chi(g) | g \in G) = \mathbb{Q}(\chi)$ of χ and

$$gA(\chi, f) = A(g\chi, f) \quad \text{for} \quad g \in G(\mathbb{Q}(\chi)/\mathbb{Q}). \square$$

Remark. The eigenvalues of a matrix of finite order are roots of unity. Hence $\mathbb{Q}(\chi)$ is an abelian field. □

One has the following reductions of Stark's conjecture and special cases for which the conjecture is true:

Theorem 5.10. *If the main conjecture is true for one choice of f and S, then it is true for all choices of f and S.*

If the conjecture is true for normal extensions $L/\mathbb{Q}$, it is true in general.

If the conjecture is true for one dimensional characters, it is true in general.

⊠

Theorem 5.11. *The main conjecture is true for rational characters, i.e. characters with values in* $\mathbb{Q}$. ⊠

Theorem 5.12. *The main conjecture is true if* $r(\chi) = 0$. ⊠

Now we restrict ourselves to abelian extensions L/K. Furthermore we assume that S contains beside S_∞ all places which are ramified in L/K and at least one place which is totally decomposed in L/K. Then Stark (1975a) announced the following conjecture which is stronger than the main conjecture: Let v be a totally decomposed place of S and w a place of L above v. Then there exists a unit $\varepsilon \in L$ such that

$$L'(0, \chi) = -\frac{1}{e} \sum_{g \in G} \chi(g) \log|g\varepsilon|_w \tag{5.10}$$

for all characters $\chi \in X(G)$.

We denote this conjecture in the following by $\mathrm{St}(L/K, S)$. (5.10) determines ε up to a root of unity in L.

Since the Artin L-series for one dimensional characters χ are equal to the L-series of the corresponding character of the class group of L/K, $\mathrm{St}(L/K, S)$ is a non expected contribution to Hilbert's 12. problem concerning the construction of class fields by means of special values of transcendental functions (Chap. 2.2). In some cases we have $L = K(\varepsilon)$: Let G be cyclic and suppose that S contains only one place totally decomposed in L. Then $L'(0, \chi) \neq 0$ for all faithful characters χ of G. Let $h \in G$ with $h\varepsilon = \varepsilon$. Then

$$L'(0, \chi) = -\frac{1}{e} \sum_{g \in G} \chi(gh) \log|gh\varepsilon|_w = \chi(h)L'(0, \chi),$$

hence $\chi(h) = 1$, which implies $h = 1$.

The conjecture was proved by Stark for $K = \mathbb{Q}$ and K an imaginary-quadratic number field, using the fact that in this cases we have constructions of class fields by special values of transcendental functions (Chap. 2.2). The conjecture is also true if S contains two totally ramified places: If $|S| \geqslant 3$, then $L'(0, \chi) = 0$ for all $\chi \in X(G)$ and the conjecture is trivial. If $|S| = 2$, then $L'(0, \chi) = 0$ for $\chi \neq \chi_0$ and the conjecture is easily proved. The group of S-units in K has rank 1 and ε can be taken as a power of the fundamental unit of K. Hence the interesting case is the case that there is only one totally ramified place in S.

Example 4 (Stark (1980)). The polynomial

$$x^3 - x^2 - 9x + 8 = (x - 1)(x + 3)(x - 3) - 1$$

has three real zeroes $\beta = \beta_1 > \beta_2 > \beta_3$, $\beta = 3{,}079\ldots$. The field $K = \mathbb{Q}(\beta)$ has discriminant $7^2 \cdot 53$ and its ring of integers is $\mathbb{Z}[\beta]$. The elements $\beta - 1, \beta + 3$ form a fundamental system of units and the class group has order 3 and is generated by $\mathfrak{p}_2 = (\beta, 2)$.

Furthermore let $\theta := 2\cos\frac{2\pi}{7}$ and

$$\delta = \frac{(\beta + 1)}{2} - \frac{1}{2}\sqrt{(\beta + 1)^2 - 4},$$

$F := K(\theta)$, $L := K(\theta, \delta)$, $L' := K(\delta)$.

$\mathbb{Q}(\theta)/\mathbb{Q}$ is only ramified for 7, $\mathbb{Q}_7(\theta)$ and $\mathbb{Q}_7(\beta)$ are cyclic extensions of degree 3. Since the tame ramification group is always cyclic, F/K is unramified at finite places. L'/K is unramified at finite places, too, because the discriminant of δ is the unit $(\beta + 1)^2 - 4 = (\beta + 3)(\beta - 1)$. K has three infinite places $\infty, \infty_2, \infty_3$ corresponding to the embeddings $\beta \to \beta$, $\beta \to \beta_2$, $\beta \to \beta_3$ of K in R. It is easy to see that ∞ decomposes in L'/K, but ∞_2 and ∞_3 are ramified. Let $S := \{\infty, \infty_2, \infty_3\}$, then Stark has numerically verified the conjecture St$(L/K, S)$. For more information and examples see Stark (1975a)–(1980) and Tate (1984). ⊠

There is a p-adic analogue of Stark's conjecture due to Gross (1981) for $s = 0$ and to Serre (1978a) for $s = 1$ (Tate (1984), Chap. 6).

Beilinson (see Rapoport e.a. (1988)) generalized Stark's conjectures to L-functions of varieties.

§2. Galois Module Structure and Artin Root Numbers

(Main reference: Fröhlich (1983))

Let K be an algebraic number field and L/K a finite normal extension with Galois group G. Then L considered as $K[G]$-module is isomorphic to $K[G]$ considered as $K[G]$-module (L/K has a normal basis). It arises the natural question what one can say about $\mathfrak{O}_L$ considered as an $\mathfrak{O}_K[G]$-module. Leopoldt (1959) gave a complete answer to this question in the case that $K = \mathbb{Q}$ and $L/\mathbb{Q}$ is an abelian extension (Chap. 4.1). But in general it seems to be hopeless to get any reasonable answer since $\mathfrak{O}_L$ in general is far from belonging to some simple type of $\mathfrak{O}_K[G]$-modules.

Therefore one looks for a more moderate question which one can hope to answer: One asks for conditions on L which imply that $\mathfrak{O}_L$ has a *normal integral basis* (NIB), i.e. there exist $\alpha_1, \ldots, \alpha_m \in \mathfrak{O}_L$, $m := [K:\mathbb{Q}]$, such that $\{g\alpha_i | g \in G, i = 1, \ldots, m\}$ is a basis of $\mathfrak{O}_L$. Of course one is mainly interested in the case $K = \mathbb{Q}$, but the general situation is useful for technical reasons, e.g. induction arguments.

It is easy to see from the theory of the different (Chap. 1) that a normal local extension has a NIB if and only if it is tame (Noether's theorem) Hence tame ramification is a necessary condition for the existence of a NIB for $\mathfrak{O}_L$. Already Hilbert (1895), Satz 132, proved that for abelian extensions $L/\mathbb{Q}$ this condition is also sufficient. Martinet showed that this is also true for extensions $L/\mathbb{Q}$ with dihedral group of order $2l$, l an odd prime. But if $G(L/\mathbb{Q})$ is the quaternion group

H_8 of order 8, Martinet (1971) showed that this is wrong. In the case of the group H_8 we have a nice situation since there are only two isomorphy classes of $\mathbb{Z}[H_8]$-modules M with $M \otimes_{\mathbb{Z}} \mathbb{Z}_p \cong \mathbb{Z}_p[H_8]$ for all primes p. Serre had the idea to connect this question with the Artin root number ε_0 of the unique irreducible representation of $G(L/\mathbb{Q})$ of dimension 2. He conjectured that $\mathfrak{O}_L$ has NIB if and only if $\varepsilon_0 = 1$ (For a quaternion extension $L/\mathbb{Q}$ one has $\varepsilon_0 = \pm 1$). This was proved by Fröhlich (1972) on the basis of his results about central extensions of abelian fields (Chap. 3.2.6). In the following this was generalized mainly by Fröhlich. We give here a description of essential parts of Fröhlich's theory and the main result due to Taylor (1981).

2.1. The Class Group of $\mathbb{Z}[G]$. Let G be a finite group. A locally free $\mathbb{Z}[G]$-module is a finitely generated $\mathbb{Z}[G]$-module M such that $M_p := M \otimes_{\mathbb{Z}} \mathbb{Z}_p$ is a free $\mathbb{Z}_p[G]$-module for all primes p. The rank $rk(M)$ is defined as the rank of the free $\mathbb{Q}[G]$-module $M \otimes_{\mathbb{Z}} \mathbb{Q}$. This is also the rank of M_p as free $\mathbb{Z}_p[G]$-module for all p.

The class group $CL(\mathbb{Z}[G])$ is the factor group of the free group generated by the isomorphy classes of finitely generated locally free $\mathbb{Z}[G]$-modules to the group generated by the modules of the form $M_1 \oplus M_2 - M_1 - M_2$ and by the free $\mathbb{Z}[G]$-modules. Let $[M]$ be the class in $CL(\mathbb{Z}[G])$ of the module M. Then $[M_1] = [M_2]$ and $rk(M_1) = rk(M_2)$ if and only if M_1 and M_2 are stably isomorphic, i.e. $M + \mathbb{Z}[G] \cong M_2 + \mathbb{Z}[G]$. If the ranks are > 1, then this implies actually $M_1 \cong M_2$.

Let $R(G)$ be the group of virtual representations of G (§1.6) and let $E/\mathbb{Q}$ be a finite extension such that the absolute irreducible representations of G can be realized over E. Then $G(E/\mathbb{Q})$ acts on $R(G)$ by

$$(\omega\rho)(g) = \omega(\rho(g)) \quad \text{for} \quad \rho \in R(G), \omega \in G(E/\mathbb{Q}), g \in G.$$

We want to give an interpretation of $CL(\mathbb{Z}[G])$ as a factor group of $\mathrm{Hom}_{G(E/\mathbb{Q})}(R(G), J_E)$, where J_E denotes the idele group (Chap. 1.5.5), and we need for this purpose the following homomorphism

$$\mathrm{Det}\colon \prod_p (\mathbb{Z}_p[G])^\times \to \mathrm{Hom}_{G(E/\mathbb{Q})}(R(G), U_E),$$

where U_E denotes the group of unit ideles of E and the product runs over all places p of $\mathbb{Q}$ (we put $\mathbb{Z}_\infty = \mathbb{R}$). Det is defined by the corresponding local maps Det_p: Let $\rho\colon G \to GL_n(E)$ be an irreducible representation of G. Then we extend ρ to an algebra homomorphism $\rho\colon \mathbb{Q}_p[G] \to M_n(E \otimes_{\mathbb{Q}} \mathbb{Q}_p)$. We put

$$(\mathrm{Det}_p\, x_p)(\rho) := \det(\rho x_p) \qquad \text{for } x_p \in \mathbb{Q}_p[G].$$

If $x_p \in (\mathbb{Z}_p[G])^\times$, then $\det(\rho x_p)$ lies in the group of units $\prod_{\mathfrak{p}|p} U_{\mathfrak{p}}$ of $E \otimes_{\mathbb{Q}} \mathbb{Q}_p = \prod_{\mathfrak{p}|p} E_{\mathfrak{p}}$.

Theorem 5.13. 1. *Let M be a locally free rank one $\mathbb{Z}[G]$-module. Choose a free generator v of $M \otimes_{\mathbb{Z}} \mathbb{Q}$ over $\mathbb{Q}[G]$ and for each prime p choose a free generator v_p of $M \otimes_{\mathbb{Z}} \mathbb{Z}_p$ over $\mathbb{Z}_p[G]$. Then*

$$v_p = v\lambda_p \text{ with } \lambda_p \in (\mathbb{Q}_p[G])^\times$$

and the class $f(M)$ of $\prod_p \mathrm{Det}_p \lambda_p \in \mathrm{Hom}_{G(E/\mathbb{Q})}(R(G), J_E)$ with respect to

$$D := \mathrm{Hom}_{G(E/\mathbb{Q})}(R(G), E)\mathrm{Det}\left(\prod_p (\mathbb{Z}_p[G])^\times\right)$$

only depends on the isomorphism class of M.

2. *There is a unique isomorphism*

$$CL(\mathbb{Z}[G]) \cong \mathrm{Hom}_{G(E/\mathbb{Q})}(R(G), J_E)/D$$

such that for every locally free rank one module M the class $[M]$ in $CL(\mathbb{Z}[G])$ is mapped onto $f(M)$. ⊠

2.2. The Galois Module Structure of Tame Extensions. Now we are able to generalize the connection between Galois module structure and Artin root numbers to arbitrary normal tame extensions L of the algebraic number field K. Only symplectic representation appear in this connection: A representation ρ of the finite group G is called symplectic (resp. orthogonal) if it factors through $G \to \mathrm{Sp}_n(\mathbb{C})$ (resp. through $G \to \mathrm{O}_n(\mathbb{C})$). Every irreducible representation with real valued character is either symplectic or orthogonal. Moreover it is easy to see that the Artin root number of such a representation is $+1$ or -1.

Theorem 5.14. *Let L/K be a normal tame extension and ρ a symplectic representation of $G := G(L/K)$. Then the Artin root number $\varepsilon_0(\rho)$ satisfies*

$$\varepsilon_0(\rho) = \varepsilon_0(\omega\rho) \quad \text{for } \omega \in G(\overline{\mathbb{Q}}/\mathbb{Q}).$$

We associate to L/K an element $\varepsilon_0(L/K)$ of $CL(\mathbb{Z}[G])$ as follows:

Let $\varepsilon_0'(L/K)$ be the element of $\mathrm{Hom}_{G(E/\mathbb{Q})}(R(G), J_E)$ which is given on irreducible representations ρ of G by

$$\varepsilon_0'(L/K)(\rho)_\mathfrak{p} = \varepsilon_0(\rho) \quad \text{for } \mathfrak{p} \text{ finite}$$

$$\varepsilon_0'(L/K)(\rho)_\mathfrak{p} = 1 \quad \text{for } \mathfrak{p} \text{ infinite}$$

if ρ is symplectic,

$$\varepsilon_0'(L/K)(\rho) = 1$$

if ρ is not symplectic.

Then $\varepsilon_0(L/K)$ is the element of $CL(\mathbb{Z}[G])$ corresponding to $\varepsilon_0'(L/K)$ according to Theorem 5.13. ⊠

Theorem 5.15. *Let L/K be a finite normal tame extension with Galois group G. The class of the $Z[G]$-module $\mathfrak{O}_L$ in $CL(\mathbb{Z}[G])$ is equal to $\varepsilon_0(L/K)$.* ⊠

The theorem implies that $\mathfrak{O}_L$ is a free $\mathbb{Z}[G]$-module if G has no irreducible symplectic representations and that $\mathfrak{O}_L \oplus \mathfrak{O}_L$ is always a free $\mathbb{Z}[G]$-module.

2.3. Further Results on Galois Module Structure. There is a vast literature about Galois module structure for subgroups of the additive or multiplicative group of local or global fields mostly with very restrictive conditions for the Galois group, e.g. cyclic or dihedral groups, and the base field. We mention here only a few results of a more general nature.

Let K be a $\mathfrak{p}$-adic field with residue class characteristic p and let L/K be a finite normal extension with Galois group G. Krasner (1936) and Iwasawa (1955) studied the structure of $L^\times$ as G-module in the case that L/K is tamely ramified (Chap. 3.2.4). The case of wild ramification was studied by Borevich and some of his pupils (see Borevich (1967) and further references mentioned there). Let L/K be a p-extension, let $\hat{L}^\times$ be the pro-p-completion of $L^\times$, let F be the Galois group of the maximal p-extension of K, and let R be the kernel of the natural map $F \to G$. Then the G-module $\hat{L}^\times$ is isomorphic to $R/[R, R]$ by class field theory. Hence one can derive information about $\hat{L}^\times$ by means of the known structure of F (Chap. 3.2.3). Borevich gives a description of $\hat{L}^\times$ by means of generators and relations. His further results concern the structure of the group of principal units U_1 of L, which can be considered as a subgroup of $\hat{L}^\times$. In the case that K does not contain the p-th roots of unity he gives a full description of the G-module structure of U_1.

Now let K be an algebraic number field and let L/K be a finite normal extension with Galois group G. The structure of the group E of units of L as G-module was first studied by Herbrand (1931a) who showed that E contains a G-submodule of finite index with a structure which can be easily described (§1.12). Recently Fröhlich (1989) began the investigation of E in the case of tame extensions L/K in analogy to his additive theory explained in § 2.2.

Appendix 1
Fields, Domains, and Complexes

(Main reference: van der Waerden (1971))

1.1. Finite Field Extensions (van der Waerden (1971), Chap. 6). Let K be a field and L a finite extension of K, i.e. L contains K and has finite dimension $n = [L:K]$ considered as K-vector space. Let $\alpha \in L$ and A_α the endomorphism of L defined by $A_\alpha(\xi) = \alpha\xi$ for $\xi \in L$. The characteristic polynomial of A_α will be denoted by f_α. If $f_\alpha(t) = t^n + a_1 t^{n-1} + \cdots + a_n$, then $\mathrm{tr}_{L/K}(\alpha) = \mathrm{tr}(\alpha) := -a_1$ is called the *trace* and $N_{L/K}(\alpha) = N(\alpha) := (-1)^n a_n$ is called the *norm* of α. Furthermore $D(\alpha) := f_\alpha'(\alpha)$ is called the *different* of α. $D(\alpha) \neq 0$ if and only if $L = K(\alpha)$ and L/K is separable.

Let $\alpha_1, \ldots, \alpha_n \in L$, then $d(\alpha_1, \ldots, \alpha_n) := \det(\mathrm{tr}(\alpha_i\alpha_j))_{ij})$ is called the *discriminant* of $\alpha_1, \ldots, \alpha_n$. In particular $d(\alpha) := N(D(\alpha)) = (-1)^{n(n-1)/2} d(1, \alpha, \ldots, \alpha^{n-1})$ is called the *discriminant* of α.

If $\alpha, \beta \in L$, then $\mathrm{tr}(\alpha + \beta) = \mathrm{tr}(\alpha) + \mathrm{tr}(\beta)$, $N(\alpha\beta) = N(\alpha)N(\beta)$.

$d(\alpha_1, \ldots, \alpha_n) \neq 0$ if and only if L/K is separable and $\alpha_1, \ldots, \alpha_n$ is a basis of L/K, i.e. a basis of the K-vector space L.

Let L/K be separable and $\omega_1, \ldots, \omega_n$ a basis of L/K. Then there exists a uniquely determined basis $\kappa_1, \ldots, \kappa_n$ of L/K such that $\mathrm{tr}(\omega_i \kappa_j) = \delta_{ij}$ for $i, j \in \{1, \ldots, n\}$. In fact, the system of linear equations

$$\mathrm{tr}(\omega_i(\xi_{j1}\omega_1 + \cdots + \xi_{jn}\omega_n) = \delta_{ij}, \qquad i = 1, \ldots, n,$$

for $\xi_{j1}, \ldots, \xi_{jn} \in K$ has determinant $d(\omega_1, \ldots, \omega_n)$.

$\kappa_1, \ldots, \kappa_n$ is called the *complementary basis* of $\omega_1, \ldots, \omega_n$.

1.2. Galois Theory (van der Waerden (1971), Chap. 8). Let K be a field and L a finite extension of L. We denote by $G(L/K)$ the group of automorphisms of L which fix the elements of K. Then $|G(L/K)| \leqslant [L:K]$ and L/K is called *separable and normal* (or *galois*) if $|G(L/K)| = [L:K]$. The group $G(L/K)$ is called the *Galois group* of L/K.

Theorem A1.1. *Let L/K be a finite, separable, normal extension with Galois group G.*

a) *There is a one to one correspondence ϕ between the intermediate fields F of L/K and the subgroups of G given by $\phi(F) = G(L/F)$. If H is a subgroup of G, then $\phi^{-1}(H)$ is the field $\{x \in L \mid gx = x \text{ for } g \in H\}$, called the fixed field of H.*

b) *L contains an element θ such that $\{g\theta, g \in G\}$ is a basis of the vector space L/K, called a normal basis of L/K.* ⊠

Let L/K be a finite separable extension. Then there exists a finite separable, normal extension N/K with $L \subset N$. There are exactly $n := [L:K]$ isomorphisms $g_1, \ldots, g_n$ of L in N. If $\alpha \in L$, then $f_\alpha(t) = \prod_{i=1}^n (t - g_i\alpha)$. Hence $\mathrm{tr}(\alpha) = \sum_{i=1}^n g_i\alpha$, $N(\alpha) = \prod_{i=1}^n g_i\alpha$. $g_1\alpha, \ldots, g_n\alpha$ are called the *conjugates* of α (in N).

In particular let K be an algebraic number field, i.e. K is a finite extension of $\mathbb{Q}$ contained in $\mathbb{C}$. Then there are $n = [K:\mathbb{Q}]$ isomorphisms of K in $\mathbb{C}$. Such an isomorphism g is called *real* if $g(K) \subset \mathbb{R}$ and *complex* if $g(K) \not\subset \mathbb{R}$. In the latter case the complex conjugate $\bar{g}$ of g is defined by $\bar{g}(\alpha) = \overline{g(\alpha)}$, where the bar means complex conjugation in $\mathbb{C}$. Hence complex isomorphisms occur always pairwise. K is called *totally real* (*complex*) if all isomorphisms g are real (complex).

1.3. Domains (van der Waerden (1967), Chap. 17). A domain R is a commutative ring with unit element and without zero divisors. Such a ring is contained in a field. The smallest field containing R is uniquely determined up to isomorphism and is called the quotient field of R. It will be denoted by $Q(R)$.

Let L be a field containing R. Then $\alpha \in L$ is called integral over R if there is a polynomial $f(x) = x^m + a_1x^{m-1} + \cdots + a_m \in R[x]$ such that $f(\alpha) = 0$. α is integral with respect to R if and only if the minimal polynomial of α with respect to $Q(R)$ has coefficients in R.

The elements of L which are integral over R form a ring S called the *integral closure* of R in L. The ring R is called *integrally closed* if the integral closure of R in $Q(R)$ is equal to R. The integral closure of R in L is integrally closed.

Proposition A1.2. *Let z be a variable over L. Then $S[z]$ is the integral closure of $R[z]$ in $L(z)$.* □

We put $K := Q(R)$. Let $\alpha \in S$. Then the characteristic polynomial $f_\alpha(z)$ with respect to K has coefficients in R since $f_\alpha(z)$ is a power of the minimal polynomial. In particular $\mathrm{tr}_{L/K}\alpha$, $N_{L/K}\alpha$, $d_{L/K}\alpha \in R$, $D_{L/K}\alpha \in S$.

Let $f(z)$, $g(z) \in L[z]$ be polynomials with highest coefficients 1 and suppose that $g(z)$ is a divisor of $f(z)$ in $L[z]$. Then $f(z) \in S[z]$ implies $g(z) \in S[z]$. In particular $\alpha \in S$ is a divisor of $N_{L/K}\alpha$ in S since $z - \alpha$ divides $f_\alpha(z)$.

1.4. Complexes (Main reference: Cartan, Eilenberg (1956)). A *complex* $\mathfrak{K} = (E, d)$ is a family $\{E^i | i \in \mathbb{Z}\}$ of abelian groups together with homomorphisms d^i of E^i into E^{i+1} such that $d^{i+1}d^i = 0$. A morphism φ of the complexes $\mathfrak{K} = (E, d)$, $\mathfrak{K}' = (E', d')$ is a family $\{\varphi_i | i \in \mathbb{Z}\}$ of homomorphisms of E^i into E'^i such that $d'^i\varphi^i = \varphi^{i+1}d^i$. We transfer the notions of kernel, image, and exact sequence in obvious manner from the category of abelian groups to the category of complexes.

The *i-th cohomology group* $H^i(\mathfrak{K})$ of the complex $\mathfrak{K} = (E, d)$ is defined as

$$H^i(\mathfrak{K}) = \mathrm{Ker}\, d^i/\mathrm{Im}\, d^{i-1}.$$

A morphism $\mathfrak{K} \to \mathfrak{K}'$ induces a homomorphism $H^i(\mathfrak{K}) \to H^i(\mathfrak{K}')$. Furthermore an exact sequence

$$0 \to \mathfrak{K}_1 \underset{\varphi_n}{\to} \mathfrak{K}_2 \underset{\varphi_2}{\to} \mathfrak{K}_3 \to 0$$

induces the long exact sequence

$$\cdots \to H^i(\mathfrak{K}_1) \underset{\varphi_1^*}{\to} H^i(\mathfrak{K}_2) \underset{\varphi_2^*}{\to} H^i(\mathfrak{K}_3) \underset{\Delta^i}{\to} H^{i+1}(\mathfrak{K}_1) \to \cdots$$

where the *connecting homomorphism* Δ^i is defined as follows.

Let $\bar{x} \in H^i(\mathfrak{K}_3)$, i.e. $x \in E_3^i$, $d_3^i x = 0$. We take a $y \in E_2^i$ such that $\varphi_2^i y = x$. Then $\varphi_2^{i+1}(d_2^i y) = d_3^i \varphi_2^i y = 0$, hence there is a $z \in E_1^{i+1}$ with $\varphi_1^{i+1} z = d_2^i y$. since

$$\varphi_1^{i+2} d_1^{i+1} z = d_2^{i+1} \varphi_1^{i+1} z = d_2^{i+1} d_2^i y = 0,$$

we have $d_1^{i+1} z = 0$. The class $\bar{z} \in H^{i+1}(\mathfrak{K}_1)$ is independent of the choice of y, z. We put $\Delta^i \bar{x} = \bar{z}$. The homomorphism Δ^i is functorial in the sense that a commutative diagram

$$\begin{array}{ccccccccc} 0 & \to & \mathfrak{K}_1 & \to & \mathfrak{K}_2 & \to & \mathfrak{K}_3 & \to & 0 \\ & & \downarrow & & \downarrow & & \downarrow & & \\ 0 & \to & \mathfrak{K}_1' & \to & \mathfrak{K}_2' & \to & \mathfrak{K}_3' & \to & 0 \end{array}$$

with exact rows induces connecting homomorphisms Δ^i, Δ'^i such that

$$\begin{array}{ccc} H^i(\mathfrak{K}_3) & \underset{\Delta^i}{\to} & H^{i+1}(\mathfrak{K}_1) \\ \downarrow & & \downarrow \\ H^i(\mathfrak{K}_3') & \underset{\Delta'^i}{\to} & H^{i+1}(\mathfrak{K}_1') \end{array}$$

is commutative.

If the family (E, d) is given only for indices $i = 0, 1, \ldots$, we put $E_i = 0$, $d_i = 0$ for $i = -1, -2, \ldots$, and get a complex in the above sense.

Appendix 2 Quadratic Residues

(Main reference: Vinogradov (1965))

Let p be an odd prime. An integer $a \not\equiv 0 \pmod p$ is called quadratic residue mod p if there is an integer x with $a \equiv x^2 \pmod p$.

By definition the *Legendre symbol* $\left(\frac{a}{p}\right)$ is equal to 1 if a is a quadratic residue mod p and it is equal to -1 if a is not a quadratic residue mod p. Obviously $\left(\frac{a}{p}\right)$ depends only on the residue class of a mod p and it defines a homomorphism of $(\mathbb{Z}/p\mathbb{Z})^\times$ onto the group $\mathbb{Z}^\times = \{\pm 1\}$.

Proposition A2.1 (Euler's criterion).

$$\left(\frac{a}{p}\right) \equiv a^{(p-1)/2} \pmod p.$$

Proof. Let g be a primitive root mod p, i.e. g is a generator of the cyclic group $(\mathbb{Z}/p\mathbb{Z})^\times$. Then $g^\nu \equiv 1 \pmod p$ if and only if $\nu \equiv 0 \pmod{p-1}$, $g^\nu \equiv -1 \pmod p$ if and only if $\nu \equiv (p-1)/2 \pmod{p-1}$. □

The first interesting question about quadratic residues is the following: If a is fixed, what can be said about the dependence of $\left(\frac{a}{p}\right)$ on p. The answer is given by the *quadratic reciprocity law*:

Theorem A2.2 (Gauss). *Let p, q be distinct odd primes.*

a)
$$\left(\frac{p}{q}\right)\left(\frac{q}{p}\right) = (-1)^{(p-1)(q-1)/4},$$

b)
$$\left(\frac{-1}{p}\right) = (-1)^{(p-1)/2},$$

c)
$$\left(\frac{2}{p}\right) = (-1)^{(p^2-1)/8}.$$

We sketch a proof of this theorem in order to show the use of algebraic numbers in elementary number theory: b) follows immediately from Proposition A2.1. For the proof of a) we put

$$\tau_p := \sum_{i=1}^{p-1} \left(\frac{i}{p}\right) \zeta^i,$$

where ζ is a primitive p-th root of unity (Chap. 1.3.9). Then

$$\tau_p^2 = \sum_{i,j=1}^{p-1} \left(\frac{ij}{p}\right) \zeta^{i+j} = \sum_{k=1}^{p-1} \left(\frac{k}{p}\right) \sum_{i=1}^{p-1} \zeta^{i(1+k)} = \left(\frac{-1}{p}\right) p = (-1)^{(p-1)/2} p.$$

Hence

$$\tau_p^q = (-1)^{(p-1)(q-1)/4} p^{(q-1)/2} \tau_p \equiv (-1)^{(p-1)(q-1)/2} \left(\frac{p}{q}\right) \tau_p (\text{mod } q).$$

On the other hand

$$\tau_p^q \equiv \sum_{i=1}^{p-1} \left(\frac{i}{p}\right) \zeta^{iq} \equiv \left(\frac{q}{p}\right) \tau_p (\text{mod } q).$$

Therefore

$$(-1)^{(p-1)(q-1)/4} \left(\frac{p}{q}\right) \equiv \left(\frac{q}{p}\right) (\text{mod } q).$$

This congruence implies equality since both sides are equal to ± 1.

For the proof of c) we put $\tau_2 := \zeta + \zeta^{-1}$ with a primitive 8-th root of unity ζ. Then $\tau_2^2 = 2$ and

$$\tau_2^p = 2^{(p-1)/2} \tau_2 \equiv \left(\frac{2}{p}\right) \tau_2 \equiv \zeta^p + \zeta^{-p} = (-1)^{(p^2-1)/8} \tau_2 \ (\text{mod } p). \qquad \square$$

Let $a \neq 0$ be an integer, let $b > 0$ an integer with $(2a, b) = 1$, and let $b = p_1 \dots p_s$ be the decomposition of b into the product of primes. The *Jacobi symbol* $\left(\frac{a}{b}\right)$ is defined by $\left(\frac{a}{b}\right) := \prod_{i=1}^{s} \left(\frac{a}{p_i}\right)$.

The symbol $\left(\frac{a}{b}\right)$ defines a homomorphism of $(\mathbb{Z}/b\mathbb{Z})^\times$ onto $\{\pm 1\}$ and Theorem A2.2 is valid if p and q are relatively prime odd integers.

It follows that if $a \equiv 1 \ (\text{mod } 4)$, then $\left(\frac{a}{b}\right)$ only depends on the class of b mod a, and in general $\left(\frac{a}{b}\right)$ only depends on the class of b mod $4a$.

Appendix 3
Locally Compact Groups

3.1. Locally Compact Abelian Groups (Main reference: Bourbaki (1967), Chap. 2). Let G be a locally compact abelian group. A character χ of G is a continuous homomorphism of G into $\mathbb{C}^\times$ with $|\chi(g)| = 1$ for $g \in G$. We define a group structure in the set $X(G)$ of all characters of G by

$$\chi_1\chi_2(g) := \chi_1(g)\chi_2(g) \quad \text{for } \chi_1, \chi_2 \in X(G).$$

Furthermore we define a topology in $X(G)$ by means of the following family $\{U(\varepsilon, K)\}$ of neighborhoods of the zero element of G:

$$U(\varepsilon, K) := \{\chi \in X(G) \,|\, |\chi(g) - 1| < \varepsilon \text{ for } g \in K\},$$

where ε is a positive number and K is a compact subset of G. With this topology $X(G)$ becomes a locally compact group. $X(G)$ is called the *character group* or *dual group* of G.

Theorem A3.1 (Pontrjagin's duality theorem). a) *Let H be a closed subgroup of G. We identify $X(H)$ with the factor group $X(G)/\{\chi/\chi(H) = \{1\}\}$. Then $H \to X(H)$ is a one to one correspondence between the subgroups of G and the factor groups of $X(G)$.*

b) *Let φ be the homomorphism of G into $X(X(G))$ with*

$$\varphi(g)(\chi) = \chi(g) \text{ for } \chi \in X(G).$$

Then φ is a topological isomorphism of G onto $X(X(G))$. ⊠

In the following we identify G with $X(X(G))$.

Example A3.1. The dual group of $\mathbb{R}^+$ is $\mathbb{R}^+$, the dual group of $\mathbb{R}^+/\mathbb{Z}^+$ is $\mathbb{Z}^+$. □

Example A3.2. The dual group of a compact group is a discrete group. □

Let μ be a Haar measure on G. As usual, $L^m(G)$ denotes the space of complex-valued measurable functions f on G with $\int_G |f(g)|^m \, d\mu(g) < \infty$.

The *Fourier transform* $\hat{f}$ of a function $f \in L^1(G)$ is the continuous function on $X(G)$ with

$$\hat{f}(\chi) := \int_G f(g)\chi(-g)\, d\mu(g).$$

Theorem A3.2 (inversion theorem). *Let G be a locally compact abelian group and μ a Haar measure of G. Then there exists a unique Haar measure $\hat{\mu}$ on $X(G)$ such that for any function $f \in L^1(G)$ with $\hat{f} \in L^1(X(G))$ the inversion formula*

$$f(g) = \hat{\hat{f}}(-g)$$

holds. □

Theorem A3.3 (Poisson's summation formula). *Let H be a closed subgroup of G, let μ be a Haar measure on G, let ν be a Haar measure on H, let γ be the induced Haar measure on G/H, let $\hat{\gamma}$ be the Haar measure on $X(G/H)$ which corresponds to γ in the sense of Theorem* A3.2.

Furthermore let f be a continuous function in $L^1(G)$ such that

a) *for all $g \in G$ the function $h \to f(g + h)$ is integrable on H,*

b) *$\int_H f(g + h)\, d\nu(h)$ is a continuous function on G,*

c) *$\hat{f}(y)$ is integrable on $X(G/H)$.*

Then

$$\int_H f(h)\,dv(h) = \int_{X(G/H)} \hat{f}(y)\,d\hat{\gamma}(y). \boxtimes$$

3.2. Restricted Products. Let $\{G_v | v \in V\}$ be a family of locally compact groups and let S be a finite subset of V such that for $v \notin S$ there exists an open compact subgroup H_v of G_v, which will be fixed in the following.

The *restricted product* of $\{G_v | v \in V\}$ with respect to $\{H_v | v \notin S\}$ is the subgroup G of the direct product $\prod_{v \in V} G_v$ consisting of all elements $\prod_v g_v$ such that $g_v \in H_v$ for almost all v. We introduce in G the topology induced from the product topology of the subgroup $G_S := \prod_{v \in S} G_v \times \prod_{v \notin S} H_v$. Then G is a locally compact group.

We choose Haar measures μ_v for $v \in V$ such that $\mu_v(H_v) = 1$ for $v \in S$. Then the product measure on G_S is well defined and induces a measure on G, which will be called the product measure.

Appendix 4
Bernoulli Numbers

(Main references: Borevich, Shafarevich (1985) Chap. 5, § 8, Washington (1982), Chap. 4, 5)

The *Bernoulli numbers* $B_0, B_1, \ldots$ are defined by the generating series

$$\sum_{n=0}^{\infty} B_n \frac{t^n}{n!} := \frac{t}{e^t - 1}.$$

Since $B_1 = -\frac{1}{2}$ and since $\frac{t}{e^t - 1} + \frac{t}{2}$ is an even function one has $B_n = 0$ for odd $n > 1$.

Let χ be a primitive Dirichlet character with conductor f (Chap. 4.1). The *generalized Bernoulli numbers* $B_{n,\chi}$ are defined by

$$\sum_{n=0}^{\infty} B_{n,\chi} \frac{t^n}{n!} := \sum_{i=1}^{f} \frac{\chi(i) t e^{it}}{e^{ft} - 1}$$

If $\chi = 1$, we have

$$\sum_{n=0}^{\infty} B_{n,1} \frac{t^n}{n!} = \frac{te^t}{e^t - 1} = \frac{t}{e^t - 1} + t,$$

hence $B_{n,1} = B_n$ except for $n = 1$, where $B_{1,1} = \frac{1}{2}$. If $\chi \neq 1$, then $B_{0,\chi} = 0$. Furthermore in this case the generating function for $B_{n,\chi}$ is even if χ is even and odd if χ is odd. Hence $B_{2n+1,\chi} = 0$ if χ is even, $B_{2n,\chi} = 0$ if χ is odd.

We introduce the *Bernoulli polynomials* $B_n(x)$ by means of

$$\sum_{n=0}^{\infty} B_n(x)\frac{t^n}{n!} := \frac{te^{xt}}{e^t - 1}.$$

Since

$$\sum_{n=0}^{\infty} B_n(x)\frac{t^n}{n!} = \left(\sum_{i=0}^{\infty} B_i\frac{t^i}{i!}\right)\left(\sum_{j=0}^{\infty} x^j\frac{t^j}{j!}\right),$$

one has

$$B_n(x) = \sum_{i=0}^{n} B_i\binom{n}{i}x^{n-i}.$$

On the other hand, the generalized Bernoulli numbers are easily expressed by the Bernoulli polynomials:

Proposition A4.1. *Let F be a multiple of the conductor f. Then*

$$B_{n,\chi} = F^{n-1}\sum_{i=1}^{F} \chi(i)B_n\left(\frac{i}{F}\right). \boxtimes$$

In particular

$$B_{1,\chi} = \frac{1}{f}\sum_{i=1}^{f} \chi(i)i \quad \text{for} \quad \chi \neq 1.$$

The Bernoulli numbers B_n are rational numbers whose denominator is given by the following *theorem of von Staudt-Clausen*:

Theorem A4.2. *Let $n > 0$ be even, then*

$$B_n + \sum_{(p-1)|n} \frac{1}{p} \in \mathbb{Z},$$

where the sum runs over all primes p such that $p - 1$ divides n. ⊠

Tables

There is a growing amount of computations of data (e.g. class numbers, class groups, regulators) of number fields in the literature, mostly in the journal "Mathematics of computation". We give here only very few examples of tables resulting from such computations. Our first two tables can be established by hand or with a personal computer using the computational method explained in Chap. 1.1.4, 1.1.7.

Table 1. Class number h of imaginary-quadratic fields $\mathbb{Q}(\sqrt{-a})$ with a square-free, $1 \leqslant a < 500$

a	h	a	h	a	h	a	h	a	h	a	h	a	h
1	1	71	7	143	10	215	14	287	14	365	20	434	24
2	1	73	4	145	8	217	8	290	20	366	12	435	4
3	1	74	10	146	16	218	10	291	4	367	9	437	20
5	2	77	8	149	14	219	4	293	18	370	12	438	8
6	2	78	4	151	7	221	16	295	8	371	8	439	15
7	1	79	5	154	8	222	12	298	6	373	10	442	8
10	2	82	4	155	4	223	7	299	8	374	28	443	5
11	1	83	3	157	6	226	8	301	8	377	16	445	8
13	2	85	4	158	8	227	5	302	12	379	3	446	32
14	4	86	10	159	10	229	10	303	10	381	20	447	14
15	2	87	6	161	16	230	20	305	16	382	8	449	20
17	4	89	12	163	1	231	12	307	3	383	17	451	6
19	1	91	2	165	8	233	12	309	12	385	8	453	12
21	4	93	4	166	10	235	2	310	8	386	20	454	14
22	2	94	8	167	11	237	12	311	19	389	22	455	20
23	3	95	8	170	12	238	8	313	8	390	16	457	8
26	6	97	4	173	14	239	15	314	26	391	14	458	26
29	6	101	14	174	12	241	12	317	10	393	12	461	30
30	4	102	4	177	4	246	12	318	12	394	10	462	8
31	3	103	5	178	8	247	6	319	10	395	8	463	7
33	4	105	8	179	5	249	12	321	20	397	6	465	16
34	4	106	6	181	10	251	7	322	8	398	20	466	8
35	2	107	3	182	12	253	4	323	4	399	16	467	7
37	2	109	6	183	8	254	16	326	22	401	20	469	16
38	6	110	12	185	16	255	12	327	12	402	16	470	20
39	4	111	8	186	12	257	16	329	24	403	2	471	16
41	8	113	8	187	2	258	8	330	8	406	16	473	12
42	4	114	8	190	4	259	4	331	3	407	16	474	20
43	1	115	2	191	13	262	6	334	12	409	16	478	8
46	4	118	6	193	4	263	13	335	18	410	16	479	25
47	5	119	10	194	20	265	8	337	8	411	6	481	16
51	2	122	10	195	4	266	20	339	6	413	20	482	20
53	6	123	2	197	10	267	2	341	28	415	10	483	4
55	4	127	5	199	9	269	22	345	8	417	12	485	20
57	4	129	12	201	12	271	11	346	10	418	8	487	7
58	2	130	4	202	6	273	8	347	5	419	9	489	20
59	3	131	5	203	4	274	12	349	14	421	10	491	9
61	6	133	4	205	8	277	6	353	16	422	10	493	12
62	8	134	14	206	20	278	14	354	16	426	24	494	28
65	8	137	8	209	20	281	20	355	4	427	2	497	24
66	8	138	8	210	8	282	8	357	8	429	16	498	8
67	1	139	3	211	3	283	3	358	6	430	12	499	3
69	8	141	8	213	8	285	16	359	19	431	21		
70	4	142	4	214	6	286	12	362	18	433	12		

Table 2. Class number h and fundamental unit $\varepsilon > 1$ of real-quadratic fields $\mathbb{Q}(\sqrt{a})$ with a square-free, $2 \leqslant a \leqslant 101$, $\omega = (1 + \sqrt{a})/2$ if $a \equiv 1 \pmod 4$, $\omega = \sqrt{a}$ if $a \equiv 2, 3 \pmod 4$

a	h	ε	$N(e)$	a	h	ε	$N(e)$
2	1	$1+\omega$	-1	53	1	$3+\omega$	-1
3	1	$2+\omega$	$+1$	55	2	$89+12\omega$	$+1$
5	1	ω	-1	57	1	$131+40\omega$	$+1$
6	1	$5+2\omega$	$+1$	58	2	$99+13\omega$	-1
7	1	$8+3\omega$	$+1$	59	1	$530+69\omega$	$+1$
10	2	$3+\omega$	-1	61	1	$17+5\omega$	-1
11	1	$10+3\omega$	$+1$	62	1	$63+8\omega$	$+1$
13	1	$1+\omega$	-1	65	2	$7+2\omega$	-1
14	1	$15+4\omega$	$+1$	66	2	$65+8\omega$	$+1$
15	2	$4+\omega$	$+1$	67	1	$48842+5967\omega$	$+1$
17	1	$3+2\omega$	-1	69	1	$11+3\omega$	$+1$
19	1	$170+39\omega$	$+1$	70	2	$251+30\omega$	$+1$
21	1	$2+\omega$	$+1$	71	1	$3480+413\omega$	$+1$
22	1	$197+42\omega$	$+1$	73	1	$943+250\omega$	-1
23	1	$24+5\omega$	$+1$	74	2	$43+5\omega$	-1
26	2	$5+\omega$	-1	77	1	$4+\omega$	$+1$
29	1	$2+\omega$	-1	78	2	$53+6\omega$	$+1$
30	2	$11+2\omega$	$+1$	79	3	$80+9\omega$	$+1$
31	1	$1520+273\omega$	$+1$	82	4	$9+\omega$	-1
33	1	$19+8\omega$	$+1$	83	1	$82+9\omega$	$+1$
34	2	$35+6\omega$	$+1$	85	2	$4+\omega$	-1
35	2	$6+\omega$	$+1$	86	1	$20405+1122\omega$	$+1$
37	1	$5+2\omega$	-1	87	2	$28+3\omega$	$+1$
38	1	$37+6\omega$	$+1$	89	1	$447+106\omega$	-1
39	2	$25+4\omega$	$+1$	91	2	$1574+165\omega$	$+1$
41	1	$27+10\omega$	-1	93	1	$13+3\omega$	$+1$
42	2	$13+2\omega$	$+1$	94	1	$2143295+221064\omega$	$+1$
43	1	$3482+531\omega$	$+1$	95	2	$39+4\omega$	$+1$
46	1	$24335+3588\omega$	$+1$	97	1	$5035+1138\omega$	-1
47	1	$48+7\omega$	$+1$	101	1	$9+2\omega$	-1
51	2	$50+7\omega$	$+1$				

Table 3 (Williams, Broere (1976)). Class number h of real-quadratic fields $\mathbb{Q}(\sqrt{a})$ with a square-free, $2 \leqslant a \leqslant 150000$. The second column of the table shows the number $f(h)$ of fields with class number h, the third column shows the smallest a for which $\mathbb{Q}(\sqrt{a})$ has class number h

h	$f(h)$	d	h	$f(h)$	d	h	$f(h)$	d
1	20574	2	28	324	5626	55	1	106537
2	26427	10	29	16	49281	56	38	39999
3	2677	79	30	113	11665	57	2	41617
4	18573	82	31	4	97753	58	7	27226
5	943	401	32	397	15130	60	18	78745
6	3453	235	33	11	55339	61	1	126499
7	462	577	34	47	19882	62	3	68179
8	6898	226	35	8	25601	63	1	57601
9	311	1129	36	165	18226	64	23	71290
10	1237	1111	37	7	24337	66	3	87271
11	176	1297	38	33	19834	68	12	53362
12	2434	730	39	6	41614	70	5	56011
13	124	4759	40	179	16899	72	11	45511
14	563	1534	41	1	55966	74	1	38026
15	115	9871	42	30	47959	76	7	93619
16	1970	2305	43	3	14401	78	1	136159
17	62	7054	44	82	11026	80	3	94546
18	385	4954	45	7	32401	84	3	77779
19	48	15409	46	14	49321	86	2	110926
20	788	3601	47	1	78401	87	2	90001
21	43	7057	48	92	21610	88	3	56170
22	163	4762	49	1	70969	94	2	99226
23	20	23593	50	8	54769	96	4	50626
24	838	9634	51	1	69697	100	2	131770
25	30	24859	52	28	23410	108	1	140626
26	110	13321	53	1	69694	110	1	125434
27	20	8761	54	8	49834	116	1	116554

Table 4 (Ennola-Turunen (1985)). Totally real cubic fields with discriminant $d < 500000$. The table shows the number of fields with given class number and discriminant in the bounds given at the top of each column

Class number	1 50000	50001 100000	100001 150000	150001 200000	200001 250000
1	2023	2169	2204	2204	2258
2	109	181	193	199	230
3	112	155	143	162	163
4	11	10	17	20	29
5	6	12	12	11	11
6	1	6	11	12	17
7	1	6	—	3	3
8	—	1	4	3	2
9	—	1	4	4	2
10	—	—	2	—	—
11	—	—	—	2	—
12	—	—	—	—	—
13	—	—	1	—	—
Total no of fields	2263	2541	2591	2620	2715

Class number	250001 300000	300001 350000	350001 400000	400001 450000	450001 500000
1	2261	2244	2278	2309	2270
2	216	232	238	229	232
3	156	199	155	154	193
4	26	24	26	29	32
5	12	14	14	12	21
6	10	16	9	17	11
7	5	5	1	2	6
8	1	3	2	3	2
9	3	3	3	3	2
10	—	3	1	3	1
11	—	1	—	—	—
12	1	1	—	5	3
13	—	—	1	—	1
14	—	—	1	—	1
15	—	1	—	—	—
16	—	—	—	—	1
Total no of fields	2691	2746	2731	2766	2776

Table 5 (Ennola-Turunen (1985)). Discriminants <500000 of totally real cubic fields with N associated non conjugate fields for $N > 1$

$N = 2$

3969	8281	13689	17689	29241	37300	38612	45684
46548	47089	55700	61009	66825	67081	69012	77841
83700	90601	92340	110889	113940	115668	138996	148372
149769	155412	157300	162324	162409	164052	168372	173556
181300	182329	182868	185652	186516	189972	191700	208980
213300	215700	215892	219961	223668	231361	235224	238140
248724	255636	257556	259700	261121	262964	263277	275700
278964	284148	296325	299700	301401	302292	305809	312481
323028	327668	331425	334260	340200	346921	348948	359700
363609	367956	370548	372276	374868	379700	391284	393012
393492	395604	399924	419796	428436	431325	431649	435348
441396	442260	452925	456948	457652	458325	460377	460404
461041	465588	470988	473300	489300	494209		

$N = 3$

22356	28212	31425	41332	47860	54324	57588	58077
62004	62644	63028	65908	77844	82484	86485	86828
89073	95992	97844	98132	99860	101876	105192	108729
109396	119604	122300	123860	129164	136628	138388	144212
144532	146452	150164	152212	153981	156244	161844	177741
180549	189777	198045	202932	205748	210708	214925	215796
217012	223540	223604	224084	225716	226580	235953	236277
239124	239476	240692	263196	270292	270405	275604	279284
293876	295284	302612	303220	304925	305268	313492	313620
314577	314772	317300	321364	323956	324308	325620	325809
326381	326516	327537	335732	339348	344568	344884	345716
350612	354772	358425	360948	378228	380884	383668	384404
392468	394292	397300	405965	407528	408244	410913	414708
418324	419688	424148	425493	428212	430228	438484	439124
444756	444852	448092	448929	452084	456425	456980	458260
458336	459892	462537	463988	464212	469233	469773	470569
471325	476820	477981	478521	486708	492212	492700	493925
498428							

$N = 4$

32009	42817	62501	72329	94636	103809	114889	130397
142097	151141	152949	153949	172252	173944	184137	189237
206776	209765	213913	214028	214712	219461	220217	250748
252977	255973	259653	265245	275881	282461	283673	298849
320785	321053	326945	333656	335229	341724	342664	358285
363397	371965	384369	390876	400369	412277	415432	422573
424236	431761	449797	459964	460817	468472	471057	471713
476124	476152	486221	486581	494236			

References*

Abrashkin, V.A. (1974): The determination of imaginary-quadratic fields with class number two and even discriminant by Heegner's method. Mat. Zametki *15*, No. 2, 241–246. English transl.: Math. Notes *15*, 137–139 (1974). Zbl.295.12002

Adleman, L.M., Heath-Brown, D.R. (1985): The first case of Fermat's last theorem. Invent. Math. *79*, No. 2, 409–416. Zbl.557.10034

Albert, A.A. (1932): A construction of non-cyclic normal division algebras. Bull. Am. Math. Soc. *38*, 449–456. Zbl.5,6

Amitsur, S.A., Rowen, L.H., Tignol, J.P. (1979): Division algebras of degree 4 and 8 with involution. Isr. J. Math. *33*, No. 2, 133–148. Zbl.422.16010

Andozhskij, I.V. (1975): On some classes of closed pro-p-groups. Izv. Akad. Nauk SSSR, Ser. Mat. *39*, No. 4, 707–738. Zbl.318.12007. English transl.: Math. USSR, Izv. *9*, 663–691 (1976)

Artin, E. (1923): Über eine neue Art von L-Reihen. Abh. Math. Semin. Univ. Hamb. *3*, 89–108. Jbuch 49,123

Artin, E. (1927): Beweis des allgemeinen Reziprozitätsgesetzes. Abh. Math. Semin. Univ. Hamb. *5*, 353–363. Jbuch 53,144

Artin, E. (1951): Algebraic Numbers and Algebraic Functions I. Inst. for Math. and Mech., New York: New York University Press, 345pp. Zbl.54,21

Artin, E., Hasse, H. (1928): Die beiden Ergänzungssätze zum Reziprozitätsgesetz der l^n-ten Potenzreste im Körper der l^n-ten Einheitswurzeln. Abh. Math. Semin. Univ. Hamb. *6*, 146–162. Jbuch.54,191

Artin, E., Tate, J. (1968): Class Field Theory. New York e.a.: Benjamin, 259pp. Zbl.176,335

Baker, A. (1966): Linear forms in the logarithms of algebraic numbers. Mathematika *13*, 204–216. Zbl.161,52

Baker, A. (1971): Imaginary quadratic fields with class number 2. Ann. Math., II. Ser. *94*, No. 1, 139–152. Zbl.219.12008

Barsky, D. (1977/1978): Fonctions zêta p-adiques d'une classe de rayon des corps de nombres totalement réels. Groupe Etude Anal. Ultramétrique, 5e année. Exp. 16, 23 pp. Zbl.406.12008

Bashmakov, M.J. (1968): On the problem of field embeddings. Mat. Zametki 4, No. 2, 137–140. Zbl.159,337. English transl.: Math. Notes *4*, 572–574

Belyj, G.V. (1979): On Galois extensions of a maximal cyclotomic field. Izv. Akad. Nauk SSSR, Ser. Mat. *43*, No. 2, 267–276. Zbl.409.12012. English transl.: Math. USSR, Izv. *14*, 247–256 (1980)

Belyj, G.V. (1983): On extensions of the maximal cyclotomic field having a given classical Galois group. J. Reine Angew. Math. *341*, 147–156. Zbl.515.12008

Beyer, G. (1956): Über eine Vermutung von Hasse zum Erweiterungsproblem galoisscher Zahlkörper. J. Reine Angew. Math. *196*, 205–212. Zbl.70,268

Borel, A., Casselman, W. (eds.) (1979): Automorphic forms, representations, and L-functions. Proc. Symp. Pure Math. *33*, I, II. Zbl.403.00002, Zbl.403.00003

Borel, A., Jacquet, H. (1979): Automorphic forms and automorphic representations. In: Borel, Casselman, I (1979) 189–202. Zbl.414.22020

Borevich, Z.I. (1967): On the groups of principal units of p-extensions of a local field. Dokl. Akad. Nauk SSSR *173*, No. 2, 253–255 (Russian). Zbl.155,97

Borevich, Z.I., Shafarevich, I.R. (1985): Number Theory. 3rd ed. Moscow: Nauka 503 pp. (Russian). Zbl.592.12001

Bourbaki, N. (1947): Algèbre. Paris: Hermann. Zbl.30,290

Bourbaki, N. (1965): Algèbre commutative. Chap. 7, Diviseurs. Paris: Hermann. Zbl.141,35

*For the convenience of the reader, references to reviews in Zentralblatt für Mathematik (Zbl.), compiled using the MATH database, and Jahrbuch über die Fortschritte der Mathematik (Jbuch.) have, as far as possible, been included in this bibliography.

Bourbaki, N. (1967): Théories spectrales. Paris: Hermann. Zbl.152,326

Brückner, H. (1967): Eine explizite Formel zum Reziprozitätsgesetz für Primzahlexponenten p. Algebraische Zahlentheorie. Math. Forschungsinst. Oberwolfach, Tagungsbericht, 1964. Mannheim: Bibliogr. Inst., 31–39. Zbl.201,378

Brückner, H. (1979): Explizites Reziprozitätsgesetz und Anwendungen. Vorlesungen Fachbereich Math. Univ. Essen. Heft 2, 83 pp. Zbl.437.12001

Brückner, H. (1979a): Hilbertsymbole zum Exponenten p^n und Pfaffsche Formen. Preprint Hamburg

Brumer, A. (1965): Ramification and class towers of number fields. Mich. Math. J. *12*, 129–131. Zbl.136,27

Brumer, A. (1967): On the units of algebraic number fields. Mathematika *14*, 121–124. Zbl.171,11

Brumer, A., Rosen, M. (1963): Class number and ramification in number fields. Nagoya Math. J. *23*, 97–101. Zbl.128,262

Buell, D.A. (1984): The expectation of success using a Monte-Carlo factoring method. Some statistics on quadratic class numbers. Math. Comput. *43*, No. 167, 313–327. Zbl.551.10009

Buell, D.A. (1987): Class groups of quadratic fields II. Math. Comput. *48*, No. 177, 85–93. Zbl.606.12004

Buhler, J.P. (1978): Icosahedral Galois representations. Lect. Notes Math. *654*. Berlin Heidelberg New York: Springer-Verlag, 143 pp. Zbl.374.12002

Cartan, H., Eilenberg, S. (1956): Homological Algebra. Princeton, N.J.: Univ. Press. Zbl.75,243

Cassels, J.W.S., Fröhlich, A. (1967): Algebraic Number Theory. New York-London: Academic Press. Zbl.153,74

Cassou-Noguès, P. (1979): Valeurs aux entiers négatifs des fonctions zêta et fonctions zêta *p*-adiques. Invent. Math. *51*, 29–59. Zbl.408.12015

Chatland, H., Davenport, H. (1950): Euclid's algorithm in real quadratic fields. Canad. J. Math. *2*, 289–296. Zbl.37,308

Chevalley, C. (1940): La théorie du corps de classes. Ann. Math., II. Ser. *41*, 394–418. Zbl.25,18

Chevalley, C. (1954): Class Field Theory. Nagoya: Nagoya University, 104 pp. Zbl.59,33

Cohen, H., Lenstra Jr., H.W. (1984): Heuristics on Class Groups of Number Fields. Number Theory. Noordwijkerhout 1983. In: Lect. Notes Math. *1068*. Berlin Heidelberg New York: Springer-Verlag, 33–62. Zbl.558.12002

Cohen, H., Martinet, J. (1987): Class groups of number fields: numerical heuristics. Math. Comput. *48*, No. 177, 123–137. Zbl.627.12006

Cohen, H., Martinet, J. (1990): Etude heuristique des groupes de classes des corps de nombres. J. Reine Angew. Math. *404*, 39–76. Zbl.699.12016

Curtis, C.W., Reiner, I. (1962): Representation Theory of Finite Groups and Associative Algebras. New York e.a.: Interscience Publishers, John Wiley and Sons, 685 pp. Zbl.131,256

Dade, E.C., Taussky, O., Zassenhaus, H. (1962): On the theory of orders, in particular on the semigroup of ideal classes and genera of an order in an algebraic number field. Math. Ann. *148*, 31–64. Zbl.113,265

Dedekind, R. (1894): Über die Theorie der ganzen algebraischen Zahlen. Supplement XI to Dirichlets Vorlesungen über Zahlentheorie. 4th ed., Braunschweig: Vieweg. Jbuch25,252

Delange, H. (1954): Généralisation du théorème de Ikehara. Ann. Sci. Ec. Norm. Supér., III. Sér. *71*, 213–242. Zbl.56,331

Deligne, P. (1969): Formes modulaires et représentations *l*-adiques. Sém. Bourbaki, 1968/69, No. 355, 139–172 (1971). Zbl.206,499

Deligne, P., Ribet, K. (1980): Values of abelian *L*-functions at negative integers over totally real fields. Invent. Math. *59*, 227–286. Zbl.43.12009

Demushkin, S.P., Shafarevich, I.R. (1962): The second obstruction for the embedding problem of algebraic number fields. Izv. Akad. Nauk SSSR, Ser. Mat. *26*, 911–924. Zbl.115,37. English transl.: Transl., II. Ser., Am. Math. Soc. 58, 245–260 (1966)

Deuring, M. (1958): Die Klassenkörper der komplexen Multiplikation. Stuttgart: Teubner Verlag. Zbl.123,40

Deuring, M. (1968): Algebren. 2nd ed., Berlin Heidelberg New York: Springer-Verlag. Zbl.159,42

Deuring, M. (1968a): Imaginäre quadratische Zahlkörper mit der Klassenzahl Eins. Invent. Math. *5*, 179–196. Zbl.155,380

Diaz y Diaz, F. (1978): Sur le 3-rang des corps quadratiques. Thèse 3. cycle. Publ. Math. Orsay. No. 7811, 93 pp. Zbl.405.12004

Diekert, V. (1984): Über die absolute Galoisgruppe dyadischer Zahlkörper. J. Reine Angew. Math. *350*, 152–172. Zbl.524.12007

Eichler, M. (1963): Einführung in die Theorie der algebraischen Zahlen und Funktionen. Basel: Birkhäuser. 338 pp. Zbl.152,195

Eichler, M. (1965): Eine Bemerkung zur Fermatschen Vermutung. Acta Arith. *11*, 129–131 (Errata p. 261). Zbl.135,94

Ennola, V., Turunen, R. (1985): On totally real cubic fields. Math. Comput. *44*, No. 170, 495–518. Zbl.564.12006

Faltings, G. (1983): Endlichkeitssätze für abelsche Varietäten über Zahlkörpern. Invent. Math. *73*, 349–366. Zbl.588.14026

Ferrero, B., Washington, L. (1979): The Iwasawa invariant μ_p vanishes for abelian number fields. Ann. Math., II. Ser. *109*, 377–395. Zbl.443.12001

Fouvry, E. (1985): Théorème de Brun-Titchmarsh, application au-Théorème de Fermat. Invent. Math. *79*, No. 2, 383–407. Zbl.557.10035

Fröhlich, A. (1954): On fields of class two. Proc. Lond. Math. Soc., III. Ser. *4*, 235–256. Zbl.55,33

Fröhlich, A. (1972): Artin root numbers and normal integral bases for quaternion fields. Invent. Math. *17*, 143–166. Zbl.261.12008

Fröhlich, A. (ed.) (1977): Algebraic Number Fields (*L*-functions and Galois Properties). London New York San Francisco: Academic Press. 704 pp. Zbl.339.00010

Fröhlich, A. (1983): Galois Module Structure of Algebraic Integers. Berlin Heidelberg New York: Springer-Verlag. 262 pp. Zbl.501.12012

Fröhlich, A. (1983a): Central extensions, Galois groups, and ideal class groups of number fields. Contemp. Math. *24*, 86 pp. Zbl.519.12001

Fröhlich, A. (1989): *L*-values at zero and multiplicative Galois module structure. J. Reine Angew. Math. *397*, 42–99. Zbl.693.12012

Gauss, C.F. (1863): Theoria residuorum biquadraticorum. Werke, vol. II, Göttingen, 93–198

Gelbart, S.S. (1975): Automorphic forms on adèle groups. Ann. Math. Stud. *83*, 267 pp. Zbl.329.10018

Goldfeld, D. (1985): Gauss' class number problem for imaginary quadratic fields. Bull. Am. Math. Soc., New Ser. *13*, No. 1, 23–27. Zbl.572.12004

Goldstein, L.J. (1971): Analytic Number Theory. Englewood Cliffs, N.J.: Prentice-Hall. 282 pp. Zbl.226.12001

Golod, E.S., Shafarevich, I.R. (1964): On class field towers. Izv. Akad. Nauk SSSR, Ser. Mat. *28*, 261–272. Zbl.136,26. English transl.: Transl., II. Ser. Am. Math. Soc. *48*, 91–102 (1965)

Gordeev, N.L. (1981): The infinity of the number of relations in the Galois group of the maximal *p*-extension with restricted ramification of a local field. Izv. Akad. Nauk SSSR, Ser. Mat. *45*, No. 3, 592–607. English transl.: Math. USSR, Izv. *18*, 513–524 (1982). Zbl.491.12015

Gras, G. (1977): Classes d'idéaux des corps abéliens et nombres de Bernoulli généralisés. Ann. Inst. Fourier *27*, 1–66. Zbl.336.12004

Gras, G. (1986): Théorie des genres analytique des fonctions *L* *p*-adiques des corps totalement réeles. Invent. Math. *86*, No. 1, 1–17. Zbl.571.12008

Greenberg, R. (1973): On a certain *l*-adic representation. Invent. Math. *21*, 117–124. Zbl.268.12004

Greenberg, R. (1974): On *p*-adic *L*-functions and cyclotomic fields I. Nagoya Math. J. *56*, 61–77. Zbl.315.12008

Greenberg, R. (1977): On *p*-adic *L*-functions and cyclotomic fields II. Nagoya Math. J. *67*, 139–158. Zbl.373.12007

Gross, B.H. (1981): *p*-adic *L*-series at $s = 0$. J. Fac. Sci., Univ. Tokyo, Sect. I A *28*, 979–994. Zbl.507.12010

Gross, B.H., Zagier, D.B. (1986): Heegner points and derivatives of *L*-series. Invent. Math. *84*, No. 2, 225–320. Zbl.608.14019

Grunwald, W. (1933): Ein allgemeines Existenztheorem für algebraische Zahlkörper. J. Reine Angew. Math. *169*, 103–107. Zbl.6,252

Haberland, K. (1978): Galois Cohomology of Algebraic Number Fields. Berlin: VEB Deutscher Verlag der Wissenschaften. 145 pp. Zbl.418.12004

Hall Jr., M. (1959): The Theory of Groups. New York: Macmillan Comp. 434 pp. Zbl.84,22

Hartshorne, R. (1977): Algebraic Geometry. New York Berlin Heidelberg: Springer-Verlag. 496 pp. Zbl.367.14001

Hasse, H. (1930): Die Normenresttheorie relativ-Abelscher Zahlkörper als Klassenkörpertheorie im Kleinen. J. Reine Angew. Math. *162*, 145–154. Jbuch56,165

Hasse, H. (1948): Existenz und Mannigfaltigkeit abelscher Algebren mit vorgegebener Galoisgruppe über einem Teilkörper des Grundkörpers. I, II, III. Math. Nachr. 1, 40–61, 213–217, 277–283. Zbl.32,255; Zbl.32,257; Zbl.32,258

Hasse, H. (1952): Über die Klassenzahl abelscher Zahlkörper. Berlin: Akademie-Verlag, 190 pp. Zbl.46,260

Hasse, H. (1964): Vorlesungen über Zahlentheorie. 2. Aufl. Berlin Heidelberg New York: Springer-Verlag. Zbl.123,42

Hasse, H. (1964a): Über den Klassenkörper zum quadratischen Zahlkörper mit der Diskriminante −47. Acta Arith. *9*, 419–434. Zbl.125,292

Hasse, H. (1967): History of class field theory. In: Cassels, Fröhlich (1967) 266–279. Zbl.153,76

Hasse, H. (1970): Bericht über neuere Untersuchungen und Probleme aus der Theorie der algebraischen Zahlkörper. Teil I: Klassenkörpertheorie. Teil Ia: Beweise zu Teil I; Teil II: Reziprozitätsgesetz. 3. Aufl.. Würzburg Wien: Physika-Verlag. 204 pp. Jbuch52,150; 53,143; 46,165

Hasse, H. (1979): Number Theory. Engl. transl. of the 3rd ed. of Zahlentheorie. Berlin: Akademie-Verlag, 638 pp. Zbl.423.12001

Hecke, E. (1923): Vorlesungen über die Theorie der algebraischen Zahlen. Leipzig: Geest und Portig, 265 pp. Jbuch 49,106

Hecke, E. (1970): Zur Theorie der elliptischen Modulformen. Math. Werke. Göttingen: Vandenhoeck & Ruprecht, 428–460. Zbl.205,289

Heegner, K. (1952): Diophantische Analysis und Modulfunktionen. Math. Z. *56*, 227–253. Zbl.49,162

Herbrand, J. (1931): Sur la théorie des groupes de décomposition, d'inertie et de ramification. J. Math. Pures Appl., IX. Sér. *10*, 481–498. Zbl.3,147

Herbrand, J. (1931a): Sur les unités d'un corps algébrique. C.R. Acad. Sci., Paris *192*, 24–27. Zbl.1,8

Herbrand, J. (1932): Sur les classes des corps circulaires. J. Math. Pures Appl., IX. Sér. 11, 417–441. Zbl.6,8

Hilbert, D. (1892): Über die Irreduzibilität ganzer rationaler Funktionen mit ganzzahligen Koeffizienten. J. Reine Angew. Math. *110*, 104–129. Jbuch 24,87

Hilbert, D. (1894/1895): Die Theorie der algebraischen Zahlkörper. Jahresbericht der Deutschen Mathematiker-Vereinigung IV

Hilbert, D. (1900): Mathematische Probleme. Göttinger Nachrichten, 253–298. English transl. in: Mathematical Developments Arising from Hilbert Problems. Proc. Symp. Pure Math. *28*, 1976, 1–34. Zbl.326.00002

Hoechsmann, K. (1968): Zum Einbettunsproblem. J. Reine Angew. Math. *229*, 81–106. Zbl.185,112

Hoyden-Siedersleben, G. (1985): Realisierung der Jankogruppen J_1 und J_2 als Galoisgruppen über $\mathbb{Q}$. J. Algebra *97*, 17–22. Zbl.574.12011

Hoyden-Siedersleben, G., Matzat, B.H. (1986): Realisierung sporadischer einfacher Gruppen als Galoisgruppen über Kreisteilungskörpern. J. Algebra *101*, 273–286. Zbl.598.12006

Hunt, D.C. (1986): Rational rigidity and the sporadic groups. J. Algebra *99*, 577–592. Zbl.598.12005

Ikeda, M. (1960): Zur Existenz eigentlicher galoisscher Körper beim Einbettungsproblem für galoissche Algebren. Abh. Math. Semin. Univ. Hamb. *24*, 126–131. Zbl.95,29

Ishchanov, V.V. (1976): On the semi direct embedding problem with nilpotent kernel. Izv. Akad. Nauk SSSR, Ser. Mat. *40*, No. 1, 3–25. English transl.: Math. USSR, Izv. *10*, 1–23. Zbl.372.12015

Iwasawa, K. (1953): On solvable extensions of algebraic number fields. Ann. Math., II. Ser. *58*, 548–572. Zbl.51,266

Iwasawa, K. (1955): On Galois groups of local fields. Trans. Am. Math. Soc. *80*, 448–469. Zbl.74,31

Iwasawa, K. (1969): On p-adic L-functions. Ann. Math., II. Ser. *89*, 198–205. Zbl.186,92

Iwasawa, K. (1973): On the μ-invariant of $\mathbb{Z}_l$-extensions, Number Theory, Algebraic Geometry and Commutative Algebra (in honor of Y. Akizuki). Tokyo, Kinokuniya, 1–11. Zbl.281.12005

Iyanaga, S. (ed.) (1975): The Theory of Numbers. Amsterdam: North-Holland Publ. Comp. 541 pp. Zbl.327.12001

Jacquet, H., Langlands, R.P. (1970): Automorphic Forms on GL (2). Lect. Notes Math. *114*. Berlin Heidelberg New York: Springer-Verlag. Zbl.236.12010

Jakovlev, A.V. (=Yakovlev, A.V.) (1968): The Galois group of the algebraic closure of a local field. Izv. Akad. Nauk SSSR, Ser. Mat.. *32*, 1283–1322. Zbl.174,341. English transl.: Math. USSR, Izv. *2*, 1231–1269

Jakovlev, A.V. (=Yakovlev, A.V.) (1978): Remarks on my paper "The Galois group of the algebraic closure of a local field". Izv. Akad. Nauk SSSR, Ser. Mat. *42*, 212–213. English transl.: Math. USSR, Izv. *12*, 205–206. Zbl.424.12010

Jannsen, U. (1982): Über Galoisgruppen lokaler Körper. Invent. Math. *70*, No. 1, 53–69. Zbl.534.12009

Jannsen, U., Wingberg, K. (1982): Die Struktur der absoluten Galoisgruppe p-adischer Zahlkörper. Invent. Math. *70*, No. 1, 70–98. Zbl.534.12010

Katz, N.M. (1976): An overview of Deligne's proof of the Riemann hypothesis for varieties over finite fields (Hilbert's problem 8). Mathematical developments arising from Hilbert problems. Proc. Symp. Pure Math. *28*, 275–305. Zbl.339.14013

Kawada, Y. (1954): On the structure of the Galois group of some infinite extensions. J. Fac. Sci., Univ. Tokyo, Sect. I A, 7, 1–18. Zbl.55,30

Kenku, M.A. (1971): Determination of the even discrimants of complex quadratic fields of class number 2. Proc. Lond. Math. Soc., III. Ser. *22*, No. 4, 734–746. Zbl.215,72

Klingen, H. (1962): Über die Werte der Dedekindschen Zetafunktion. Math. Ann. *145*, 265–272. Zbl.101,30

Kneser, M. (1951): Zum expliziten Reziprozitätsgesetz von I.R. Shafarevich. Math. Nachr. *6*, 89–96. Zbl.45,322

Koch, H. (1963): Über Darstellungsräume und die Struktur der multiplikativen Gruppe eines p-adischen Zahlkörpers. Math. Nachr. *26*, 67–100. Zbl.124,271

Koch, H. (1970): Galoissche Theorie der p-Erweiterungen. Berlin Heidelberg New York: Springer-Verlag. Zbl.216,47

Koch, H. (1978): On p-extensions with given ramification. Appendix 1 to Haberland, 1978, 89–126. Zbl.418.12004

Koch, H. (1986): Einführung in die klassische Mathematik I. Berlin Heidelberg New York: Springer-Verlag, 326 pp. Zbl.652.00001. English transl.: Kluver, Academic Publishers, Dordrecht, Boston, London 1991

Koch, H., Pieper, H. (1976): Zahlentheorie. Berlin: VEB Deutscher Verlag der Wissenschaften. 232 pp. Zbl.342.10001

Koch, H., Venkov, B.B. (1975): Über den p-Klassenkörperturm eines imaginär-quadratischen Zahlkörpers. Astérisque, 24–25, 57–67. Zbl.335.12021

Kolyvagin, V.A. (1979): Formal groups and norm residue symbol. Izv. Akad. Nauk SSSR, Ser. Mat. *43*, No. 5, 1054–1120. Zbl.429.12009. English transl.: Math. USSR, Izv. *15*, 289–348 (1980)

Krasner, M. (1936): Sur la représentation multiplicative dans les corps de nombres p-adique relative galoisiens. C.R. Acad. Sci., Paris *203*, 907–908. Zbl.15,150

Krasner, M. (1966): Nombre des extensions d'un degré donné d'un corps p-adique. Colloq. Int. Centre Nat. Rech. Sci. *143*, 143–169. Zbl.143,64

Kronecker, L. (1882): Grundzüge einer arithmetischen Theorie der algebraischen Größen. J. Reine Angew. Math. *92*, 1–123. Jbuch14,38

Krull, W. (1928): Galoissche Theorie der unendlichen algebraischen Erweiterungen. Math. Ann. *100*, 687–698. Jbuch54,157

Kubota, T., Leopoldt, H.W. (1964): Eine p-adische Theorie der Zetawerte I. Einführung der p-adischen Dirchletschen L-Funktionen. J. Reine Angew. Math. *214/215*, 328–339. Zbl.186,91

Kummer, E. (1975): Collected Papers (ed. by A. Weil). Vol. I, New York Berlin Heidelberg: Springer-Verlag. 957 pp. Zbl.327.01019

Kutzko, P., Moy, A. (1985): On the local Langlands conjecture in prime dimension. Ann. Math., II. Ser. *121*, 495–517. Zbl.609.12017

Kuz'min, L.V. (1981): Some remarks on the l-adic Dirichlet theorem and the l-adic regulator. Izv. Akad. Nauk SSSR, Ser. Mat. *45*, No. 6, 1203–1240. English transl.: Math. USSR, Izv. *19*, 445–478 (1982). Zbl.527.12012

Kuz'min, L.V. (1979): Some duality theorems for cyclotomic Γ-extensions of algebraic number fields of CM-type. Izv. Akad. Nauk SSSR, Ser. Mat. *43*, No. 3, 483–546. Zbl.434.12006. English transl.: Math. USSR, Izv. *14*, 441–498 (1980)

Labute, J.P. (1967): Classification of Demushkin groups. Canad. J. Math. *19*, 106–132. Zbl.153,42

Lang, S. (1970): Algebraic Number Theory. Reading, Mass., London: Addison-Wesley Publ. Comp., 354 pp. Zbl.211,384

Lang, S. (1973): Elliptic Functions. London, Reading, Mass.: Addison-Wesley Publ. Comp. 326 pp. Zbl.316.14001

Lang, S. (1990): Cyclotomic fields I and II. Combined 2nd ed., New York Berlin Heidelberg: Springer-Verlag. Zbl.704.11038

Langlands, R.P. (1970): Problems in the Theory of Automorphic Forms. Lect. Notes Math. *170*, Berlin Heidelberg New York: Springer-Verlag, 18–61. Zbl.225.14022

Lazard, M. (1965): Groupes analytiques p-adiques. Publ. Math., Inst. Haut. Etud. Sci. *26*, 389–603. Zbl.139,23

Lehmer, D.H., Lehmer, E., Vandiver, H. (1954): An application of highspeed computing to Fermat's Last Theorem. Proc. Natl. Acad. Sci. USA *40*, 25–33. Zbl.55,40

Lenstra Jr., H.W. (1977): Euclidean number fields of large degree. Invent. Math. *38*, 237–254. Zbl.328.12007

Leopoldt, H.W. (1953): Zur Geschlechtertheorie in abelschen Zahlkörpern. Math. Nachr. *9*, 351–362. Zbl.53,355

Leopoldt, H.W. (1958): Zur Struktur der l-Klassengruppe galoisscher Zahlkörper. J. Reine Angew. Math. *199*, 165–174. Zbl.82,254

Leopoldt, H.W. (1959): Über die Hauptordnung der ganzen Elemente eines abelschen Zahlkörpers. J. Reine Angew. Math. *201*, 119–149. Zbl.98,34

Leopoldt, H.W. (1961): Zur Approximation des p-adischen Logarithmus. Abh. Math. Semin. Univ. Hamb. *25*, 77–81. Zbl.99,26

Manin, Yu.I. (1968): Correspondences, motives, and monoidal transformations. Mat. Sb., Nov. Ser. *77*, No. 4, 475–507. Zbl.199,248. English transl.: Math. USSR, Sb. *6*, 439–470 (1969)

Martinet, J. (1971): Modules sur l'algèbre du group quaternionien. Ann. Sci. Ec. Norm. Supér., IV. Sér. *4*, 399–408. Zbl.219.12012

Martinet, J. (1978): Tours de corps de classes et estimations de discriminants. Invent. Math. *44*, No. 1, 65–73. Zbl.369.12007

Martinet, J. (1979): Petits discriminants. Ann. Inst. Fourier *29*, No. 1, 159–170. Zbl.387.12006

Masley, J. (1976): Solution of small class number problems for cyclotomic fields. Compos. Math. *33*, 179–186. Zbl.348.12011

Matzat, B.H. (1977): Zur Konstruktion von Funktionen- und Zahlkörpern mit vorgegebener Galoisgruppe. Math. Foschungsinst. Oberwolfach, Tagungsbericht *33*, Algebraische Zahlentheorie 8

Matzat, B.H. (1984): Konstruktion von Zahl- und Funktionenkörpern mit vorgegebener Galoisgruppe. J. Reine Angew. Math. *349*, 179–220. Zbl.555.12005

Matzat, B.H. (1985): Zum Einbettungsproblem der algebraischen Zahlentheorie mit nichtabelschem Kern. Invent. Math. *80*, 365–374. Zbl.567.12015

Matzat, B.H. (1987): Konstruktive Galoistheorie. Berlin Heidelberg New York: Springer-Verlag. 286 pp. Zbl.634.12011

Matzat, B.H. (1988): Über das Umkehrproblem der Galoisschen Theorie. Jahresber. Dtsch. Math.-Ver. *90*, No. 4, 155–183. Zbl.662.12008

Matzat, B.H., Zeh-Marschke, A. (1986): Realisierung der Mathieugruppen M_{11} und M_{12} als Galoisgruppen über $\mathbb{Q}$. J. Number Theory *23*, 195–202. Zbl.598.12007

Maus, E. (1967): Arithmetisch disjunkte Körper. J. Reine Angew. Math. *226*, 184–203. Zbl.149,294

Maus, E. (1968): Die gruppentheoretische Struktur der Verzweigungsgruppenreihen. J. Reine Angew. Math. *230*, 1–28. Zbl.165,357

Maus, E. (1972): Über die Verteilung der Grundverzweigungszahlen von wild verzweigten Erweiterungen p-adischer Zahlkörper. J. Reine Angew. Math. *257*, 47–79. Zbl.263.12007

Maus, E. (1973): Relationen in Verzweigungsgruppen. J. Reine Angew. Math. *258*, 23–50. Zbl.263.12008

Mazur, B., Wiles, A. (1984): Class fields of abelian extensions of $\mathbb{Q}$. Invent. Math. *76*, No. 2, 179–330. Zbl.545.12005

Merkuriev, A.S. (= Merkur'ev, A.S.), Suslin, A.A. (1982): K-cohomology of Severi-Brauer varieties and the norm residue homomorphism. Izv. Akad. Nauk SSSR, Ser. Mat. *46*, No. 5, 1011–1046. English transl.: Math. USSR, Izv. *21*, 307–340. Zbl.525.18008

Meyer, C. (1957): Die Berechnung der Klassenzahl Abelscher Körper über quadratischen Zahlkörpern. Berlin: Akademie-Verlag. 132 pp. Zbl.79,60

Milne, J.S. (1980): Etale cohomology. Princeton, N.J.: Princeton University Press. 323 pp. Zbl.433.14012

Milnor, J. (1970): Algebraic K-theory and quadratic forms. Invent. Math. *9*, No. 4, 318–344. Zbl.199,555

Milnor, J. (1971): Introduction to algebraic K-theory. Ann. Math. Stud. *72*, 184 pp. Zbl.237.18005

Močkoř, J. (1983): Groups of Divisibility. Prague, SNTL, and D. Reidel Publ. Comp., Dordrecht Boston Lancaster, 184 pp. Zbl.528.13001

Montgomery, H.L., Weinberger, P.J. (1973): Notes on small class numbers. Acta Arith. *24*, 529–542. Zbl.285.12004

Moy, A. (1982): Local constants and the tame Langlands correspondence. Thesis, Univ. of Chicago (see also: Am. J. Math. *108*, 863–929 (1986). Zbl.597.12019

Mumford, D. (1968): Abelian varieties. Tata Institute of Fundamental Research Lectures. Bombay, 242 pp. Zbl.223.14022

Narkiewicz, W. (1974): Elementary and Analytic Theory of Algebraic Numbers. Warszawa, PWN, 630 pp., Zbl.276.12002. 2nd ed. 1990

Neukirch, J. (1969): Klassenkörpertheorie. Mannheim: Bibliogr. Inst., 301 pp. Zbl.199,375

Neukirch, J. (1969a): Kennzeichnung der p-adischen und der endlich algebraischen Zahlkörper. Invent. Math. *6*, 296–314. Zbl.192,401

Neukirch, J. (1969b): Kennzeichnung der endlichalgebraischen Zahlkörper durch die Galoisgruppe der maximal auflösbaren Erweiterungen. J. Reine Angew. Math. *238*, 135–147. Zbl.201,59

Neukirch, J. (1979): On solvable number fields. Invent. Math. *53*, 135–164. Zbl.447.12008

Neukirch, J. (1986): Class Field Theory. Berlin Heidelberg New York: Springer-Verlag, 140 pp. Zbl.587.12001

Neumann, O. (1975): On p-closed algebraic number fields with restricted ramification. Izv. Akad. Nauk SSSR, Ser. Mat. 39, No. 2, 259–271. English transl.: Math. USSR, Izv. *9*, 243–254. Zbl.352.12011

Neumann, O. (1981): Two proofs of the Kronecker-Weber Theorem "according to Kronecker and Weber". J. Reine Angew. Math. *323*, 105–126. Zbl.471.12001

Ore, O. (1939): Contribution to the theory of groups of finite order. Duke Math. J. *5*, No. 2, 431–460, Zbl.21,211

Oesterlé, J. (1984): Nombre de classes des corps quadratiques imaginaires. Sém. Bourbaki, Exp. 631, 1983–1984. Astérisque 121–122, 309–323. Zbl.551.12003

Pieper, H. (1972): Die Einheitengruppe eines zahm-verzweigten galoisschen lokalen Körpers als Galois-Modul. Math. Nachr. *54*, 173–210. Zbl.263.12009

Poitou, G. (1976): Minorations de discriminants (d'après A.M. Odlyzko), Sém. Bourbaki, Exp. 479. Lect. Notes Math. *567*, 18 pp., Zbl.359.12010

Pontrjagin, L.S. (= Pontryagin, L.S.) (1973): Topological Groups. 3rd ed., Moscow: Nauka, 520 pp. (Russian) Zbl.265.22001

Rapoport, M., Schappacher, N., Schneider, P. (eds.) (1988): Beilinson's conjecture on special values of L-functions. Perspect. Math. *4*. Academic Press, 373 pp. Zbl.635.00005

Rédei, L., Reichardt, H. (1934): Die Anzahl der durch 4 teilbaren Invarianten der Klassengruppe eines beliebigen quadratischen Zahlkörpers. J. Reine Angew. Math. *170*, 69–74. Zbl.7,396

Reichardt, H. (1937): Konstruktion von Zahlkörpern mit gegebener Galoisgruppe von Primzahlpotenzordnung. J. Reine Angew. Math. *177*, 1–5. Zbl.16,151

Ribenboim, P. (1979): 13 Lectures on Fermat's Last Theorem. New York Berlin Heidelberg: Springer-Verlag, 302 pp. Zbl.456.10006

Ribet, K. (1976): A modular construction of unramified p-extensions of $\mathbb{Q}(\mu_p)$, Invent. Math. *34*, 151–162. Zbl.338.12003

Riemann, B. (1859): Über die Anzahl der Primzahlen unter einer gegebenen Größe. Monatsber. Preuss. Akad. Wiss.

Roquette, P. (1967): On class field towers. In: Cassels, Fröhlich (1967) 231–249. Zbl.153,76

Roquette, P., Zassenhaus, H. (1969): A class rank estimate for algebraic number fields. J. Lond. Math. Soc. *44*, 31–38. Zbl.169,380

Rubin, K. (1987): Tate-Shafarevich groups and L-functions of elliptic curves with complex multiplication. Invent. Math. *89*, No. 3, 527–560. Zbl.628.14018

Rubin, K. (1990): Appendix to Lang (1990). Zbl.704.11038

Schinzel, A. (1966): On a theorem of Bauer and some of its applications. Acta Arith. *11*, 333–344. Correction 12, 425 (1967) Zbl.158,301

Schmithals, B. (1985): Kapitulation der Idealklassen und Einheitenstruktur in Zahlkörpern. J. Reine Angew. Math. *358*, 43–60. Zbl.562.12009

Schöneberg, B. (1974): Elliptic Modular Functions. An Introduction. Berlin Heidelberg New York: Springer-Verlag. 232 pp. Zbl.285.10016

Scholz, A. (1925): Über die Bildung algebraischer Zahlkörper mit auflösbarer galoisscher Gruppe. Math. Z. *30*, 332–356. Jbuch 55,89

Scholz, A. (1937): Konstruktion algebraischer Zahlkörper mit beliebiger Gruppe von Primzahlpotenzordnung. Math. Z. *42*, 161–188. Zbl.16,6

Schoof, R.J. (1983): Class groups of complex quadratic fields. Math. Comput. *41*, 295–302. Zbl.516.12002

Schoof, R.J. (1986): Infinite class field towers of quadratic fields. J. Reine Angew. Math. *372*, 209–220. Zbl.589.12011

Schrutka von Rechtenstamm, G. (1964): Tabelle der (Relativ)-Klassenzahlen der Kreiskörper, deren φ-Funktion des Wurzelexponenten (Grad) nicht größer als 256 ist. Abh. Dtsch. Akad. Wiss. Berlin, Kl. Math. Phys., No. 2, 1–64. Zbl.199,98

Seifert, H., Threlfall, W. (1934): Lehrbuch der Topologie. Leipzig: Teubner-Verlag. Zbl.9,86

Sen, S. (1972): Ramification in p-adic Lie extensions. Invent. Math. *17*, No. 1, 44–50. Zbl.242.12012

Serre, J.P. (1958): Classes des corps cyclotomiques (d'après K.Iwasawa). Sémin. Bourbaki, Exp. *174*, 11pp. Zbl.119,276

Serre, J.P. (1960): Sur la rationalité des représentations d'Artin. Ann. Math., II. Ser. *72*, 405–420. Zbl.202,328

Serre, J.P. (1962): Corps locaux. Paris: Hermann, 243 pp. Zbl.137,26

Serre, J.P. (1963): Structure de certains pro-p-groupes. Sémin. Bourbaki, Exp. *252*, 11 pp. Zbl.121,44

Serre, J.P. (1964): Cohomologie galoisienne. Lect. Notes Math. *5*. Berlin Heidelberg New York: Springer-Verlag. Zbl.128,263

Serre, J.P. (1965): Sur la dimension cohomologique des groupes profinis. Topology *3*, 413–420. Zbl.136,274

Serre, J.P. (1967): Local class field theory. In: Cassels, Fröhlich (1967): 129–162. Zbl.153,75

Serre, J.P. (1967a): Représentations linéaires des groupes finis. Paris: Hermann. Zbl.189,26

Serre, J.P. (1970): Cours d'arithmétique. Paris: Presses Universit. de France. 188 pp. Zbl.225.12002

Serre, J.P. (1977): Modular forms of weight one and Galois representations. In: Fröhlich (1977) 193–268. Zbl.366.10022

Serre, J.P. (1978): Une "formule de masse" pour les extensions totalement ramifiées de degré donné d'une corps local. C.R. Acad. Sci., Paris, Sér. A *286*, No. 22, 1031–1036. Zbl.388.12005

Serre, J.P. (1978a): Sur le résidu de la fonction zêta p-adique d'un corps de nombres. C.R. Acad. Sci., Paris, Sér. A *287*, 183–188. Zbl.393.12026

Serre, J.P., Tate, J. (1968): Good reduction of abelian varieties. Ann. Math., II. Ser *88*, 492–517. Zbl.172,461

Shafarevich, I.R. (1947): On *p*-extensions. Mat. Sb., Nov. Ser. 20 (62), 351–363, Zbl.41,171. English transl.: Transl., II. Ser., Am. Math. Soc. *4*, 59–72, and in: Shafarevich, Collected Mathematical Papers. Berlin Heidelberg New York: Springer-Verlag 1989, 6–19, Zbl.669.12001

Shafarevich, I.R. (1950): A general reciprocity law. Mat. Sb., Nov. Ser. *26* (68), 113–146, Zbl.36,159. English transl.: Transl., II. Ser., Am. Math. Soc. *4*, 73–106 (1956), and in: Shafarevich, Collected Mathematical Papers. Berlin Heidelberg New York: Springer-Verlag 1989, 20–53, Zbl.669.12001

Shafarevich, I.R. (1951): A new proof of the Kronecker-Weber theorem. Tr. Mat. Inst. Steklova *38*, 382–387, Zbl.53,355. English transl. in: Shafarevich, Collected Mathematical Papers. Berlin Heidelberg New York: Springer-Verlag 1989, 54–58. Zbl.669.12001

Shafarevich, I.R. (1954a): On the construction of fields with a given Galois group of order l^a. Izv. Akad. Nauk SSSR, Ser. Mat. *18*, 261–296, Zbl.56,33. English transl.: Transl., II. Ser., Am. Math. Soc. *4*, 107–142 (1956), and in: Shafarevich, Collected Mathematical Papers. Berlin Heidelberg New York: Springer-Verlag 1989, 62–97, Zbl.12001

Shafarevich, I.R. (1954b): On the imbedding problem for fields. Dokl. Akad. Nauk SSSR *95*, No. 3, 459–461, Zbl.55,31. English transl. in: Shafarevich, Collected Mathematical Papers. Berlin Heidelberg New York: Springer-Verlag 1989, 59–61. Zbl.669.12001

Shafarevich, I.R. (1954c): Construction of algebraic number fields with given solvable Galois group. Izv. Akad. Nauk SSSR, Ser. Mat. *18*, 525–578, Zbl.57,274. English transl.: Transl., II. Ser., Am. Math. Soc. *4*, 185–237 (1956), and in: Shafarevich, Collected Mathematical Papers. Berlin Heidelberg New York: Springer-Verlag 1989, 139–191, Zbl.669.12001

Shafarevich, I.R. (1964): Extensions with given ramification points. Publ. Math., Inst. Hautes Etud. Sci. *18*, 295–319, Zbl.118,275. English transl.: Transl., II. Ser., Am. Math. Soc. *59*, 128–149 (1966): and in: Shafarevich, Collected Mathematical Papers. Berlin Heidelberg New York: Springer-Verlag 1989, 295–316, Zbl.669.12001

Shafarevich, I.R. (1969): The Zeta-function. Moscow: Moscow University Press, 148 pp. (Russian)

Shimura, G. (1971): Introduction to the arithmetic theory of automorphic functions. Tokyo, Iwanami Shoten, and Princeton, Princeton University Press. 267 pp. Zbl.221.10029

Shimura, G., Taniyama, Y. (1961): Complex multiplication of abelian varieties and its applications to number theory. Publ. Math. Soc. Japan No. 6, Tokyo, 159 pp. Zbl.112,35

Siegel, C.L. (1969): Berechnung von Zetafunktionen an ganzzahligen Stellen. Nachr. Akad. Wiss. Göttingen, II. Math. -Phys. Kl., 87–102. Zbl.186,88

Sinnott, W. (1980): On the Stickelberger ideal and the circular units of an abelian field. Invent. Math. *62*, 181–234. Zbl.465.12001

Skula, L. (1970): Divisorentheorie einer Halbgruppe. Math. Z. *114*, 113–120. Zbl.177,32

Speiser, A. (1919): Die Zerlegungsgruppe. J. Reine Angew. Math. *149*, 174–188. Jbuch 47,92

Springer, G. (1957): Introduction to Riemann surfaces. Reading, Mass.: Addison-Wesley Publ. Comp. 230 pp. Zbl.78.66

Stark, H.M. (1967): A complete determination of the complex quadratic fields of class-number one. Mich. Math. J. *14*, 1–27. Zbl148,278

Stark, H.M. (1969): On the "gap" in a theorem of Heegner. J. Number Theory *1*, No. 1, 16–27. Zbl.198,377

Stark, H.M. (1971): Values of *L*-functions at $s = 1$. I. *L*-functions for quadratic forms. Adv. Math. 7, 301–343. Zbl.263.10015

Stark, H.M. (1975): On complex quadratic fields with class-number two. Math. Comput. *29*, 289–302. Zbl.321.12009

Stark, H.M. (1975a): *L*-functions at $s = 1$. II. Artin *L*-functions with rational characters. Adv. Math. *17*, 60–92. Zbl.316.12007

Stark, H.M. (1976): *L*-functions at $s = 1$. III. Totally real fields and Hilbert's twelfth problem. Adv. Math. *22*, 64–84. Zbl.348.12017

Stark, H.M. (1977): Hilbert's twelfth problem and *L*-series. Bull. Am. Math. Soc. *83*, 1072–1074. Zbl.378.12007

Stark, H.M. (1977a): Class fields and modular forms of weight one. In: Modular functions of one variable V, Bonn 1976. Lect. Notes Math. *601*. Berlin Heidelberg New York: Springer-Verlag, 278–287. Zbl.363.12010

Stark, H.M. (1980): L-functions at $s = 1$. IV. First derivatives at $s = 0$. Adv. Math. *35*, 197–235. Zbl.475.12018

Stark, H.M. (1982): Values of zeta and L-functions. Abh. Braunschw. Wiss. Ges. *33*, 71–83. Zbl.509.12012

Sullivan, D. (1970): Geometric Topology. Cambridge, Mass.: MIT. 432 pp. Zbl.366.57003

Takagi, T. (1920): Über eine Theorie des relativ-abelschen Zahlkörpers. J. Coll. Sci. Imp. Univ. Tokyo *41*, No. 9, 1–133. Jbuch 47,147

Tate, J. (1952): The higher-dimensional cohomology groups of class field theory. Ann. Math., II. Ser. *56*, 294–297. Zbl.47,37

Tate, J. (1963): Duality theorems in Galois cohomology over number fields. Proc. Int. Congr. Math., Stockholm 1962, 288–295. Zbl.126,70

Tate, J. (1967): Global class field theory. In: Cassels, Fröhlich (1967) 163–203. Zbl.153,75

Tate, J. (1967a): Fourier analysis in number fields and Hecke's zeta-functions. In: Cassels, Fröhlich (1967) 305–347. Zbl.153,75

Tate, J. (1977): Local constants. In: Fröhlich (1977) 89–131. Zbl.425.12019

Tate, J. (1979): Number theoretic background. In: Borel, Casselman, II (1979), Proc. Symp. Pure Math. *33*, 3–26. Zbl.422.12007

Tate, J. (1981): On Stark's conjecture on the behavior of $L(s, \chi)$ at $s = 0$. J. Fac. Sci. Univ. Tokyo, Sect. I A *28*, No. 3, 963–978. Zbl.514.12013

Tate, J. (1984): Les conjectures de Stark sur les fonctions L d'Artin en $s = 0$. Boston-Basel-Stuttgart: Birkhäuser. Prog. Math. *47*, 143 pp. Zbl.545.12009

Taussky, O. (1932): Über eine Verschärfung des Hauptidealsatzes für algebraische Zahlkörper. J. Reine Angew. Math. *168*, 193–210. Zbl.6,8

Taylor, M.J. (1981): On Fröhlich's conjecture for rings of integers of tame extensions. Invent. Math. *63*, No. 1, 41–79. Zbl.469.12003

Thompson, J.G. (1984): Some finite groups which appear as Gal(L/K), where $K \subseteq \mathbb{Q}(\mu_n)$. J. Algebra *89*, No. 2, 437–499. Zbl.552.12004

Tunnell, J. (1981): Artin's conjecture for representations of octahedral type. Bull. Am. Math. Soc., New Ser. *5*, No. 2, 173–175. Zbl.475.12016

Uchida, K. (1976): Isomorphisms of Galois groups. J. Math. Soc. Japan *28*, 617–620. Zbl.329.12013

Uchida, K. (1982): Galois groups of unramified solvable extensions. Tôhoku Math. J., II. Ser. *34*, 311–317. Zbl. Zbl.502,12020

van der Waall, R.W. (1977): Holomorphy of quotients of zeta functions. In: Fröhlich (1977) 649–662. Zbl.359.12013

van der Waerden, B.L. (1967, 1971): Algebra. Berlin Heidelberg New York: Springer-Verlag, I: 8th ed. 1971, II: 5th ed. 1967. Zbl.67,5 and Zbl.192,330

Vinogradov, I.M. (1965): Basic Number Theory. 7th ed., Moscow: Nauka, 172 pp. (Russian). Zbl.124,25

Vostokov, S.V. (1978): The explicit form of the reciprocity law. Izv. Akad. Nauk SSSR, Ser. Mat. *42*, No. 6, 1288–1321 Zbl.408.12002. English transl.: Math. USSR, Izv. *13*, 557–588 (1979)

Vostokov, S.V., Fesenko, I.B. (1983): The Hilbert symbol for formal Lubin-Tate groups II. Zap. Nauchn. Semin. Leningr. Otd. Math. Inst. Steklova *132*, 85–96. English transl.: J. Sov. Math. *30*, 1854–1862 (1985). Zbl.574.12016

Wagstaff, S. (1978): The irregular primes to 125000. Math. Comput. *32*, 583–591. Zbl.377.10002

Wang, Sh. (1950): On Grunewald's theorem. Ann. Math., II. Ser. *51*, 471–484. Zbl.36,158

Washington, L. (1979): The non-p-part of the class number in a cyclotomic $\mathbb{Z}_p$-extension. Invent. Math. *49*, No. 1, 89–97. Zbl.403.12007

Washington, L. (1982): Introduction to Cyclotomic Fields. New York Berlin Heidelberg: Springer-Verlag. 389 pp. Zbl.484.12001

Weber, H. (1886, 1887): Theorie der Abelschen Zahlkörper I, II. Stockholm. I: Acta Math. *8* (1886) 193–263, II: Acta Math. *9* (1887) 105–130. Jbuch 18,55

Weber, H. (1891): Elliptische Funktionen und algebraische Zahlen. Braunschweig: Vieweg. 504 pp. Jbuch 23,455

Weber, H. (1897–1898): Über Zahlengruppen in algebraischen Körpern I, II, III. I: Math. Ann. *48* (1897) 433–473, II: Math. Ann. *49* (1897) 83–100, III: Math. Ann. 50 (1898) 1–26. Jbuch 28,83

Weil, A. (1967): Basic Number Theory. Berlin Heidelberg New York: Springer-Verlag. 296 pp. Zbl.176,336

Weil, A. (1952): Sur les "formules explicites" de la théorie des nombres premiers. Meddel. Lunds Univ. Mat. Semin., 252–265. Zbl.49,32

Weil, A. (1972): Sur les formules explicites de la théorie des nombres. Izv. Akad. Nauk SSSR, Ser. Mat. *36*, No. 1, 3–18. Zbl.245.12010; reprinted in: Math. USSR, Izv. *6*, 1–17 (1973)

Weil, A. (1983): Number Theory. An Approach Through History. From Hammurapi to Legendre. Boston-Basel-Stuttgart: Birkhäuser, 375 pp. Zbl.531.10001

Weiss, E. (1963): Algebraic Number Theory. New York: Mc Graw Hill, 275 pp. Zbl.115,36

Weissauer, R. (1982): Der Hilbertsche Irreduzibilitätssatz. J. Reine Angew. Math. *334*, 203–220. Zbl.477.12029

Weyl, H. (1940): Algebraic theory of numbers. Ann. Math. Studies *1*, 223 pp. Zbl.178,381

Wiles, A. (1978): Higher explicit reciprocity laws. Ann. Math., II. Ser. *107*, 235–254. Zbl.378.12006

Williams, H.C., Broere, J. (1976): A computational technique for evaluating $L(1, \chi)$ and the class number of a real quadratic field. Math. Comput. *30*, No. 136, 887–893. Zbl.345.12004

Wingberg, K. (1982): Der Eindeutigkeitssatz für Demushkinformationen. Invent. Math. *70*, No. 1, 99–113. Zbl.534.12011

Wisliceny, J. (1981): Zur Darstellung von pro-p-Gruppen und Lieschen Algebren durch Erzeugende und Relationen. Math. Nachr. *102*, 57–78. Zbl.498.17011

Zelvenskij, I.G. (1972): On the algebraic closure of a local field for $p = 2$. Izv. Akad. Nauk SSSR, Ser. Mat. *36*, 933–946. Zbl.252.12012. English transl.: Math. USSR, Izv. *6*, 925–937 (1973)

Zolotarev, G. (1880): Sur la théorie des nombres complexes. J. Math. Pures Appl., III. Sér. *6*, 51–85, 115–129, 145–167. Jbuch 12,125

Author Index

Subject Index